ENVIRONMENTAL DATA MANAGEMENT

Water Science and Technology Library

VOLUME 27

ENVIRONMENTAL DATA MANAGEMENT

edited by

NILGUN B. HARMANCIOGLU
Civil Engineering Department,
Dokuz Eylul University,
Izmir, Turkey

VIJAY P. SINGH
Department of Civil and Environmental Engineering,
Louisiana State University,
Baton Rouge, Louisiana, U.S.A.

and

M. NECDET ALPASLAN
Environmental Engineering Department,
Dokuz Eylul University,
Izmir, Turkey

KLUWER ACADEMIC PUBLISHERS
DORDRECHT / BOSTON / LONDON

A C.I.P. Catalogue record for this book is available from the Library of Congress.

ISBN 0-7923-4857-5

Published by Kluwer Academic Publishers,
P.O. Box 17, 3300 AA Dordrecht, The Netherlands.

Sold and distributed in the U.S.A. and Canada
by Kluwer Academic Publishers,
101 Philip Drive, Norwell, MA 02061, U.S.A.

In all other countries, sold and distributed
by Kluwer Academic Publishers,
P.O. Box 322, 3300 AH Dordrecht, The Netherlands.

Printed on acid-free paper

Printed in the Netherlands

TABLE OF CONTENTS

CHAPTER III

CHAPTER IV

CHAPTER V

CHAPTER VI

Environmental Data Management: Storage, Handling and Retrieval **123**
A. E. Hindrichs

CHAPTER VII

Data Analysis **141**
N. B. Harmancioglu, S. D. Ozkul and O. Fistikoglu

CHAPTER VIII

CHAPTER IX

CHAPTER X

PREFACE

The diverse nature of environmental problems mankind has encountered within the last decade has developed a new understanding of the nature of environmental processes. Currently, the environment is considered as a continuum of air, soil and water as the vital components for sustaining life on earth. The interactive nature of these components requires that the environment is managed and protected as a cohesive whole. This can only be accomplished through an integrated approach to environmental management. Besides the concept of environmental continuum, prospects for sustainable development of natural resources and the recent recognition of global climate change impacts have also necessitated such an integrated approach to environmental management.

Two basic tools for integrated management of the environment are modeling and environmental data. Both tools were available and valid in the past; however, the recent requirements for integrated environmental management have also led to a significant evolution of both modeling procedures and data management systems. Regarding these advances, current literature provides vast amounts of studies on modeling of different environmental processes. However, issues related to data management systems are barely touched in a comprehensive framework. Data requirements and data availability are mentioned merely as subtopics in most environmental studies although it is well recognized that data constitute the basis for all environmental management activities. In particular, there is no book as yet published that focuses exclusively on data management systems. In this respect, the present book fills in an important gap by covering various aspects of environmental data management in a systematic approach.

The contents of the book are organized in order of basic steps that constitute an environmental data management system. These steps cover in sequence: collection of environmental data, their storage, handling and retrieval, reliability issues in available data, transfer of data into information via data analysis and environmental modeling, and finally the use of available data in decision making for environmental management.

The introductory chapter sums up basic aspects of data management systems and discusses current requirements imposed on these systems by integrated approaches to environmental management. Chapters 2, 3, and 4 focus on data collection systems related to air, water and solid waste as the basic sources of environmental pollution. Chapter 5 discusses the storage, handling and retrieval of collected data as the second step in a data management system. Before stored data can be used for various purposes, their reliability must be assessed as another step in the sequence. This issue is considered in Chapter 6 for the case of water quality data. Chapters 7, 8, and 9 relate to transfer of data into information via data analyses and modeling procedures of both deterministic and stochastic types. Finally, Chapter 10 focuses on management and use of data in the decision making process for integrated environmental data management.

With its contents, the book addresses a wide community including planners, managers, scientists, engineers and graduate students working in the area of environmental management in general and of data management in particular.

The editors would like to express their deep appreciation to the contributors of the book who have generously devoted their efforts and time towards the realization of this book.

July, 1997

Nilgun B. Harmancioglu
Vijay P. Singh
M. Necdet Alpaslan
Editors

LIST OF CONTRIBUTORS

M. Necdet ALPASLAN
Dokuz Eylul University
Faculty of Engineering
Environmental Engineering Department
Bornova 35100 Izmir, TURKEY

W. CURI
Universidade Federal Paraiba
Campina Grande, Paraiba, BRAZIL

Grant C. EDWARDS
University of Guelph
School of Engineering
Guelph, Ontario, N1G 2W1 CANADA

Okan FISTIKOGLU
Dokuz Eylul University
Faculty of Engineering
Civil Engineering Department
Bornova 35100 Izmir, TURKEY

Ian C. GOULTER
University of Central Queensland
Rockhampton, Queensland
AUSTRALIA 4702

Nilgun B. HARMANCIOGLU
Dokuz Eylul University
Faculty of Engineering
Civil Engineering Department
Bornova 35100 Izmir, TURKEY

Albert E. HINDRICHS
Louisiana Dept. of Environmental Quality
Office of Water Resources
Post Office Box 82215
Baton Rouge, Louisiana 70884-2215 USA

Agus KUSMULYONO
University of Central Queensland
Rockhampton, Queensland
AUSTRALIA 4702

Edward A. MCBEAN
Associate, Conestoga-Rovers &
Associates
651 Colby Drive
Waterloo, Ontario, N2V 1C2 CANADA

P. K. MISRA
Ministry of Environment and Energy
2 St. Clair Avenue West, 14th Floor
Toronto, M4V 1L5 CANADA

Sevinc D. OZKUL
Dokuz Eylul University
Faculty of Engineering
Civil Engineering Department
Bornova 35100 Izmir, TURKEY

K. PONNAMBALAM
Department of Systems Design
Engineering
University of Waterloo
Waterloo, Ontario, N2L 3G1 CANADA

Vijay P. SINGH
Louisiana State University
Department of Civil and
Environmental Engineering
Baton Rouge, Louisiana 70803-6405 USA

CHAPTER 1

NEEDS FOR ENVIRONMENTAL DATA MANAGEMENT

N.B. Harmancioglu, M. N. Alpaslan, and V.P. Singh

1.1. Changing Attitudes Towards Environmental Management

Within the last two decades, there has been a gradual change in our conceptualization of the environment. This change has occurred due to a parallel evolution in the nature and scale of environmental problems experienced. Until the second half of the 70's, the majority of the problems we faced were of a local nature; thus, we could then be content with local solutions and remedies. In time, environmental pollution adopted a spatial nature, ranging from regional problems to the recent ones of a global nature.

On the other hand, each problem had often been considered to be specific to a particular component of the environment (i.e., air, water, soil, etc.) so that these components were treated independently of each other, often by independent organizations and disciplines. Recently, however, environment is recognized to be a "continuum" of air, soil, and water components, which are vital for sustaining life on earth. These components are not independent of each other; they interact in a number of complex ways. Thus, any intervention imposed on one of these components produces effects that propagate to the others. It is this interactive nature of environmental processes that led to the development of a new approach to environmental management, namely that the environment must be managed and protected as a cohesive whole or system (Singh, 1995). This means that all components of the environment and their interactions must be accounted for, considering both their temporal and spatial dimensions. As such, management of the environment requires multidisciplinary and inter-organizational approaches. These requirements set the basis for what is recently described as "integrated environmental management".

Besides the recognition of the "environmental continuum" concept and changes in the scale of problems experienced, there have been other developments in terms of environmental pollution that necessitated an integrated approach to management. Due to rapid population growth, urbanization, industrial and agricultural development, water scarcity has become a significant problem in most parts of the world. The quality of surface waters, aquifers and coastal zones is continuously degrading throughout the world. There is significant concern over the slow rate of progress towards the sustainable use and development of water resources for health, food production, and income generation. Land resources reflect a similar situation; overuse or misuse of these resources have resulted in land degradation, particularly in the form of deforestation and desertification. All these problems have gradually led to physical or ecological degradation of physical habitat for biodiversity. There are further environmental challenges due to ozone depletion and climate change which affect various components in a number of interactive ways (Tyson, 1995).

N.B. Harmancioglu et al. (eds.), Environmental Data Management, 1-12.

All the above problems have stemmed from unsustainable exploitation of living and nonliving resources, and the result is the alarming environmental crisis we live in today. An important feature of this crisis is that it has grown in dimensions from local levels to regional, international and global scales. These developments have eventually led to the consideration of "sustainable development" as the basic philosophy to be adopted in exploitation and management of natural resources. At first, "sustainability" had indeed been a philosophy until the U.N. Conference on Environment and Development held in Rio de Janeiro in 1992. At this Earth Summit, the integration of environmental issues into economic and developmental decision making was foreseen, and "sustainability" was formulated as a management strategy rather than a mere philosophical concept.

One of the major outputs of UNCED is Agenda 21, which is described as "an action plan for the 1990s and 21st century, elaborating strategies and integrated programme measures to halt and reverse the effects of environmental degradation and to promote environmentally sound and sustainable development in all countries" (U.N., 1992).

As foreseen in Agenda 21, environmental sustainability is "becoming a core commitment of governments … and agencies" (Clark and Gardiner, 1994). For instance, in the case of water resources management, the UK Local Government Management Board (LGMB, 1992) has reformulated the basic concepts of Agenda 21 as the following:

- "Water resources must be planned and managed in an integrated and holistic way to prevent shortage of water, or pollution of water resources, from impeding development…".
- "By the year 2000 all states should have national action programmes for water management, based on catchment basins or sub-basins, and efficient water use programmes. These could include integration of water resource planning with land use planning and other development and conservation activities, demand management through pricing or regulation, conservation, reuse and recycling of water".

Similarly, environmentally sound and sustainable development of natural resources is in the agenda of all governments, particularly in developing countries, which undergo rapid socioeconomic development.

With respect to sustainable development, the key term is "integrated". In other words, the sustainable management of natural resources requires an "integrated" approach to:

a) promote the conservation and sustainable use of natural resources;
b) assure compability with long-term economic growth; and
c) enhance productive capacity, which is equitable and environmentally acceptable.

An "integrated" approach implies multidisciplinary management practices that allow for the consideration of all components and processes in the environment (e.g., water, soil and biotic resources), their interactions and correlations, coupled with social, economic, political and legal impacts.

1.2. Information Needs for Integrated Environmental Management

1.2.1. CURRENT STATUS OF INFORMATION SYSTEMS

Agenda 21 of UNCED has officially stated the new outlook towards environmental management, namely that the environment should be managed by an integrated approach in respect of sustainability. It was further emphasized in Agenda 21 that effective management relies essentially on reliable and adequate information on how the environment behaves under natural and man-made impacts. Yet, Agenda 21 and several other similar reports have also recognized that current systems of information production, i.e., data management systems, do not fulfill the requirements of environmental management and decision making. This is a highly unfortunate situation in view of the rapidly growing environmental problems. At a time when we need informational support the most, we find that our data management systems experience a declining trend (WMO, 1994). Recognition of this trend has brought focus to current monitoring systems, databases, and data use. Accordingly, major efforts have been initiated at regional and international levels to improve the status of existing information systems.

Several examples may be cited on activities toward assessment and revision of data management systems. In view of the significant deficiencies in the available environmental information in the European Community, European Environment Agency (EEA) is now assigned "the task of supplying those concerned with the Community environmental policy with reliable and comparable information" (Santos, 1997). For the case of water resources in particular, a World Hydrological Cycle Observing System (WHYCOS) was proposed by WMO and the World Bank in 1993 (WMO, 1994). Several programs such as MEDI (IOC), GRID and GEMS (UNEP) and EDMED (European Communities - MAST programme) have taken the task of identifying the existence of environmental data archives (Geerders, 1997). Efforts have been initiated by WMO, FAO, UNEP, IOC, and ICSU to develop a common data and information plan for GCOS/GOOS/GTOS by establishing a joint data and information management panel (Oliounine, 1997). All these and similar examples reflect the emphasis placed on development of sound and adequate informational systems to support integrated management of the environment.

1.2.2. BASIC TOOLS OF INTEGRATED MANAGEMENT

At the technical level, integrated management of the environment relies on two basic tools: modeling and data. Relevant technologies regarding these tools have long been developed at an advanced level to investigate and solve various discrete problems in the management of environmental resources. Realization of integrated approaches to management imply that such technologies will also need to be integrated as a system of tools towards decision making in environmental management.

In essence, the adoption of an integrated approach to environmental management has imposed new requirements on both tools. With respect to modeling, simpler models based on system balance (e.g., water balance) to assess discrete components of the environment are no longer sufficient. Rather, we now see more of comprehensive integrated type of models at regional (e.g., watershed models) or global scales (GCM's), comprising soil and land processes, water quantity and quality components, and other processes in the ecosystem (Harmancioglu *et al.*, 1997a).

With respect to data, the problems that must be addressed today require interdisciplinary approaches and hence much more sharing of data and information than in the past. As noted in section 1.2.1, "Agenda 21 has emphasized that the priority activities for environmental management should include establishment and integration of existing data on physical, biological, demographic and user conditions into a database; maintenance of these databases as part of the assessment and management databases; promotion of exchange of data and information with a view to the development of standard intercalibrated procedures, measuring techniques, data storage, and management capabilities" (Oliounine, 1997). The requirements imposed on data imply that, again, integrated approaches should be applied in managing environmental data. In this respect, data management should be considered as an activity for handling data so that they are available where they are needed, when they are needed, and have with them all the supporting information that is necessary for the user to understand and use the data at their full potential (Oliounine, 1997; Harmancioglu *et al.*, 1997a).

1.3. Environmental Data Management Systems

1.3.1. NEED FOR DATA

Assessment of natural resources requires knowledge and full understanding of environmental processes. Apart from considerations related to management of the environment, an increasing concern has developed in all communities over the impact of pollution (primarily surface water, coastal zone, air and solid waste pollution) on public health and general environmental conditions. Consequently, besides project-makers, the society itself stresses the need for a better understanding of how environmental processes evolve under natural and man-made conditions. Thus, information on environmental processes is needed with respect to natural resources management in general and to pollution control in particular. Retrieval of such information requires collection of data; basically, the purpose of data collection practices is to produce the information needed for efficient management of the environment. Thus, there is a significant pressure in all communities to monitor the environment, and this pressure is recognized by policy-makers, scientists, practitioners, and the society itself (Harmancioglu, 1997).

The general trend until the second half the 70's in management of natural resources has been to gather and use information on environmental variables for purposes of planning, design, and operation of particular schemes and treatment facilities. Thus, most attempts at procurement of information on environmental processes have been problem, project, or rather user-oriented. Recently, however, the accelerated growth of environmental problems, both in their extent and scale, has put broader needs on information availability.

In general, environmental data are needed to delineate (Harmancioglu *et al.*, 1992; Whitfield, 1988):

a) the general nature and trends in characteristics of environmental processes for a better understanding of these processes;
b) the effects of natural and man-made factors upon the general trends in environmental processes;
c) the effectiveness of pollution control measures;
d) the compliance of environmental quality characteristics with established quality standards for eventual purposes of enforcing quality control measures.

Furthermore, data are the essential inputs for:

a) environmental impact assessment;
b) assessment of general quality conditions over a wide area or "general surveillance";
c) modeling of environmental processes.

The crucial point in all of the above issues are evidently the availability of appropriate and adequate environmental data and the full extraction of information from collected data.

The significance of environmental data lies in the fact that they are our only means of being informed about the environment. Data constitute the link between the actual process and our understanding, interpretation, and assessment of the highly complex environmental processes. Therefore, data collection and information production is the most crucial activity an man's side with respect to all management and control efforts. Adequate and reliable data may serve to increase our knowledge on environmental processes and hence reduce the uncertainties; whereas lack of such data may lead to erroneous interpretations and decisions (Harmancioglu *et al.*, 1992).

The above discussion basically emphasizes the significance of data in environmental management. Another point to be stressed is the fact that data needs undergo changes in time. Environmental problems become more and more varied as the impact of man on the environment changes. Accordingly, information expectations also vary, leading to changes in the nature and types of data needed. As denoted in section 1.1, environmental problems had previously been more of a local nature; thus, it was often sufficient to collect data at a single point in space. Recently, however, such problems reflect a significant spatial component so that environmental processes have to be evaluated in both the time and the space dimensions. Accordingly, data to be collected are expected to reflect the spatial variations of environmental processes as well as the temporal changes.

Another significant development described in section 1.1 is the recognition of the environmental continuum. This new outlook at the environment has also changed data needs. Environmental data have to be collected in such a way as to properly account for all components of the environment and their interactions. In other words, data on different components of the environment should be integrated to eventually produce complete information about the environmental continuum.

It follows from the above that, as the complexity of environmental problems increase, information expectations and hence data needs become more varied and complicated.

1.3.2. NEED FOR DATA MANAGEMENT

As pointed out in the previous section, data availability is not a sufficient condition to produce the required information about the environment. It is the utility or usefulness of data that contributes to production of information. In the past, the primary concern was to conceive what available data showed about prevailing conditions of the environment. The question nowadays is whether the available data convey the expected information. Data collection systems have indeed become sophisticated with new methods and technologies. However, when it comes to utilizing collected data, no matter how numerous they may be, one often finds that available samples fail to meet specific data requirements foreseen for the solution of a certain problem. In this case, the data lack utility and cannot be transferred into the required information. This is one of the reasons why we need to manage our data systems; that is data management is required to produce an efficient information system where data utility is maximized (Harmancioglu, 1997).

Another aspect of the problem lies in the cost considerations. Data collection and dissemination are costly procedures; they require significant investments which have to be amortized by versatile uses of data. Even in the developed countries, a data collection system has to be realized under the constraints of limited financial sources, sampling and analysis facilities, and manpower. If the output of this system, or the data, do not fulfill information expectations, the investment made on the system cannot be amortized so that the result will inevitably be economic loss. Cost considerations do not only relate to costs of monitoring; they are also reflected in the eventual decision making process. If available data produce the required information, decisions are made more accurately, and the smaller the chances are of underdesign and overdesign. Proper decisions minimize economic losses and lead to an overall increase in the benefit/cost ratio. Thus, a data collection system has to be cost-effective and efficient to avoid economic losses both in the monitoring system itself and in the eventual design based on the information produced by this system (Harmancioglu & Alpaslan, 1992 and 1994).

The transfer of data into information involves several activities in sequence as summarized in Fig. 1.1. Each of these activities contribute to retrieval of the required information. Thus, all of these steps must be efficient to maximize data utility. To respect the condition of cost-effectiveness, again each step has to be economically optimized. Thus, these activities have to be managed to ensure the efficiency and cost-effectiveness of the whole information system.

At present, a further requirement is imposed on data management systems, namely that they should be evaluated via integrated approaches. This issue was stressed at a recent workshop where an international and multidisciplinary group of experts delineated the needs underlying an integrated approach as the following (Harmancioglu *et al.*, 1997a and b):

a) "There is a significant gap between information needs on the environment and information produced by current systems of data collection and management. The presence of this gap contradicts the nature of the *Information Age* we live in. That is, we now have developed the most sophisticated means of collecting, processing, storing and communicating data; yet, we still suffer from poor information when we attempt to use the available data. This gap can be filled in by appropriate monitoring and management of data. In view of numerous problems encountered in monitoring and information production, the adoption of integrated approaches to data management appears to be the only means by which the existing gap can at least be minimized".
b) "Various programmes on environmental management, e.g., World Climate (WRCP) and Geosphere-Biosphere (IGBP) programmes, Cooperative Programme for Monitoring and Evaluation of the Long-Range Transmission of Air Pollutants in Europe (EMEP), Global Environmental Facility (GEF), United Nations Environment Programme (UNEP), World Weather Watch (WWW), and the similar, have a multidisciplinary regional or global character. They need strengthening of collaboration between data management activities of different organizations to ensure

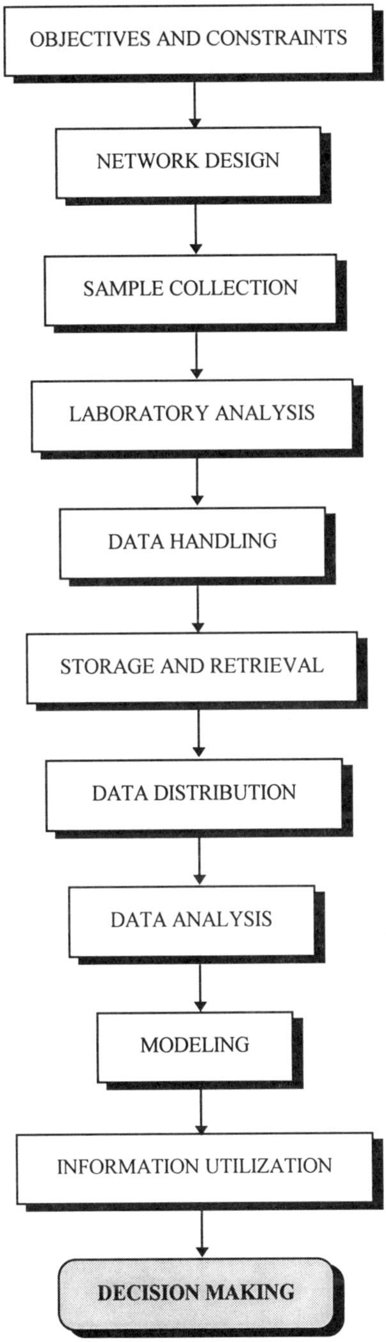

Figure 1.1. Basic steps in environmental data management

proper coordination of environmental data collection, data flow, and archiving and to avoid duplication of efforts on both national and international levels. Such collaboration can only be realized by integrated approaches to data management".

c) "The solution to environmental problems often requires data exchange at local, national, and global (international) levels. Such an exchange may be needed for:
   1) data of the same type, e.g., water quality data collected by different methods;
   2) data of different types of one discipline, e.g., marine physical, chemical, biological, and other oceanographic data types; and
   3) data of different disciplines, e.g., oceanographic, meteorological, geophysical, or demographic data.

   We live in a decade when computer and communication technologies have made significant advances in terms of technical capability and connectivity. Such advances facilitate data exchange on various levels; however, they also impose significant demands on our capacity to handle environmental data so that information flow can be properly realized at local, regional and global levels. The development of computer and communication technologies have changed fundamentally the way in which data and information can be managed and made available. These demands imply the requirement for integrated approaches to data handling".

1.3.3. BASIC ELEMENTS OF DATA MANAGEMENT SYSTEMS

Environmental data management systems comprise the basic steps outlined in Fig. 1.1. Here, the ultimate goal of the system is decision making for environmental management. The key to proper management decisions is information on environmental processes, and retrieval of this information relies on data to be collected, analyzed and evaluated.

Figure 1.1 shows that the two basic tools of integrated environmental management, i.e., modeling and data, can be integrated in the data management system. In essence, modeling is the stage where data are transferred into information for the eventual decision making process. Thus, it constitutes a significant component of the environmental data management system.

On the other hand, production of the desired information from available data is a difficult task; it is subject to numerous uncertainties and problems in the collection, processing, handling, analysis, and interpretation of data. Thus, management of the system of activities shown in Fig. 1.1 has become an end in itself apart from the management of the environment.

The major difficulty associated with the current data management systems relates to deficiencies in defining specific objectives for monitoring. Constraints in the form of social, legal, economic, and administrative factors complicate this step further (Alpaslan, 1997). Essentially, lack of clearly stated objectives implies failure to define information expectations so that, eventually, the data management system cannot produce the information required for decision making. In this case, one may consider not to collect any data for which the objective is not specified.

With respect to the design of data collection programs, there are yet no standard guidelines to be followed in the design of monitoring programs. Basic problems relate to the selection of sampling sites, frequencies, variables and sampling duration. When these network features are not properly selected, the efficiency of the monitoring network is significantly reduced (Harmancioglu *et al.*, 1992; Harmancioglu and Alpaslan, 1994).

The major difficulty in physical sampling relates to realization of representative sampling. Furthermore, the selection of proper tools and equipment for sampling may complicate the problem particularly in case of equipment failures. Sampling has to be followed by proper preservation of sampling, and timely and safe transport to the laboratories. These activities, if not appropriately realized, may lead to poor samples (Alpaslan, 1997).

Laboratory analyses result in significant uncertainties due to lack of standardization among laboratories with respect to analysis methods and units used. There is a significant need for reference laboratories. Furthermore, laboratory analyses must include quality control/quality assurance of available samples, which are not properly realized in most laboratories. This issue significantly hinders exchange of data on local, regional, and global levels (Timmerman *et al.*, 1996).

With respect to storage of data, most developed countries have well-established databases which can be accessed easily by the users. The main problem here is that data banks have been filled up with huge amounts of data; and there is the question of what should be done with too many data. Developing countries either have no data banks or have poor databases which are hardly accessible by the users. The main problem related to data banks is the appropriateness of formats with which the data are stored. Again, there is a need for harmonization or standardization in development of databases so that data exchange can be facilitated on regional and global levels (Alpaslan, 1997).

Data analysis is the initial step of transferring data into information. There are numerous analysis methods proposed by different researchers. The problem is to select the best one among them. Modeling, as a means of data analysis, has its own uncertainties and complexities. Models often prove to be unsatisfactory when the underlying mechanisms of environmental processes are not fully and reliably perceived. Another difficulty related to data analyses is that the messy character of environmental data require special treatment via modified or new techniques. These methods have been developed, but they have not yet been validated to the fullest extent (Alpaslan, 1997).

It follows from the above that each step of the data management system has its own difficulties and uncertainties such that the resulting data are often of a messy character with deficiencies in both quantity and quality. Actually, each task in the system contributes to data utility; problems in any one step reduces the reliability of the output information. Thus, to improve the status of existing data management systems, these problems should be solved, or at least minimized. Second, the system should be viewed as a cohesive whole since the output of one step constitutes the input to the next step. Coordination of data flow among these steps is often difficult since each task is performed by a different discipline. Thus, agreement should be established between multidisciplinary approaches if current data management systems are to be improved.

1.4. Purpose and Scope of This Book

The above sections have emphasized two basic factors that underline current practices of environmental management, namely that:

a) the environment should be managed via integrated approaches, for which a reliable and adequate information system is a prerequisite;
b) current environmental data management systems reflect a declining trend, i.e., there are numerous problems and uncertainties to be resolved in each step of the system.

These two factors essentially indicate the need to assess and revise existing data management systems. It is this need that gave impetus to the current work. It is intended herein to contribute to the ongoing efforts on local, regional and global levels towards improvement of environmental data management systems.

As described in the previous sections, recent requirements for integrated environmental management have led to the evolution of both modeling procedures and data management systems. Regarding these advances, current literature provides substantial work on modeling and data analysis relevant to different environmental processes. However, issues related to data management as a system are barely touched in a comprehensive framework. Data requirements, data availability and diverse modeling approaches are covered as specific topics in most environmental studies. Thus, the need remains for more publications that focus exclusively on data management systems. The present book intends to fill an important gap in this respect by covering various aspects of environmental data management in a systematic approach. Current status of data management systems are reviewed in order of the basic steps summarized in Fig. 1.1. Basic approaches employed and difficulties encountered in each step are presented for air, solid waste and water, which constitute the major areas of environmental pollution.

Chapter 2,3 and 4 concentrate on collection of data on air quality, solid waste and water quality, respectively. The remaining chapters focus on basic steps of data management in a sequential order. These chapters concentrate primarily on water quality as water is essentially the largest area where environmental pollution occurs. Furthermore, basic approaches and methods employed for water quality data management are valid in general for all other environmental processes. Specific conditions that pertain to such processes are noted wherever appropriate within the text. A separate section for soil and land resources is not presented; these environmental components interact closely with water so that they are covered as part of data management practices for water quantity and quality. It must also be noted that groundwater is not considered exclusively in the text as this environmental component has unique features that have to be evaluated by specific approaches and methodologies. Nevertheless, some aspects of data management are common to both surface and ground waters, and these are referred to in appropriate sections of the text. Essentially, the last chapter on decision making for environmental management is covered in a general framework that represents the environment as a continuum of different components.

References

Alpaslan, M.N. (1997) Prevailing problems in environmental data management, in N.B. Harmancioglu, M.N. Alpaslan, S.D. Ozkul and V.P. Singh (eds.), *Integrated Approach to Environmental Data Management Systems*, Proceedings of the NATO Advanced Research Workshop on Integrated Approach to Environmental Data Management Systems, September 16-20, 1996, Izmir, Turkey, Kluwer Academic Publishers, NATO ASI Series, 2. Environment, vol. 31, pp. 15-22.

Clark, M. J.; Gardiner, J. (1994): Strategies for handling uncertainty in integrated river basin planning, in C. Kirby & W.R. White (eds.), *Integrated River Basin Development*, John Wiley & Sons, pp. 437-445.

Geerders, P. J.F. (1997) Nature's data and data's nature, in N.B. Harmancioglu, M.N. Alpaslan, S.D. Ozkul and V.P. Singh (eds.), *Integrated Approach to Environmental Data Management Systems*, Proceedings of the NATO Advanced Research Workshop on Integrated Approach to Environmental Data Management Systems, September 16-20, 1996, Izmir, Turkey, Kluwer Academic Publishers, NATO ASI Series, 2. Environment, vol. 31, pp. 49-60.

Harmancioglu, N.B. (1997) The need for integrated approaches to environmental data management, in N.B. Harmancioglu, M.N. Alpaslan, S.D. Ozkul and V.P. Singh (eds.), *Integrated Approach to Environmental Data Management Systems*, Proceedings of the NATO Advanced Research Workshop on Integrated Approach to Environmental Data Management Systems, September 16-20, 1996, Izmir, Turkey, Kluwer Academic Publishers, NATO ASI Series, 2. Environment, vol. 31, pp. 3-14.

Harmancioglu, N.B.; Alpaslan, M.N. and Ozkul, S.D. (1997a) Conclusions and recommendations, in N.B. Harmancioglu, M.N. Alpaslan, S.D. Ozkul and V.P. Singh (eds.), *Integrated Approach to Environmental Data Management Systems*, Proceedings of the NATO Advanced Research Workshop on Integrated Approach to Environmental Data Management Systems, September 16-20, 1996, Izmir, Turkey, Kluwer Academic Publishers, NATO ASI Series, 2. Environment, vol. 31, pp. 423-436.

Harmancioglu, N.B.; Alpaslan, M.N.; Ozkul, S.D. and Singh, V.P. (eds.) (1997b) *Integrated Approach to Environmental Data Management Systems*, Proceedings of the NATO Advanced Research Workshop on Integrated Approach to Environmental Data Management Systems, September 16-20, 1996, Izmir, Turkey, Kluwer Academic Publishers, NATO ASI Series, 2. Environment, vol. 31, 546 p.

Harmancioglu, N.B. and Alpaslan, N. (1994) Basic approaches to design of water quality monitoring networks, Elsevier, *Water Science and Technology* **30-10**, 49-56.

Harmancioglu, N.B.; Alpaslan, N.; Alkan, A.; Ozkul, S.; Mazlum, S. and Fistikoglu, O. (1994) *Design and Evaluation of Water Quality Monitoring Networks for Environmental Management* (in Turkish), Report prepared for the Research Project granted by TUBITAK (Scientific and Technical Research Council of Turkey), Project code: DEBAG-23, January 1994, Izmir.

Harmancioglu, N.B.; Alpaslan, N. and Singh, V.P. (1992) Design of water quality monitoring networks, in R. N. Chowdhury (ed.), *Geomechanics and Water Engineering in Environmental Management*, ch. 8, pp. 267-296.

Harmancioglu, N.B.; and Alpaslan, N. (1992) Water quality monitoring network design: a problem of multi-objective decision making, *AWRA, Water Resources Bulletin* **28-1**, 179-192.

LGMB (1992) *Agenda 21: A Guide for Local Authorities in the UK*, The Local Government Management Board, England.

Oliounine, I. (1997) Integrated approach - A key to solving global problems, in N.B. Harmancioglu, M.N. Alpaslan, S.D. Ozkul and V.P. Singh (eds.), *Integrated Approach to Environmental Data Management Systems*, Proceedings of the NATO Advanced Research Workshop on Integrated Approach to Environmental Data Management Systems, September 16-20, 1996, Izmir, Turkey, Kluwer Academic Publishers, NATO ASI Series, 2. Environment, vol. 31, pp. 61-66.

Santos, M. (1997) Data management and the European Union information policy, in N.B. Harmancioglu, M.N. Alpaslan, S.D. Ozkul and V.P. Singh (eds.), *Integrated Approach to Environmental Data Management Systems*, Proceedings of the NATO Advanced Research Workshop on Integrated Approach to Environmental Data Management Systems, September 16-20, 1996, Izmir, Turkey, Kluwer Academic Publishers, NATO ASI Series, 2. Environment, vol. 31, pp. 35-48.

Singh, V. P. (1995) What is environmental hydrology?, in V. P. Singh (ed.), *Environmental Hydrology*, Kluwer, Water Science and Technology Library, ch. 1, pp. 1-12.

Timmerman, J.G.; Gardner, M.J. and Ravenscraft, J.E. (1996) *Quality Assurance*, UN/ECE Task Force on Monitoring and Assessment, Working Programme 1994/1995, vol. 4, RIZA report no.: 95.067, Lelystad, January 1996, 119 p.

Tyson, J.M. (1995): Quo Vadis - Sustainability? Pergamon, *Water Science and Technology*, **32**, 5-6, pp. 1-5.

UN (1992) *Agenda 21: Programme of Action for Sustainable Development*, United Nations, New York, NY, USA.

Whitfield, P.H. (1988) Goals and data collection design for water quality monitoring, *Water Resources Bulletin, AWRA* **24**, 775-780.

WMO (1994) *Advances in Water Quality Monitoring - Report of a WMO Regional Workshop (Vienna, 7-11 March 1994)*, World Meteorological Organization, Technical Reports in Hydrology and Water Resources, No. 42, WMO/TD-NO 612, Geneva, Switzerland, 332 p.

Acronyms and Abbreviations

EDMED	European Directory of Marine Environmental Data (EC-MAST)
EEA	European Environment Agency (EC)
EMEP	Cooperative Programme for Monitoring and Evaluation of the Long-Range Transmission of Air pollutants in Europe
FAO	Food and Agriculture Organization (UN)
GCM's	General Circulation Models
GCOS	Global Climate Observing System (WMO)
GEF	Global Environmental Facility
GEMS	Global Environment Monitoring System
GOOS	Global Ocean Observing System (IOC)
GRID	Global Reference Information Database (UNEP)
GTOS	Global Terrestrial Observing System (UNEP)
ICSU	International Council of Scientific Unions
IGBP	International Geosphere-Biosphere Project
IOC	Intergovernmental Oceanographic Commission (UNESCO)
LGMB	UK Local Government Management Board
MEDI	Marine Environmental Data Inventory (IOC)
UN	United Nations
UNCED	United Nations Conference on Environment and Development
UNEP	United Nations Environment Programme
WHYCOS	World Hydrological Cycle Observing System (WMO and World Bank)
WMO	World Meteorological Organisation
WRCP	World Climate Programme
WWW	World Weather Watch (WMO)

CHAPTER 2

AIR QUALITY MODELING AND MEASUREMENT

G.C. Edwards and P.K. Misra

Abstract. Air quality modeling and measurement is a relatively new science with most of the concepts and theories having been developed over the past century and many of these over the past two decades. There have been many books and articles written on the topic. The following chapter by no stretch of the imagination is able to summarize all the knowledge on this subject. The authors therefore have attempted to provide the reader with an overview of some of the key concepts in modeling air quality and its measurement. The chapter starts out on the topic by looking at some of the fundamental atmospheric physics concepts associated with air quality modeling. This is followed by an overview of some of the important short range dispersion modeling approaches commonly in use today. The second half of the chapter is devoted to providing information on measurement methods and instrumentation typically used in support of air quality modeling.

2.1. Introduction

A fluid (gas or liquid) is an aggregation of molecules which are in constant chaotic motion. The mechanics of a fluid relates to the equations of motion of a macroscopic element of the fluid, which is large enough to consist of a large number of molecules, but small enough to be regarded as a mathematical point mass (Batchelor, 1967). It is also possible to derive the equation of motion of fluid flow by the statistics of molecular motion (Csanady, 1967). In either method, the influence of the chaotic motion of the molecules is included in the surface forces (such as the shear stresses) acting on the fluid element.

The equations describing the fluid motion including shear stresses are generally known as the Navier-Stokes equation (Currie, 1974). These equations follow Newtonian mechanics, and for most applications, the solutions are conveniently obtained when treated in an Eulerian framework; i.e., with respect to a fixed coordinate system to the observer.

One of the distinctive characteristics of the Navier-Stokes equations is that they are unstable to small disturbances (Chandrasekhar, 1968). Therefore, the amplitudes of any small disturbances introduced into the fluid flow will grow in time, eventually leading to a fully turbulent flow when the flow Reynolds number exceeds a critical value (the reader is referred to Monin and Yaglom, (1971) for a full description of turbulent flow).

N.B. Harmancioglu et al. (eds.), Environmental Data Management, 13-48.

Reynolds number is defined as:

$$R = \frac{UL}{\nu} \tag{2.1}$$

where U is a characteristic velocity, and L a characteristic length of the flow, and ν the kinematic viscosity of the fluid.

In a typical turbulent flow, energy is introduced from the mean flow at low wave number (large wave length). The nonlinear nature of the Navier-Stokes equations results in this energy to be passed down to the higher wave numbers. This continues until, at the highest wave number, fluid viscosity becomes important and the energy is dissipated into heat. This is known as the cascading of energy in a turbulent fluid flow. It is therefore plausible that turbulent energy at a range of wave numbers higher than the energy producing wave numbers, but lower than at the viscous end, will depend only on the rate at which energy is dissipated. Kolmogoroff (1935) was the first to recognize this possibility and postulated the velocity and length scales corresponding to this range of wave numbers, which is generally called the *inertial* range. Numerous experiments both in the laboratory and in the atmosphere have verified Kolmogoroff's postulates (Businger *et al.*, 1971; Kaimal *et al.*, 1994).

A typical velocity scale of the atmospheric fluid flow is 10 m s^{-1}, and a length scale near the surface is 1 km. Therefore, the Reynolds number of atmospheric boundary layer flow is typically 10^8 which is significantly larger than the critical Reynolds number at which laboratory fluid flows reach criticality for flow instability (Chandrasekhar, 1968). The atmospheric boundary layer flow can, therefore, be assumed to be always in a state of turbulence. The state of atmospheric flows in general can be said to be turbulent. However, in the free atmosphere, the nature of turbulence is different from the boundary layer.

It is well known that the random motion of the molecules diffuse particles released into a static fluid at a point. Similarly, the random nature of turbulent flow causes the particles released at a point to be diffused. The total diffusion is a combination of the molecular and turbulent diffusion. However, the magnitude of turbulent diffusion is significantly larger than the molecular diffusion so that the latter can be neglected relative to the former for most problems.

This chapter deals with the science of atmospheric turbulent diffusion. It is assumed that the reader has a basic understanding of the science of turbulent fluid flow. Details on turbulent fluid flow can be found in texts such as Monin and Yaglom (1973), Csanady (1973), Tennekes and Lumley (1972).

2.2. Conservation of Mass

Turbulent diffusion is the subject of the determination of the space-time distribution of mass of a group of particles released into a turbulent fluid medium. It is assumed that the introduction of the particles does not alter the state of the turbulent fluid. Chemical reactions, however, are allowed to occur.

It is appropriate to revisit the derivation of the conservation of mass equations in order to fully appreciate the nature of turbulent diffusion. The conservation of mass is derived in two different ways, i.e., from a fixed coordinate system (Eulerian) and a moving coordinate system (Lagrangian) to bring out the subtlety of the problem.

2.2.1. DERIVATION OF THE CONSERVATION OF MASS EQUATION: EULERIAN FRAMEWORK

Let a group of particles be distributed inside a fluid with a concentration distribution $c(\vec{r}\,,\,t)$. Let us consider a fixed volume V enclosed by a surface S inside the flow domain. Let $\hat{n}$ represent the unit normal at any point $(\vec{r}\,)$ on the surface of this volume. Let $u(\vec{r}\,,\,,t)$ be the flow velocity vector at this point. Therefore, the net mass of a scalar quantify entering the volume at a given time interval, Δt, (i.e., mass entering minus the mass exiting through the surface of the volume) is given as follows:

$$\Delta M = \Delta t . \int_s c\vec{u}.\hat{n}\, ds \tag{2.2}$$

Also,

$$\Delta M = \int_V (\Delta c)\, dV \tag{2.3}$$

Substituting Eq. (2.3) into Eq. (2.2) we get,

$$\int_V \left(\frac{\Delta c}{\Delta t}\right) dV = \int_s c\, \vec{u}.\hat{n}\, ds \tag{2.4}$$

If we make Δt arbitrarily small, then using Gauss's integral theorem on to the surface integral on the right hand side of Eq. (2.4) we get

$$\int \frac{\partial c}{\partial t} \delta V = -\int_\upsilon \nabla.\, (c\, \vec{u})\, dV \tag{2.5}$$

The negative sign on the right hand side of Eq. (2.5) renders the divergence positive and convergence negative. Since V was chosen arbitrarily, the identity of Eq. (2.5) is satisfied only if

$$\frac{\partial c}{\partial t} + \nabla.\, (C\vec{U}) = 0 \tag{2.6}$$

This is the required conservation mass equation.

2.2.2. LAGRANGIAN FORM

Assume a volume element δv inside the fluid. At a given time t, it has a mass of the particles given by the following:

$$M = \int_{\delta V} c\, dV \tag{2.7}$$

where the integral is taken over the volume δv. Since mass is conserved,

$$\frac{dM}{dt} = 0 \tag{2.8}$$

Therefore, substituting (2.8) into (2.7) we get:

$$\frac{d}{dt}\int_{\delta V} C\, dV = 0 \tag{2.9}$$

If, at some initial time t_o, the magnitude of this elemental volume was δv_o, then Eq. (2.9) can be written as:

$$\frac{d}{dt}\int_{\delta V_o} C\,(\frac{\partial V}{\partial V_0})\, d\,V_0 = 0 \tag{2.10}$$

where

$$\frac{\partial V}{\partial V_o} = (\frac{\partial x}{\partial x_o}\ \frac{\partial y}{\partial y_o}\ \frac{\partial z}{\partial z_0})$$

is called the Jacobian J in coordinate transformation:

$$\frac{d\,J}{\partial\, t} = (\nabla.\ \vec{u})\ J \tag{2.11}$$

Substituting (2.11) into (2.10), we obtain the following:

$$\int_{\delta V}\left\{\frac{\partial\, c}{\partial\, t} + c(\nabla.\vec{u})\right\} dV = 0 \tag{2.12}$$

Since δv is chosen arbitrarily, Eq. (2.12) is an identity only if

$$\frac{\partial\, c}{\partial\, t} + c(\nabla.\ \vec{u}) = 0 \tag{2.13}$$

This equation is the same as (2.6).

2.3. Concentration Distribution in a Turbulent Fluid

Equation (2.6) or (2.13) correctly describes the variation of the concentration field of particles inside a fluid and, with suitable initial and boundary conditions solutions to (2.6) or (2.13), will provide the distribution of concentration in space and time c($\vec{r}$, t). C($\vec{r}$, t) will depend on the nature of the velocity field $\vec{u}$. Inside a turbulent fluid, $\vec{u}$ exhibits random fluctuations, the details of which are not easily obtained either through measurements or through solutions to the momentum equations. For most practical problems, it is sufficient to express the velocity fields by their various moments. Often, only the mean value of $\vec{u}$ (i.e., its first moment) and its variance σ_u^2 (i.e., its second moment) provide adequate information to describe the statistical state of the velocity field.

Fluctuations in $\vec{u}$ inevitably lead to fluctuations in the concentration field $C(\vec{r}\,,\,t)$. If one makes a continuous measurement of the particle concentrations inside a turbulent fluid at a fixed location in space, then the resulting values will show peaks and valleys randomly distributed on a temporal chart. A convenient way to organize these concentration variations in time is to determine the various momenta of the concentration field. Towards this end, the following statistical parameters are defined:

$$\bar{c}(\,\vec{r},\,t) \;=\; \frac{1}{t}\,\int_o^t c(\tau)\,d\tau \tag{2.14}$$

$$c \;=\; \bar{c} \;+\; c' \tag{2.15}$$

$$\sigma_c^n \;=\; \frac{1}{t}\,\int_o^t (\,c - \bar{c}\,)^{\,n}\;d\tau \tag{2.16}$$

Here, an overbar refers to the mean value, and σ_c^n is the nth moment of C about its mean value.

We notice from the above definitions that $\bar{c}$ and σ_c^n are functions of the averaging time t. In order for these quantities to be useful in a statistical sense, we require them to be 'stable' in the sense that they do not vary significantly with our choice of the averaging time. This is possible in atmospheric flows only for a narrow range of conditions where the flow is stationary and homogeneous. In these situations, one can assume that, for an averaging time t, sufficient number of 'equivalent' conditions would have been summed so that increasing the averaging time further would not substantially alter the resulting values $\bar{c}$ and σ_c^n. It is important to understand the term 'equivalent' conditions in our averaging procedures.

In a sense, we are trying to determine the 'ensemble' average by simulating these conditions. Although there is still some debate on the exact nature of the solutions to the Navier-Stokes equations when the fluid is turbulent, it is assumed that, in the turbulent phase of the fluid flow, a multitude of solutions are possible for the same initial conditions of flow variables. Thus, carrying out the averaging over all feasible solutions for the same initial conditions, one obtains the ensemble averages of flow variables, which are stable and can be represented in some relations to each other.

An explanation of the condition where the initial conditions are kept fixed is in order here. One can conceptualize an infinite number of physical variables which can adequately describe a state. In order for this state to be repeated, for example in an experiment, to obtain the various possible solutions, one must clearly define all these variables. Also, all these variables must be measurable for the reproducibility of the initial condition. In reality, this is rarely the case. Experiments carried out in a control environment of a wind tunnel, perhaps, allow this possibility to exist. In the atmospheric flows, however, no two events separated in space and/or time can be assumed to be identical. Therefore, within the averaging time period 't' as defined in Eqs. (2.14) to (2.16), if one obtains 'n' events, then they do not necessarily belong to the same ensemble, i.e., they are not necessarily solutions resulting from the same initial conditions.

The concept of stationarity and homogeneity can only be approximately valid for atmospheric flows if at all. This has important implications in obtaining solutions to (2.6) or (2.13) and employing the statistical definitions (2.14) to (2.16). In spite of these difficulties, however, we will employ these statistics with a view that the resulting solutions are at least approximately correct.

2.4. Solution to Mass Conservation Equation in a Turbulent Fluid Medium

Using the definitions of (2.14), (2.15) and (2.16) let us simplify Eq. (2.6) into an equation for the ensemble mean $\bar{c}$. We decompose the velocity vector $\vec{u}$ as follows:

$$\vec{u} = \bar{\vec{u}} + \vec{u}' \qquad (2.17)$$

Here, $\bar{\vec{u}}$ is the ensemble mean value of the velocity vector and $\vec{u}$ its fluctuation.

Substitution of Eqs. (2.15) and (2.17) into (2.6) yields the following:

$$\frac{\partial \bar{c}}{\partial t} + \frac{\partial c'}{\partial t} + \nabla.(\bar{\vec{u}}\,\bar{c} + \bar{\vec{u}}\,c' + c\,\vec{u}' + \overline{\vec{u}'c'}) = 0 \qquad (2.18)$$

If we take the ensemble average of Eq. (2.18), we get the following:

$$\frac{\partial \bar{c}}{\partial t} + \nabla.(\bar{\vec{u}}\,\bar{c}) + \nabla.(\overline{\vec{u}'c'}) = 0 \qquad (2.19)$$

It is apparent from (2.19) that turbulence leads to an extra flux term $\overline{\vec{u}'c'}$ which must be included in the equation for the ensemble average concentration field $\bar{c}$.

One of the properties of turbulent flow is that, when the conservation equations are derived for the various momenta of the appropriate variables, the equation for the nth moment of the variable contains terms higher than the nth moment. For instance, the equation for $\bar{c}$ as shown in (2.19) contains the second order terms $\overline{\vec{u}'c'}$. Therefore, the equations for the various momenta cannot be written in a closed form. In other words, we need to define $\overline{\vec{u}'c'}$ as functions of $\bar{c}$ in order to be able to solve (2.19) in a closed form.

If turbulent flows are visualized as consisting of eddies of various sizes, and the eddies are assumed to behave as molecules in an ordinary fluid, then one can extend the molecular mass transfer concepts to turbulent mass transfer. This leads to the following parameterization of $\overline{\vec{u}'c'}$:

$$\overline{\vec{u}'\, c'} \;=\; -\, k\, \nabla\, \bar{c} \tag{2.20}$$

Here, k is the 'eddy diffusivity' and the above parameterization is termed Fickian diffusion (Csanady, 1973). The negative sign on the right hand side of (2.20) ensures that diffusion proceeds from the higher values of $\bar{c}$ to lower values.

In Fickian diffusion, it is assumed that dispersion of particles in a turbulent medium depends only on the local gradients of the mean concentration. It is not immediately obvious why this should be the case. Further, as will be shown later, the rate of diffusion of particles at a point $\{\vec{r}\}$ in space depends on the points of origin $\{\vec{r}_o\}$ of these particles. Therefore, k as defined in (2.20) will depend on $\{\vec{r}_o\}$. This presents a conceptual difficulty in the definition of k for seeking solution to $\bar{c}$.

In spite of these difficulties, some insight into the turbulent dispersion processes can be gained by using (2.20) in (2.19). Substituting (2.20) into (2.19) we get:

$$\frac{\partial \bar{c}}{\partial t} + \nabla .\, (\vec{u}\, \bar{c}) \;=\; \nabla .\, (k\, \nabla\, \bar{c}) \tag{2.21}$$

Let us consider the case of a continuous point source where particles are emitted at a rate of Q g/s into a homogeneous and stationary atmosphere with a constant mean wind speed U along the X direction. Let us also assume that k is a constant and C goes to zero as y and z go to infinity.

For this problem, the solution to (2.21) is given as follows (Pasquill and Smith, 1983):

$$\bar{c} \;=\; \frac{Q}{2\pi u\, \sigma_y \sigma_z}\; exp\left[\; \frac{(y-\bar{y})^2}{2\, \sigma_y^z} \;\; \frac{(z-\bar{z})^2}{2\, \sigma_z^2} \; + \; \frac{(z+\bar{z})^2}{2\, \sigma_z^2} \right] \tag{2.22}$$

where

$$\sigma_y \;=\; \sigma_z \;=\; \sqrt{2kt}$$

and $\bar{y}$ and $\bar{z}$ refer to the mean position of the source.

It is emphasized that the solution resembles a Gaussian distribution where the standard deviations σ_y and σ_z are equal to $\sqrt{2kt}$. The shape of the plume of the particles, therefore, grows in the cross-wind directions in time. Thus, the mean concentration $\bar{c}$ will decrease as we move away from the source. This is an important conclusion as there is no possibility of the particles to concentrate and create an increase in $\bar{c}$ at any point in space away from the source.

Several authors (Venkatram and Wyngaard, 1988) have attempted to find solution to Eq. (2.20) for various assumptions on spatial dependence of k. It is not beneficial to pursue this line of investigation in view of the conceptual difficulties with k.

For the sources located very close to earth's surface for atmospheric flows, however, it has been found that eddy diffusivity parameterization as in (2.20), and with suitable variation in k with height, can yield useful solutions. This will be discussed later in the chapter.

2.5. Statistical Treatment

Taylor (1921) provided the basis for the statistical treatment of turbulent diffusion. This is reproduced here to clarify the statistical methods which will be dealt with in this section.

Consider a point source of particles at a point in space, which is taken to coincide with the origin of a coordinate system for convenience. Each particle will move away from the point of release with the velocity field $\bar{u}\ (\bar{r}, t)$ of the flow.

The trajectory of a particle inside the flow domain is therefore given by the following:

$$x(t) = \int_o^t u_x\ (t')\, d\,t' \tag{2.23}$$

$$y(t) = \int_o^t u_y\ (t')\, d\,t' \tag{2.24}$$

$$z(t) = \int_o^t u_z(t')\, d\,t' \tag{2.25}$$

Here, subscripts x, y, z refer to the x, y and z components of the velocity field. Let us consider only one component x. The derived results can be easily extended to the other components.

The trajectory of a single particle is obviously a random path. Therefore, $x(t)$, $y(t)$, and $z(t)$ are random variables.

Now

$$u_x\ (t) = \frac{dx(t)}{dt} \tag{2.26}$$

Therefore,

$$\frac{dx(t)}{dt}\ .\ x(t) = u_x(t) \int_o^t u_x(t')\, d\,t' \tag{2.27}$$

The right hand side of (2.27) can be written as

$$\int_o^t u_x\ (t)\ u_x(t')\ d\,t'$$

Therefore, we get:

$$x(t\,)^2 \;=\; 2 \int_o^t \int_o^\tau u_x(t)\ u_x(t')\ d\,t'\ d\tau \tag{2.28}$$

If we take the ensemble averaging of both sides of (2.28), we get:

$$\overline{x\,(t)}^2 \;=\; z \int_o^t \int_o^\tau \overline{u_x(t)\ u_x(t')}\ d't\ d\tau \tag{2.29}$$

If the turbulent flow under consideration is stationary and homogeneous, then the right hand side of (2.29) can be expressed in a simplified form. For example, let us define the auto-correlation function of u_x as follows:

$$R_{u_x}\ (t,t') \;=\; \frac{\overline{u_x(t)\ u_x(t')}}{\sigma_{u_x}{}^2} \tag{2.30}$$

Here $\sigma_{ux}{}^2$ is the variance $(u_x - \bar{u}_x\ )^2$ of u_x. Then, for a stationary and homogeneous turbulent field, we can assume that the auto-correlation function will be independent of the origin, and will depend only on the difference $(t\text{-}t')$. Further, $R_{ux}\ (t\text{-}t')$ goes to unity when $(t\text{-}t')$ goes to zero, and it goes to zero when $(t\text{-}t')$ goes to infinity.

If we denote $\{\ \overline{x\ (t)}^2\ \}$ as σ_x^2, we note that for very small travel times,

$$\sigma_x^2 \approx\ \sigma_{Ux}^2\ t^2 \tag{2.31}$$

and for very large travel times,

$$\sigma_x^2 \approx\ \sigma_{Ux}^2\ t \tag{2.32}$$

It is noted that

$$\int_o^\infty R\,u_x\ (t-t')\ d\,(t-t') \;=\; T \tag{2.33}$$

where T is the time scale. Therefore, small travel times refer to, $t << T$, and large travel times refer to $t >> T$.

As noted earlier, similar expressions for σ_y^2 and σ_z^2 can be obtained by using the above mathematical steps. The expressions for σ_x^2, σ_y^2 and σ_z^2 imply that they monotonically increase with time. Therefore, concentrations of particles released at a point into a turbulent fluid will always decrease with time.

The asymptotic behavior of σ_x^2, σ_y^2 and σ_z^2 are always valid. In atmospheric turbulence, however, stationarity of the flow is valid only for a short time period. Therefore, the $t^{1/2}$ variation of the variance of particle position for large travel times is usually not observed for atmospheric flows.

As mentioned earlier, Kolmogoroff's energy cascading arguments pointed to the existence of an inertial range of turbulent energy, where turbulent scales are determined by the energy dissipation rate. Batchelor (1967) argued that the mean shape σ of a group of particles released at an instant of time into a turbulent flow, as it is advected downstream, will increase initially as t. As the shape of the particles becomes comparable to the eddy sizes belonging to the inertial range, these eddies will be the most effective in diffusing it. Using dimensional arguments, he showed that σ in this stage will increase as $t^{3/2}$ and, with increasing time, σ will increase as $t^{1/2}$.

The diffusion of a group of particles, also called a puff of particles being advected downstream with the mean flow, is called instantaneous diffusion. When we average over a large number of these puffs over a long time period, we get a plume. The diffusion of a plume is called continuous diffusion.

For a puff, σ will increase initially as t, then as $t^{3/2}$ at an intermediate stage when the inertial range eddies are dominant, and as $t^{1/2}$ for large elapsed time. For a continuous plume, σ will increase initially as t and as $t^{1/2}$ for large elapsed time. A continuous plume does not experience the influence of the eddies in the inertial range in the same way as an instantaneously released puff.

The usefulness of the above analyses lies in the formulation of the space-time variation of $\bar{c}$. For instance, if the probability distribution function of the position of particles at a distance (x, y, z) from their point of release is $p(x,y,z)$, then the mean concentration distribution is given by the following expression:

$$\bar{c}\,(x,y,z) = \frac{p(x,y,z)\,Q}{U} \tag{2.34}$$

where Q is the release rate of the particles and U the mean velocity of the flow. If $p(x,y,z)$ can be expressed as a function of σ_x, σ_y, σ_z alone, then $\bar{c}\,(x,y,z)$ is easily defined. It is then a matter of defining σ_x, σ_y, σ_z as suitable functions of $t = (x/u)$ for the problem at hand. Typically, experimental data are used to define σ_x, σ_y and σ_z. This is conceptually more appealing than solving the mass conservation equation as it avoids any ad-hoc assumptions on the value of the eddy diffusivity k.

It is not easy to define $p(x,y,z)$ from theoretical considerations (Monin and Yaglom, 1973). In some cases, it may be linked to the probability density function of the velocity. For example, for near-field diffusion, the probability of finding a particle at a point X after its release from a point x_0 can be assumed to be linked to the probability density function $p_u\,(x_0)$ of the velocity at point x_0 by the following equation:

$$p(x) = p_u\,(x_o)\left|\frac{dy}{dx}\right| \tag{2.35}$$

where x is a single valued function of u.

It is a clear that the probability density functions p and the associated parameters σ_x, σ_y, σ_z will depend on the state of turbulence of the fluid medium. In the next section, a brief description of the state of turbulence of the atmospheric flows is given.

2.6. Atmospheric Boundary Layer

An active and ongoing area of research is the study of the Atmospheric Boundary Layer (ABL). Information on its characteristics and behavior have been primarily gathered over the last three decades. Most of what is known of the ABL is confined to the simplest state, that is its structure and characteristics over open and flat terrain. For more in-depth information than will be offered here on what is known about the ABL, the reader is referred to several other monographs: Kaimal and Finnigan (1994), Fleagle and Businger (1980), Venkatram and Wyngaard (1988), and Sorbjan (1989).

For atmospheric flows, it is generally accepted that the boundary layer is defined as the domain next to earth's surface with significant three dimensional turbulence. The height of this domain is a function of solar radiation and wind speed and shows a strong diurnal variation. For example, in the temperate climate zones, this height varies between 1 - 2 km in the spring and summer months.

Turbulent energy within this layer is generated by both thermal and frictional forces. The thermal effect is defined by the heat flux, at the surface, which can be positive (upward from a heated surface) or negative (downward towards a cold surface) depending on the solar radiation and ground cover. The effect of surface friction is to create shear between successive layers of the atmosphere upward from the surface. It will be shown later that it is generally convenient to classify the state of the turbulence within the atmospheric boundary layer on the basis of the ratio of shear generated turbulence to thermally generated turbulence.

Another characteristic of the ABL are its stability states. These states are typically described by classes; i.e., unstable stability when heat flux is positive and thermally generated turbulence is dominant; stable when the heat flux is negative and turbulence is generated by shear and destroyed by negative heat flux; and neutral when surface heat flux is negligible and shear generated turbulence is dominant. Other schemes are available, for example atmospheric stability was divided into 7 classes by Gifford (1976). In any case, these stability states are used as indicators of the dispersive capacity of the ABL.

Other parameters often used in relation to the ABL are velocity and length scales associated with the prevalent turbulent state. Shear results in gradients in the wind speed. These gradients tend to be stronger at the surface since the no-slip boundary condition applies there. In the layer close to the surface, the velocity scale is called u_* for the shear generated turbulence and is defined from the surface shear stress τ_o to as follows:

$$p u_*^2 = \tau_o \tag{2.36}$$

where p is the density of the air. Above this layer, a velocity scale is defined based on the thermally generated turbulence (also called convective turbulence) and is denoted as w_*. It is defined mathematically as follows:

$$W_* = \left(\frac{g\, Q_O\; zi}{p\, c_p\, T} \right)^{1/3} \tag{2.37}$$

where Q_o. is the surface heat flux, g the acceleration due to gravity, c_p the specific heat of air, zi, the height of the boundary layer, and T an average temperature within the atmospheric boundary layer. In the surface layer, the wind goes to zero at the roughness height zo. Externally defined length sides for the ABL are zo at the surface and zi inside the boundary layer away from the surface.

With both shear and buoyant forces at work in the ABL simultaneously, it is clear that the relative importance of these two forces will be an important indicator of the dispersive condition of the ABL, and in turn its stability dispersive. The two established parameters indicating the selective importance of these two forces are the Richardson number Ri, and the Monin-Obukhov length L. These are defined as follows:

$$L = -\frac{\rho\, c_p\, T u_*^2}{k\, g\, Q_0} \tag{2.38}$$

$$Ri = \frac{\left(\frac{g}{T}\frac{\partial T}{\partial z}\right)}{\left(\frac{\partial u}{\partial z}\right)^2} \tag{2.39}$$

The ratio of the height z and the Monin-Obukhov length L (i.e., z/L) is now widely used as the stability indicator for the surface layer.

It has been shown that experimental data for atmospheric turbulence, when organized using these velocity and length scales, show consistent patterns (Kaimal, 1973). It is important that the parameters σ_x, σ_y, and σ_z, should also be expressed using the above mentioned length and velocity scales to obtain consistent formulations. From the asymptotic behavior of σ_x, σ_y, and σ_z, one can express these quantities as follows:

$$\sigma_y = \sigma_u\, t\, f_v\, (t/T) \tag{2.40}$$

$$\sigma_x = \sigma_u\, t\, f_u\, (t/T) \tag{2.41}$$

$$\sigma_z = \sigma_w t\, f_w\, (t/T) \tag{2.42}$$

where f_u, f_v and f_w, are functions of (t/T). The exact mathematical nature of these functions can be obtained from experimental data. It is evident that these functions should behave as $(t/T)^{-1/2}$, and as $(t/T) \rightarrow \infty$. Very often for atmospheric flows, the stationarity assumption breaks down before the limiting condition, $t/T \rightarrow \infty$ is achieved. This may cause the functions f_u, f_v, and f_w, when derived from experimental data, to not show the asymptotic limit $(t/T)^{-1/2}$ when

$(t/T) \to \infty$. Examples of recent formulations for σ_c, σ_y, and σ_z based on experimental data can be found in Venkatram and Wyngaard (1988).

The concepts introduced here describing the state of the ABL are central to modeling pollutant releases. The next section provides some examples of air pollution modeling in the ABL.

2.7. Examples of Model Applications

Mathematical models are used to describe the meteorological transport and dispersion processes that are active in the ABL and thus the fate of pollutants released into it. Since there are many possible dispersion scenarios that can be active in the ABL, there are therefore necessarily many different models needed to describe them. This chapter will attempt to provide the reader with some insight into these models and their scope.

As our knowledge of the ABL has increased over the past 30 years, so has our ability to model its behavior. As well, with the advent of more powerful desk top computing and the ability to obtain, inexpensively, accurate source and meteorological data to drive models, our ability to assess pollutant impact with modeling has increased. These factors have led to the development of better modeling approaches for use in regulatory applications (Weil *et al.*, 1992).

In this section, several modeling approaches are discussed, their relevance, application, and limitations. This is by no means a comprehensive discussion of modeling. Here, we focus primarily on short range modes where distance scales are less than 30 km, and travel times typically less than 3 hours.

2.7.1. PASQUILL-GIFFORD METHODS

The Gaussian plume model forms the basis for estimating concentrations from pollutant sources. The model, as its name suggests, is based on the key assumption that the locus of molecules describing the pollutant concentration in the lateral and vertical directions are distributed normally (bell-shaped), that is the turbulence is random in all directions. The model is robust in its application often because the shape of the distribution is not the most critical condition to be met in its use. What is most important in its application is the accurate assessment of the standard deviations of the plume concentration σ_y and σ_z in the two directions y and z respectively.

The model also assumes, in its simplest form, that the pollutant does not undergo transformation and that the pollutant reaching the ground or the top of the ABL as it grows is reflected back toward the plume center line. It typically assumes that the atmospheric stability is uniform throughout the layer the pollutant disperses in.

The model is commonly described mathematically when $z = 0$ (i.e., at ground level) as follows for a point source:

$$C(x,\ y,\ 0,\ H) = \frac{Q}{\pi\ u\sigma_y\sigma_z}\left[exp\left[\frac{1}{2}\left(\frac{y}{\sigma_y}\right)^2\right]\right]\left[exp\left[-\frac{1}{2}\left(\frac{H}{\sigma_z}\right)^2\right]\right] \tag{2.43}$$

where $C(x, y, 0, H)$ is the downwind concentration at ground level (g/m^3), Q is the emission rate of the pollutant (g/s), σ_y and σ_z are the standard deviations of the plume concentrations in these directions, u is a representative wind speed (m/s), and H is the effective height of pollutant emission (e.g., emission height plus plume rise).

The model says that the continuous emission from a point is diluted by the ABL in the direction of the wind at a rate inversely proportional to the wind speed. The downwind concentration is empirically related to the plume spreads σ_y and σ_z. These spreads are a function of the atmospheric turbulence (stability) and the distance downwind from the source.

Pasquill and Gifford were one of the first to be involved in the measurement of these parameters during the 60's. As such, their data have been the most widely used until recently in regulation of air pollutants. Their data were collected over flat uniform terrain for averaging times of ten minutes. The use of these data has been widely abused by extrapolation and application to inappropriate situations.

Experimental data collected over the past two decade (Venkatram and Wyngaard (1988); Weil *et al.*, (1992); Irwin (1983)) provide more accurate methods for estimating σ_y and σ_z. The Gaussian plume model is highly sensitive to good plume spread data. The same is true for the plume height *'H'* estimates. Considerable effort during the 70's went into the study of plume rise in the ABL. As a result, there are models available which will describe the plume rise quite well. These models perform well because they contain representative descriptions of the physics of plume rise. The reader is referred to the work of Briggs (1975), Csanady (1967) and Slawson and Csanady (1971) for more detail on this component of the Gaussian plume model.

2.7.2. SIMILARITY MODELS

The concept of similarity is as follows. If the variables, both dependent and independent, are expressed in non-dimensional forms utilizing appropriate length, velocity and time scales, then the solutions for the dependent variables as functions of the independent variables are similar for all situations, where only the empirical constants may vary. The nature of the mathematical functions representing these solutions can be derived either empirically by fitting the experimental results or analytically where possible. One such example is given in this chapter.

Analytical Considerations in Similarity Models

It has been pointed out earlier that mass conservation equations, where diffusion of particles is parameterized by an eddy diffusivity, yield useful solutions only for limited situations. Near the surface where eddy sizes are limited by the height above the ground, solutions to the mass conservation equation with suitable assumptions on the functional representation of the eddy diffusivity have been found to be very useful.

Nieuwstadt and Van-Ulden (1978) assumed that the eddy diffusivity K near the surface can be expressed as a function of height and atmospheric boundary layer stability as follows:

$$K = k\, u_* z \,/\, \phi_h \,(z / L) \tag{2.44}$$

where k is von-Karman's constant, u_* friction velocity, z is height above the surface, and L is Obukhov length.

The functions ϕ_h *(z/L)* are obtained empirically (Businger, 1973) and are given as follows:

$$\phi_h = 0.74(1 + 6.3\, z / L) \quad for\ 1/L \geq 0, \tag{2.45}$$

$$\phi_h = 0.74(1 - 9\, z / L)^{-1/2} \quad for\ 1/L \leq 0, \tag{2.46}$$

With this expression for the eddy diffusivity, the solutions to the mass conservation equation is given as follows for a continuous source at coordinates $x = 0$ and $z = h$:

$$\frac{c\,\bar{u}}{Q} = A\ exp\ (-\left\{\frac{Bz}{\bar{z}}\right\}^2) \tag{2.47}$$

where Q (kg m s^{-1}) is the source strength.

$$\bar{z} \equiv \frac{\int_0^\infty z\,c\,dz}{\int_0^\infty c\,dz} \tag{2.48}$$

is the mean height of particles that have traveled a distance x and

$$\bar{z} \equiv \frac{\int_0^\infty z\,c\,dz}{\int_0^\infty c\,dz} \tag{2.49}$$

is the mean horizontal velocity of these particles.

The shape factor s is related to the powers m and n by:

$$s = 2 + m - n \tag{2.50}$$

A and B depend on s as follows:

$$A \equiv s\,\Gamma(2/s) \;\; / \;\; [\Gamma(1/s)]^2 \tag{2.51}$$

$$B \equiv \Gamma(2/s) \; / \; \Gamma(1/s) \tag{2.52}$$

where (Gamma) is the gamma function. The concentration profile is thus determined by the mean particle height, the mean horizontal particle velocity, and the shape factor.

The comparison of the predictions of concentration fields using the above equation is shown in Nieuwstadt and Van-Ulden (1978). The predictions of the concentration fields using this equation are in excellent agreement with observations.

Equation (2.52) is valid when the pollutants are released into the surface layer. Empirical expressions for these parameters have been determined by several researchers (Venkatram and Wingaard, 1988).

2.7.3. DISPERSION OF POLLUTANTS INSIDE A CONVECTIVE BOUNDARY LAYER: NATURE OF THE CONVECTIVE BOUNDARY LAYER

A convective boundary layer is characterized by the presence of large scale three dimensional eddies, often called 'updrafts' and 'downdrafts' (Deardorff and Willis, 1984; Lamb, 1978; Caughey, 1982). These eddies have a typical length scale equivalent to the height of the convective boundary layer z_i. A velocity scale inside a convective boundary layer is w_*. The time scale is, therefore, equal to z_i/w_*.

Dispersion of particles, being a Lagrangian process, is conveniently described by the average position of particles in space after a defined travel time. For homogeneous steady state situations, travel time 't' is approximately equal to x/u, where x is the down wind distance and u, the mean wind speed. Therefore, a logical definition of the non-dimensionalized downwind distance $\bar{x}$ is expressed as:

$$\bar{x} = \left(\frac{x\, w_*}{u\, z_i}\right) \tag{2.53}$$

which is the ratio of the horizontal travel time of the particles to the time scale inside a convective boundary layer. The dispersion of particles inside a convective boundary layer is, therefore, expressed as functions of $\bar{x}$.

A typical value of z_i is 1 km in the temperate climate zones. W_* has a typical value of 1 ms^{-1} during summer months. This translates to a typical value of the time-scale (z_i/w_*) to be on the order of 15-20 minutes. This implies that portions of a pollutant plume, which are caught in an updraft or a downdraft, travel relative to the vertical motions within these eddies for a time period of 15-20 minutes. This is a long enough time period to shear these plume segments giving the appearance of a ‘looping plume’. The segments which are caught in the downdrafts are brought to the ground causing high ground level concentrations of the chemical species inside these segments.

In addition, the average vertical velocities inside an updraft is significantly greater than the vertical velocities inside a downdraft. The cross-sectional area of a downdraft is correspondingly larger than that of an updraft. The ratio of the cross-sectional areas of the downdraft to the updraft varies with height, being about 3:2 in the middle of the boundary layer. For most tall 'stacks', therefore, the fraction of the plume which would disperse into the downdrafts is 60 percent compared to 40 percent into the updrafts. This skewness adds to the higher ground level concentrations caused by a looping plume inside a convective boundary layer.

The highest value of the ground level concentration occurs relatively close to the stack inside a convective boundary layer. Although the probability density function (pdf) describing the position of particles at a given time after their release inside a turbulent boundary layer evolves with time, for short travel times, it can be assumed that this pdf will be related to the pdf of the 3-dimensional velocity field at the point of release.

Misra (1982) assumed that the maximum ground level concentrations for non-buoyant particles dispersing inside a convective boundary layer can be simulated by invoking the above short travel time hypothesis. Towards this end, he assumed that the pdf of the particle position *p(z)* on a vertical plane located at a downwind distance *x* from the point of release is given by the following:

$$p(z) = p(w) \left| \frac{dw}{dz} \right| \tag{2.54}$$

where *p(w)* is the pdf of the vertical velocity at the point of release.

The above relationship assumes that the particle position *z* is a single-valued function of the vertical velocity *w*. Further, for homogeneous, steady-state situations,

$$\frac{dz}{dx} = w / u \tag{2.55}$$

The relationship between the concentration field and *p*(*z*) is given as follows:

$$c^y = \frac{Q\, p(z)}{u} \tag{2.56}$$

where c^y is the cross-wind integrated concentration field.

Misra (1982) assumed p(w) to be a bi-gaussian form given as follows:

$$p(w) = \frac{f_1(z_s)}{\sqrt{2\pi}\, \sigma_w(z_s)} \exp\left\{ -\frac{1}{2\, \sigma_w^2(z_s)} [a f_1(z_s) - \overline{w}_d(z_s)]^2 \right\} \tag{2.57}$$

This has also been used by others (Weil *et al.*, 1992). Using this form of *p(w)*, the derived equations for the ground level concentrations were shown to reproduce the maximum ground level concentrations reasonably well. Weil *et al.* (1992) has extended these concepts for the buoyant plumes.

The purpose of this section has been to introduce the readers to the concepts of dispersion of particles inside a convective boundary layer. For more details on this subject, the reader is referred to Weil *et al.* (1992), Venkatram and Wyngaard (1988).

2.7.4. STABLE LAYERS

The ABL is stably stratified when the density of the atmosphere decreases with height sufficiently, or the temperature increases rapidly enough such that, if one displaces a parcel of air in this layer and then releases it, it will tend to restore itself to its original position. Stable layers will form near the surface at night when the air is in contact with a cool surface as well as aloft above the ABL.

Turbulent motions in stable layers have special characteristics. Vertical diffusion is reduced, and long-wave radiation and gravity waves play a significant role. Because of the complexity of stable layers, models often restrict themselves to dealing with just the turbulent processes effecting pollutant dispersion.

Stable conditions can be defined under the following conditions:

$$Q_o \preceq 5\, w/m^2 \qquad (2.58)$$

or,

$$\frac{z_i}{L} \succeq 1 \qquad (2.59)$$

One example of a model used to estimate pollutant dispersion in stable layers is that of van Ulden (1978). It is a surface similarity model used when the effective release height is very near the surface (i.e., $H \preceq L/50$). In this case, the cross-wind integrated ground level downwind concentration is given by:

$$\bar{c}_o^y = 0.73 \frac{Q}{\bar{u}\, \bar{z}} \qquad (2.60)$$

where z is given by:

$$x = 0.74\, \bar{z}\, ((\ ln\ (0.6 \frac{\bar{z}}{z_o}) + 4.9 \frac{\bar{z}}{L})\ (1 + 4.9\, \bar{z}\, L)\ +\ (1.2\ \frac{\bar{z}}{L})\ k^2 \qquad (2.61)$$

and $\bar{u}$ is defined by:

$$\bar{u} = u_*\ (\ ln\ (0.6\ \frac{\bar{z}}{z_o}) + 4.7\ \frac{\bar{z}}{L})\ /\ k \qquad (2.62)$$

The crosswind dispersion of the plume is parameterized in terms of a Gaussian distribution.

$$C(x, y, z = 0) = \frac{\bar{C}_0^y\ exp\ (\ -y^2 / 2\ \sigma_y^2)}{(2\pi)^{1/2}\ \sigma_y} \qquad (2.63)$$

The lateral dispersion parameter σ_y is calculated using the assumption that the initial plume spread σ_{y_0} (which may be due to plume induced spread, wake induced spread, or to the initial height of an area source) and the ambient turbulence induced spreading σ_{y_a} are independent. Hence,

$$\sigma_y = (\sigma_{y_0}^2 + \sigma_{y_a}^2)^{1/2} \qquad (2.64)$$

where

$$\sigma_{y_a} = \frac{1.3\, u_*\, x}{u} (1 - z_e / H)\, f_y(x) \qquad (2.65)$$

The form of the universe function $f_y(x)$ is taken from Pasquill (1976) and Irwin (1979):

$$f_y(x) = (1 + (x / 2500)^{1/2})^{-1}, \text{ if } x < 10 \text{ kilometers} \qquad (2.66)$$

$$f_y(x) = 33\, x^{-1/2})^{-1}, \text{ otherwise.} \qquad (2.67)$$

The vertical spread of the plume is assumed to follow an exponential form. Thus,

$$C(x, y, z) = C(x, y, z = 0) \exp (- (z / 1.52\, \bar{z}\ )^{1.5}) \qquad (2.68)$$

If the effective release height z_e, or the mean plume height $\bar{z}_0$, is non-zero, then the above calculations are modified in the same manner as the convective near surface release model. Given the effective source height or the mean plume height, Eq. (2.61) is solved for the virtual ground level source location, x_0, assuming $\bar{z} = z_e$, or $\bar{z} = \bar{z}_0$ respectively. $\bar{u}$ is calculated using (2.62) for the height $\bar{z}\,(x + x_0)$.

In some cases, a near surface release with $z_e > 0$ may be considered as an elevated source when viewed at receptor locations very near the source. For modeling purposes, such a source is considered elevated by all receptors located at downwind distances (x) which are less than x_0. The modeling of concentration at receptors for which the source appears as an elevated release proceeds in the manner described below.

Elevated releases into the stable boundary layer

The downwind concentration due to an elevated release may be approximated using the Gaussian plume formulation. The lateral dispersion parameter, σ_y, may be calculated in the same manner as was employed in the stable, near surface release model using (2.64). The ambient turbulence induced spread, σ_{y_a}, may be determined using (2.65) and (2.66). The initial plume spread, σ_{y_0}, for the elevated release is given by $\sigma_y = \Delta h/4$ which describes the initial dispersion of the plume due to its own turbulence during plume rise. The vertical dispersion parameter, σ_z, may be estimated using an approximation for the standard deviation of the turbulent vertical velocity fluctuations.

$$\sigma_w = 1.3\, u_* (1 - z_e / H) \qquad (2.69)$$

where

$$\sigma_{z_a} = \frac{\sigma_w\, x}{u} f_z(x) \qquad (2.70)$$

The form of the universal function $f_z(x)$ was selected, based on Irwin's (1983) review of various parameterizations, to be

$$f_z(x) = (1 + 0.9\,(x / 50\,u)^{1/2})^{-1} \tag{2.71}$$

Including the initial plume spread, σ_{z_0}, due to the internal plume turbulence during plume rise gives:

$$\sigma_z = (\sigma_{z_0}^2 + \sigma_{z_a}^2)^{1/2} \tag{2.72}$$

σ_z is limited to a maximum value of $\sigma_z = H/(2\pi)^{1/2}$ corresponding to complete vertical mixing of the plume within the stable boundary layer.

The downwind concentrations due to an elevated release of gaseous emissions are determined by employing the effective plume height and the above parameterizations of σ_z and σ_y in the Gaussian plume formulation:

$$C(x, y, z) = \frac{fQ}{2\pi\,\sigma_y \sigma_z u}\left\{ exp\left[\frac{-(z - z_e)^2}{2\,\sigma_{zsup}2}\right] + exp\left[\frac{-(z - z_e)^2}{2\sigma_z^2}\right]\right\} exp\left[\frac{-y^2}{2\,\sigma_{zsup}2}\right] \tag{2.73}$$

2.7.5. COMPLEX TERRAIN AND BUILDING WAKE MODELS

The ideal dispersion conditions described in section 2.7.1 rarely occur in real applications. Most of the earth's surface is not flat, and often, nearby structures will influence the dispersion of pollutant emissions. For many applications, however, the departures from the ideal are not sufficient to require the use of a different modeling approach. For those situations that do, then, usually, physical modeling approaches are invoked rather than mathematical ones.

Some attempts have been made to define complex terrain in the context of a regulatory framework. For example, a site would be considered to be a complex terrain if:

1. The pollutant release height is less than two times the maximum terrain height. The maximum terrain height *dh* in this case is defined as the difference between the highest level (including tree tops) and the lowest level within the larger of twenty times the stack height or 1 km from the source complex.
2. In addition, the gradient of the terrain height with distance, *dx*, from the source complex (i.e., *dh/dx*) must be greater than 1/5.

Similarly, there are criteria which distinguish when building induced turbulence may influence plume dispersion. Some examples are:

a) the dispersion of material released on or near a building or structure is assumed to be influenced by structure-induced turbulence if the release height is less than two times the building height *Hb*.

b) if the release height is less than two times the building height, but is equal to or greater than ten times the building width, the pollutant emission will be considered to be outside the building-induced turbulence.
c) for a source upwind of a building, the influence of the structure on the plume dispersion will extend upwind a distance of 1.3 times the building height.

In most circumstances, the effect of the enhanced turbulence levels around a building will be to increase plume dispersion over the situation where the building is not there, leading to lower glc's downwind. In the near field, however, these situations create the potential for elevated concentrations impacting at locations on or near a building.

Considerable effort in the last decade has been directed towards developing a better understanding of how terrain and structures affect dispersion. Nonetheless, the present understanding does not allow for the recommendation of a single model which will be applicable to the wide range of complex dispersion scenarios.

Attempts to incorporate the effect of terrain in models have been made and generally are based on derating the plume rise by some empirical parameterization. This certainly results in a higher glc estimate; however, the validity of this approach is questionable as there is no theoretical or physical justification for it.

In some complex terrain situations, mathematical models which are based on sound physical and theoretical arguments have been developed. One example of this type of complex terrain model is one that describes plume behavior over and around a simple hill. The influence of a simple hill on a plume impacting it is to depress the plume centerline due to converging streamlines and thus decreasing σ_z, σ_y is increased by streamline divergence. Increased wind speeds decrease concentration, while increased turbulence counteracts the decrease in σ_z.

To understand a model describing this behavior, the dividing streamline height concept must be described. Mathematically, we consider the balance of potential and kinetic energy as follows:

$$1/2\, U^2\, (Hc) = N^2(z)(H-z)dz \tag{2.74}$$

where N^2 is the Brunt-Vaisalla frequency defined by: $N^2 = g/To\; dT/dz$. The LHS of Eq. (2.74) is the kinetic energy of the fluid at Hc. The RHS is the potential energy of the fluid in rising through the height *H-Hc* to the top of the hill at $z=H$. This energy equation suggests that the plume in the fluid below *Hc* will not have sufficient energy to go over the hill and, therefore, will tend to impact the hill and encompass the hill laterally. For the part of the plume in the fluid above *Hc*, it will tend to flow over the hill.

For constant U and N^2 Eq. (2.74) reduces to:

$$Hc = H(1-Fr) \tag{2.75}$$

where the Froude number is defined by

$$Fr = U/NH \tag{2.76}$$

To mathematically describe the effect of the simple hill on the plume, two models have been proposed by Venkatram and Wyngaard (1988). These models, one to describe the lift component and the other, the wrap component, are good examples of theoretically based approaches.

In the case where a structure influences plume behavior, some simple models are also available. These models also tend not to represent the physics nor the theory very well and are empirically set to provide conservative estimates of the glc.

Recently, Weil (1988) has proposed an approach to modeling these situations, which are more physically and theoretically based. Weil's model uses Taylor's (1921) statistical theory as a guide, where the glc is divided into short and long time regimes. For the short-time component, a probability density function (pdf) approach is adopted in which particles are traced or followed from the source to the downwind locations, given the pdf's of the lateral and vertical velocity fluctuations. This follows a similar approach successfully applied to dispersion of tall stack plumes in the convective boundary layer. In the wake model, the horizontal inhomogeneity of the wake turbulence statistics and the downwind variation of the wake mean velocity deficit are additional complications that are treated. Another feature of his approach is a mean negative vertical velocity in the wake due to (1) the flow deformation about the building, and (2) the presence of axial vortices on either side of the building. This mean velocity has a Gaussian pdf of turbulent velocities superimposed on it, with the pdf being used to track particles in the wake. For the long-time regime, an eddy-diffusion approach is used to predict the dispersion. This approach is presently being evaluated against available wind tunnel data by Lepage (1993) and Snyder and Lawson (1993).

In general, complex situations still are poorly understood, and until more data become available, physical modeling will play a prominent role in estimating the impact of sources influenced by terrain and structures.

2.7.6. COASTAL FUMIGATION

The convective boundary layer described in the previous sections dealt with homogeneous and steady state situations. The convective boundary layer existing over land near a land-water interface is non-homogeneous for on-shore winds.

During spring and summer, solar radiation causes the land to heat up faster than the water. As a result, for on-shore winds, a stable boundary layer forms over water. As this stable boundary layer is advected over land, surface heating causes a convective boundary layer to develop. The height of this convective boundary layer increases with downwind distance from a near-zero or small value at the shore line. This developing convective boundary layer is also called a Thermal Internal Boundary Layer (TIBL).

The TIBL typically varies parabolically with downwind distance X. Pollutant plume from a tall stack situated at the shoreline (at the boundary of land and water) will initially disperse in a stable boundary layer and intersect the height of the boundary layer further downwind, at which point the plume will gradually be entrained into the underlying convective boundary layer (TIBL). The more vigorous nature of the turbulence inside the TIBL will disperse the plume relatively quickly to the ground, causing high ground level concentrations. This phenomenon is called continuous fumigation of the plume.

Several investigators have developed analytical formulas to determine the ground level concentration of pollutants for continuous fumigation. These include Lyons and Cole (1973), Nieuwstadt and van-Dop (1982), and Misra (1982). Of these formulations, Misra's model is described here as it includes convective boundary layer scaling more rigorously.

Misra's Model

Misra (1982) assumed a Gaussian type distribution of pollutants initially in the stable boundary layer. He assumed the dispersion inside the TIBL to proceed in an independent manner, where the source is an area source coincident with the under-surface of the TIBL.

The resulting equations describing the ground level concentrations are as follows:

$$C(x,y,o) = \frac{Q}{zih(x)} \int_o^x \frac{1}{\sigma_{ye} u} \exp\left(\frac{p^2}{z}\right) \frac{dp}{d'x} \exp\left(\frac{y^2}{2\sigma_{ye}^2}\right) d'x \qquad (2.77)$$

where

$$p(x') = \frac{h(x') \; h_e(x')}{\sigma_{z_s}(x')} \qquad (2.78)$$

and

$$\sigma_{ye}^2 = \sigma_{ys}^{\;2}(x') + \sigma_y^2 \;(x,x') \qquad (2.79)$$

In the above equation, $h(x')$ represents the height of the TIBL at x', $h_e(x')$ is the height of the plume, σ_{zs} and σ_{ys} are the dispersion parameters of th, initial Gaussian plume, and σ_y is the dispersion parameter inside the TIBL.

Misra (1982) assumed an instantaneous mixing of the plume in the vertical inside the TIBL. This can be modified by introducing a delay parameter (Willis and Deardorff, 1976).

The reader is referred to the original paper of Misra (1982) for the mathematical derivation of the equations. The following points are noteworthy in the application of these equations:

1. the wind directions inside the TIBL and the capping stable layer will generally be different. This is not included in the present models.
2. the convective velocity scale W* inside the TIBL will vary with downwind distance.

However, the surface heat flux decreases while the TIBL height increases with downwind distance. Therefore, W* is only a very weak function of downwind distance. Misra (1982) found that reasonable results are obtained when local values of W* and the TIBL height h are used in computing the ground level concentrations.

2.8. Measurement Methods

2.8.1. MEASUREMENT OBJECTIVES

There are four main purposes for atmospheric sampling and measurement: (1) environmental surveys; (2) monitoring networks; (3) compliance monitoring; (4) model validation. The nature of the sampling and measurements will vary according to these purposes and the objectives set for a program in its entirety. Practical constraints (discussed below) will generally limit the extent of sampling; thus, efficient use must be made of available resources. This will be achieved by careful planning and execution of a sampling and measurement program. In planning a program, it is essential to start with clear objectives or hypotheses to test and to precisely define the data required. This will involve selecting the species to monitor, the geographical extent and spatial density of sampling locations, the frequency and duration of monitoring, and the quality of the data. (Maher *et al.*,1994; Reid, N.,1996). The considerations that should be made when planning a sampling program are outlined in Table 2.1 and are discussed in detail in the following text.

TABLE 2.1 Considerations when planning a sampling program*

1. Has the problem/reason for sampling been clearly stated?

2. Are specific objectives:
- **(a)** Clear and concisely defined?
- **(b)** Sufficient to specify what is to be achieved?
- **(c)** Specific enough to indicate when each stage is complete?
- **(d)** Agreed between the users of the data and the collectors?

3. Has a conceptual model of the system been made explicit and agreed?
- **(a)** Have the study boundaries been agreed.
- **(b)** Has the length of study been agreed?
- **(c)** Has the scale of study been agreed?

4. Have appropriate indicators been identified?

5. Have testable hypotheses been established?
- **(a)** Will data from different sources be compatible?
- **(b)** Will data collected yield information to test hypotheses?
- **(c)** Are statistical procedures clearly identified?
- **(d)** Are the assumptions of the proposed statistics tests met?
- **(e)** Has the minimum detectable concentration been specified?

6. Have all potential sources of sampling and analytical variability been identified?
- **(a)** Are there sufficient sampling stations to accommodate variability?
- **(b)**Is replication adequate to obtain the desired level of precision in data?
- **(c)** On what basis is frequency of sampling proposed?

7. Will the sampling device collect a representative sample?
- **(a)** Does disturbance of the environment being sampled occur?
- **(b)** Does alteration of the sample occur by contact with the sampling device?
- **(c)** What are the effects of the sampling device being in contact with media other than the sample of interest?

8. Is there a quality assurance program in place?
- **(a)** How are the samples to be preserved before analysis?
- **(b)** Can the integrity of a sample be guaranteed?
- **(c)** Have sampling, instrumental and analytical procedures been written?
- **(d)** Is there a quality control program for laboratory and field instruments and analytical procedures?
- **(e)** How are problems to be rectified?
- **(f)** How is data to be recorded and reported?

*Adapted from Maher *et al*, (1994)

Environmental Surveys

Probably the most difficult programs to plan and carry out are environmental surveys. Frequently, these are "first time" measurements and may not benefit from previous experience indicating species type or concentration range according to location, time, or other variables. Thus, judgment based on experience with other geographically similar locations must usually be made in the selection of measurement techniques and instrumentation, number of measurements, and duration of sampling. Environmental surveys often have the objective of identifying and quantifying a wide selection of possible airborne species, or to provide data for a possible more extensive survey or permanent monitoring network. Thus, a number of sampling and analytical techniques will likely be required. Usually, such surveys start with a pilot program, mainly qualitative in nature, to help define the scope and objectives of the main program.

Monitoring Networks

Environmental surveys frequently lay the groundwork for a monitoring network - an ongoing monitoring effort over a particular geographical area. Often, the need to establish a monitoring network arises because of the concern over concentrations or trends, or the need to establish long-term correlations with meteorological or climatological factors or with possible sources of the contaminant. Often, a single pollutant and a single objective determine the nature of a monitoring network, but economics and efficient utilization of resources can often be achieved by utilizing or adding on to existing monitoring networks for other pollutants (Trujillo-Ventura and Ellis,1991).

There are a number of steps that must be taken to properly design and set-up a monitoring network. Firstly, existing data must be compiled. This will include species, concentration ranges, spatial variations, and time trends. Sometimes, statistical analysis of existing data will assist with program design. Models may also be available to predict concentrations where data is lacking. Secondly, the objectives of the monitoring network must be specified. Thirdly, budgets and funding commitments for setting-up and operating the network must be justified and approved. This will probably include detailed specifications for sample measurements and locations. Fourthly, equipment is purchased, installed and commissioned, and all other support arrangements are put in place. Finally, the data generated by the program are carefully reviewed, and any necessary changes to the program are made to achieve optimum use of resources and best possible data.

Permanent monitoring network sites are established in a grid within the geographical study region. The density (spacing) of monitoring sites is usually a compromise between data requirements and practical considerations of cost, availability and accessibility of suitable sites, and personnel availability to maintain sites and perform sampling and analysis.

Compliance Monitoring

Compliance monitoring generally poses specific data requirements and measurement locations. The objectives are usually to determine whether short-term peak or time-averaged concentrations exceed a regulatory limit. Thus, the required sensitivity of instruments and the accuracy and precision of data are determined by the instantaneous or time averaged concentration specified in the regulatory limit for the species subject to compliance monitoring.

Model Validation

Model validation can be simple or complex, depending on the model or elements of a model that are being validated. Generally, the species are given, the concentration ranges to be measured will be known, and the accuracy and precision required of measurements will be determined by the accuracy expected from the model. The objective will usually be to confirm, refine, or re-establish a mathematical relationship between concentrations in different compartments or locations subject to modeling. The choice will exist of the number or duration of measurements at a particular location to give a measure of natural variability, the number of measurements at different geographic locations, and the effects of different meteorological and topographical situations. For some model verifications, source-term measurements will be required. This may require instrumentation at or over the contaminant source. The source can take many configurations (e.g., single stack, multiple stacks, line source, ground area source).

Measurements made in support of model validation often aim to establish mean concentrations. However, natural variation occurs in most environmental systems modeled. This variability may span a wide range in a short to intermediate time period. This is an often overlooked aspect of model validation (Weil, J.C. *et al.*,1992). Modeling should ideally be able to indicate the likely probability bound for values about the mean, including short-term variability. Validation of such models, therefore, requires sufficient measurements to be taken and statistically analyzed.

2.8.2. MEASUREMENT CONSTRAINTS

Often, budgetary constraints will necessitate limiting measurement programs. Compromises must usually be made in the objectives set to make realistic use of fund amounts and time-frames for funding. Detailed planning is always a prerequisite to maximizing the amount and usefulness of data obtained from a measurement program.

Non-monetary constraints may also exist. These may include inaccessibility of preferred sampling locations due to lack of road access, non-cooperation of landowners, lack of electrical power supply, lack of security (risk of theft or vandalism of equipment), unavailability of suitable instrumentation, people not available, and analytical laboratory not available.

2.8.3. DATA REQUIREMENTS

The data required will be those needed to satisfy the objectives of the measurements program. Once the species to be measured have been decided, the minimum detectable concentration, the accuracy and precision requirements, the time resolution (peaks, averages, averaging time), and the measurement duration must be determined. Also, the requirements for spatial resolution (density of measurement points) and geographical extent of data must be decided. These and other considerations will be major factors in the design of a sampling program. A summary of these considerations is given in Table 2.1.

Minimum Detectable Concentration

The minimum detectable concentration, also known as detection limit and lower detectable limit, is the lowest concentration of a substance that can be distinguished with statistical confidence. It is generally defined as twice the average value of the random electronic noise associated with the output signal of a detector, or the random variations mass or other physical measurements when analyzing samples that contain none of that substance (a blank sample). The minimum detectable concentration will vary according to the analytical technique used. In general, the lower the minimum detectable concentration, the more difficult or expensive is the instrument or analytical technique. The lowest achievable minimum detectable concentration may not be necessary for a particular situation and objective.

Accuracy and Precision

An accurate measurement, or average of several measurements, is one that is close to the real value. Precision is the variation in results when a sample is analyzed repeatedly (often defined as one standard deviation about the mean). Results can be precise but inaccurate. For example, an incorrectly calibrated instrument might consistently give results from a particular sample within a 5 percent range, but the mean of these results might be 20 percent lower than the correct value. In general, the better the accuracy and precision of an instrument or technique, the more expensive it becomes. The best achievable accuracy and precision may not be necessary for a particular situation and objective.

Time Resolution

Sometimes it will be required to measure the full range of instantaneous values. For this, a "real time" or rapidly responding sampling and analysis system will be necessary. Other measurements may be required, that are averages over certain time intervals. These could range from minutes to months. The sampling program objectives will generally set the time resolution required.

Spatial Resolution

The spatial resolution is determined by the geographical area covered and the spacing of sampling locations (usually on a grid pattern). The larger the area and the denser the grid, the more sampling locations will be required and the greater will be the effort and expense of data acquisition. For some purposes, samples at various heights above ground may also be necessary. It is important to determine reasonable boundaries and scale for environmental surveys and monitoring networks. Inappropriate boundaries may focus the study away from important driving or consequential factors. Thus, if the concern is with a particular source, the effort should be directed to areas directly impacted by that source and not over an area that might include other similar sources (Maher *et al.*,1994).

Quality Assurance and Quality Control

All data generation should be accompanied by additional information to provide confidence that accuracy and precision are being maintained to acceptable levels, that sampling and measurements are being carried out at those locations and for those times specified in the program design, and that instrument failures, sample losses and analytical failures are being minimized. This is achieved through a well designed and executed quality assurance program. Quality control procedures will include the regular analysis of blank samples, and standard samples of known concentrations. Often, these are provided to laboratory personnel as blind or double blind samples from an external standards agency.

2.8.4. MEASUREMENT TECHNIQUES

There are three basic measurement methods: real time, integrating, and grab samples. Real time measurements are those in which the species concentration is measured continuously, with the results being presented as a sequence of numerical values at the time of each measurement and/or concentration-time plots (e.g., a paper chart). The time resolution (time elapsed between each measurement) will vary according to the species and instrument. In some cases, the instrument can respond instantaneously to changes in concentration. These will be true real time measurements. In other cases, the measurement instrument uses sensors that take time to respond to a change in concentration, or alternatively involves some form of averaging over a period of time. Where the averaging time is comparatively long, the instrument cannot be considered real time. Resulting measurements will be integrated and will represent an integrated or average value over a time period such as an hour or a day, depending on the instrument and how it is set-up and operated (Boubel *et al*., 1994).

For some measurements, it is impractical or unnecessary for instruments to be set up in the field. In such cases, samples are obtained at the field locations and are taken to a laboratory for analysis. Samples obtained in this way may be taken at a particular instant of time (a "grab sample"). They can also be taken over a time period to provide an integrated sample. An example of a grab sample is where a container, initially under vacuum, is opened at the field location to admit an air sample and is then immediately re-sealed. An example of an integrated sample is where air is drawn at a constant rate by pump through a trap or sorbent over a particular time period.

Real Time Air Monitoring Instruments

Real time air monitoring instruments are mainly spectroscopic. Several spectroscopic methods are used depending on species. These include absorption, fluorescence and chemiluminescence. In some methods, ambient air is drawn *in situ* through a sampling manifold and enters the monitor. The measurements are made by an automated technique which may be continuous or discrete. Sometimes, separation methods are necessary. A commonly used technique is gas chromatography. If the time taken for the species to elute from the column and be detected is a few seconds, then the measurement may be considered to be real time. Gas chromatography instruments used in this way are often set-up with automated periodic sample injection. Alternatively, manual sample collection and injection can be used. This can be considered grab sampling with a close to real time analysis.

Integrating Samplers

Several integrating sampling techniques are available depending on the species. Bubblers, in which the ambient air is pumped at a constant rate through a liquid sorbent, are used for species that can be absorbed by acid or alkaline aqueous solutions, or that undergo reaction with aqueous solutions of a particular chemical, or that are absorbed by liquid organic solvents. Considerations in using bubblers are flow rate, absorption efficiency, and capacity of solution for the species. Sometimes, a series of bubblers is used to achieve the required efficiency. The bubbler system must be selected to be appropriate for the concentration and the type of species, the volume of ambient air that must be processed, and the integration time required. Where the ambient temperature is low, heated enclosures must be provided to prevent freezing of aqueous solutions and to maintain absorption efficiency.

Solid sorbents are also used as pumped integrating sample collectors. The most common is activated charcoal. Other materials, usually in the form of granules packed in a column include a variety of proprietary sorbents and surface-coated inorganic pellets. Examples are Carbosorb and Tenax. Sometimes catalysts are used to oxidize the species of interest to more readily detected molecules which may be absorbed either *in-situ* or in a second absorption device (a bubbler or a solid sorbent) downstream.

Shallow bed granular materials, or fibre filters, impregnated with a suitable reactant, sometimes make convenient samplers for pollutant gases. Such samplers have a comparatively large surface area and enable large volumes of air to be filtered quickly. Sometimes, these samplers use multiple filters in order to determine collection efficiency. Typical applications are SO_2 collection on potassium carbonate granules, HNO_3 on nylon, and NH_3 on citric acid granules.

Cascade samplers are sometimes used to separate several species from air. 'These usually consist of an impactor at the inlet. Following this, the air stream passes through a series of denuders, tubes each packed with a different sorbent material specific to a particular pollutant species or group of species. The outlet end consists of a filter pack often with several layers of different filter/sorbent materials.

Solid sorbents may also be used in passive samplers. In these, the sorbent is placed in a container having an orifice to the ambient air. Diffusion occurs into the sorbent chamber at a fairly constant rate, and the species is absorbed. Such samplers are suitable for integration periods of days to weeks and do not require pumps or other powered air-moving components.

For particulate samples, filters or impingers are used. Filters are not generally capable of particulate size determination. Particle size is best determined by impingers in which particle inertia in an air stream is utilized to deposit particles of particular mass ranges on different surfaces.

For pumped samplers, care must be taken to obtain a representative sample. The inlet nozzle and pipework to the detector or sorbent must not alter the air concentration in any way. For reactive gaseous species, interaction with inlet surfaces must be avoided by suitable material choice, e.g., Teflon or stainless steel, as appropriate. For particulate species, special attention must be given not to deflect particles or discriminate by size or mass at the entrance to the intake pipework. This is especially important where the sample is being obtained in a flowing system or where variable air currents exist (e.g., in the outdoor environment). Pumps, flow meters, flow integrators, and control valves are usually located at the exhaust end of the sample chamber or detection instrument in order not to interfere with the measured species.

In general, the sorbent or other capture medium used in integrating samplers must be returned periodically to a laboratory for analysis. To provide continuous monitoring, it is usual to replace the sampler with a fresh bubbler and solution or other sorbent or capture system at the end of each monitoring period.

When used for quantitative analysis, it is important to determine the collection efficiency, maximum loading and, in the case of solid sorbents, the desorption efficiency of the sorbent. Desorption from solid sorbents is usually accomplished by heating the sorbent in its container and purging the species with an inert gas. Although only one shot is possible to collect the purged sample, more than one analysis may be performed on the purged sample.

Grab Samples

Grab samples are taken when it is impractical or unnecessary to set up continuous real time or integrating samplers in field locations. Samples are captured either in containers, or on filters, or solid sorbents, and they generally represent an instantaneous ambient air concentration. Samples can either be admitted to an evacuated container, or flushed by pump through a container, or collected in bags. Sometimes, a known volume of ambient air can be drawn through a sorbent either by pumping or using an evacuated chamber. These samples are analyzed in a laboratory.

For some species, sample preservation is a concern because of decomposition or other losses in the sample container due to interaction with the container material or due to inherent instability of the species. Thus, appropriate choice of sample container types and materials must be made. Care must be taken to minimize sample contamination if sample containers are reused and to avoid traces of cleaning or other contaminants that might interact with the species sampled. Shipping arrangements to the laboratory and sample storage must be appropriate for the species to be analyzed.

Usually, more than one analysis can be performed on grab samples returned to a laboratory. This allows for two or more results to be obtained in order to determine the analytical method precision. Sequential grab samples may also be taken, if desired, with which to determine how concentration changes with time.

Precipitation Samplers

Rain or snow tends to wash out some atmospheric pollutants. Analysis of collected precipitation is therefore sometimes useful in assessing the magnitude of ground deposition of airborne pollutants. In addition, some dry deposition of airborne particulate pollution may occur. Samplers usually consist of a deep bucket collector with an opening of known area. The opening is positioned above the splash zone from surrounding ground and above the maximum expected snow depth in a location well away from buildings, trees, overhead wires, etc. Often, the bucket is lined with a plastic liner which is removed along with any water collected for analysis on a periodic basis. Sometimes, only the wet deposition is of interest, for example in acid precipitation monitoring. Such samplers will likely have an automated system to open a lid over the bucket opening when precipitation is detected. In this way, undesirable materials in the collector such as insects, leaves, bird droppings and road dust are minimized. For some purposes, real time precipitation samplers are used; however, these are mostly limited to continuous measurements of rainfall rate, pH, and conductivity.

The results from precipitation collectors must generally be accumulated over a number of samplings and will likely vary according to the frequency and amount of rainfall and the meteorological conditions accompanying the precipitation occurrences.

General Sampling Precautions
Care is needed to obtain meaningful and representative samples. The choice of location, time, and meteorological conditions can be very important especially if species concentration depends strongly on these factors or vary widely for other reasons.

The sampling locations should be on a grid pattern to provide suitable areal coverage if spatial concentration profiles are the objective. Alternatively, locations should be close to people, agricultural operations, or natural ecosystems that are potentially exposed to the airborne species.

In some cases, where the species of interest can be exhaled or desorbed or otherwise emitted from the skin, clothing, or vehicles used by sampling personnel, care must be taken for such sources to be kept well away or well downwind of the sampler intake.

For some species, exhalation, desorption, or resuspension from the ground or from foliage or buildings may occur following an earlier direct exposure. This possibility should always be considered in interpreting results.

Background Concentration
Many airborne pollutant species will be present in detectable concentrations even in pristine environments well distant from industrial or other anthropomorphic sources. This has important consequences. In setting-up and calibrating instrumentation, such background concentrations must be recognized. Zeroing of instruments should be carried out only with species-free gases or with gases of known low concentrations.

For measurements made for model verification purposes, it may be necessary to determine the background concentration (i.e., the concentration of species that would exist without the source term or other inputs to the system modeled). If the background concentration is relatively significant, the use of the measured concentration without deducting the background will lead to an overestimation of the impact of the source.

2.8.5. INSTRUMENTAL AND ANALYTICAL METHODS

For a particular species, the number of analytical choices are usually quite limited. Some of the techniques used for the more commonly measured air pollutants are discussed in the following section. The detection limits for various pollutant species and detection methods are summarized in Table 2.2.

Carbon Monoxide
A commonly used instrumental method for carbon monoxide is non-dispersive infrared photometry (NDIR). This utilizes the preferential absorption of infrared radiation by carbon monoxide. Analyzers use hot filament sources of infrared radiation, sample cells, reference cells, detectors, and control electronics. Such instruments can be used in real time mode in the field or in the laboratory to analyze grab samples.

TABLE 2.2. Summary of detection methods and limits for selected airborne gaseous pollutants

Pollutant	Method	Minimum detectable concentration
NO	Chemiluminescence	5 ppt
	Single photon laser-induced fluorescence	4-240 ppt
	Lidar laser-induced fluorescence	10 ppt
	TDLS	500 ppt
	DOAS	400 ppt
N_2O	TDLS	0.1 ppb
NO_2	TDLS	0.1 ppb
NH_3	TDLS	0.1 ppb
	Tungstic acid denuder	10 ppt
	Fluorescence derivatization	0.1 ppb
	Photoacoustic detection	0.5 ppb
	FTIR (1 km pathlength)	1 ppb
	Chemiluminescence with thermal converter	1 ppb
HNO_3	Chemiluminescence	200 ppt
	Tungstic acid	70 ppt
	Denuder tubes	80 ppt
	Nylon filter	2 ppt
	TDLS	350 ppt
	FTIR	3 ppb
HCHO	DOAS (10 km pathlength)	600 ppt
	FTIR (1 km pathlength)	6 ppb
	TDLS	300 ppt
	Solid sorbent, desorb into gas chromatograph	0.3 ppb
	Wet chemistry spectrophotometric methods	40 ppb
H_2O_2	TDLS	600 ppt
SO_2	TDLS	3-12 ppb
O_3	TDLS	0.5 ppb
	FTIR	10 ppb
CO	TDLS	0.25 ppb
HONO	TDLS	10 ppb

Nitrogen Oxides

Chemiluminescence is used for measuring nitric oxide (NO) and nitrogen dioxide (NO_2). The chemiluminescent reaction between NO and ozone at high concentration (produced by an ozone generator) is detected by photomultiplier tube and electronic amplifiers producing a signal proportional to the NO concentration. NO_2 is detected by first passing the air through a converter to quantitatively change it to NO. Nox detectors (NO + NO_2) use a dual pathway through the instrument with signal-processing circuitry to give the separate NO, NO_2, and Nox concentrations. Such instruments can be used in real time mode in the field (with electrical supply). Differential optical absorption spectrometry (DOAS) is used to detect NO in long path operation (several kilometers). The UV/visible light source is viewed by a reflecting telescope, and the light passed through a grating to a photodiode array. The absorption spectrum is analyzed using a computer to give concentration values.

Nitric Acid

The tunable diode laser (TDL) has been used for the measurement of airborne nitric acid (HNO_3). This technique makes use of absorption of visible light by HNO_3 at particular frequencies. The diode laser output is "tuned" to the frequency at which maximum absorption occurs; the attenuation of the beam is proportional to the HNO_3 concentration.

Sulfur Dioxide

Several instrumental methods are used to detect atmospheric SO_2. Bubbler techniques include the West-Gaeke method which forms the basis for the U.S. Environmental Protection Agency reference method for SO_2 (Boubel *et al.*,1994). In this method, a known volume of air is bubbled through a solution of sodium or potassium tetrachloromercurate (TCM). In the laboratory, the TCM-SO_2 complex is further treated to form pararosaniline methylsulfonic acid. This is analyzed by colorimetry. The optical absorption at 548 nm is proportional to the air SO_2 concentration. Continuous methods include automated colorimetric, conductometric and colorimetric sampling and analysis methods. Gas stream methods include flame photometric detection and fluorescence spectroscopy.

Ozone

The principle method for ozone uses chemiluminescence. Ethylene is mixed with the air sample. Any ozone present reacts with the ethylene to produce electronically excited reaction products which fluoresce. The released light is detected by a photomultiplier tube and electronic amplifiers to produce a signal proportional to the ozone concentration. Such instruments can be used in real time mode in the field (with electrical supply and a cylinder of ethylene), or to analyze grab samples in the laboratory. Ozone analysis sometimes also uses UV absorption. This technique is used in long range (km) real time instruments similar to those used for NO.

Hydrogen Peroxide

The tunable diode laser (TDL) has been used for the measurement of airborne hydrogen peroxide (H_2O_2). This technique makes use of absorption of visible light by H_2O_2 at particular frequencies. The diode laser output is "tuned" to the frequency at which maximum absorption occurs; the attenuation of the beam is proportional to the H_2O_2 concentration.

Hydroxyl Radical

Airborne hydroxyl radical concentration has been measured using laser induced fluorescence and ultraviolet (UV) absorption.

Volatile Organic Compounds (VOCs)

Analysis of VOCs generally requires some chemical separation. The atmosphere contains numerous natural and man-made organic compounds. Methane is naturally occurring at about 1.6 ppm worldwide. Additional local elevated methane concentrations occur from industrial and agricultural processes. Other airborne organic compounds which may be present include various aliphatic and aromatic hydrocarbons, organic acids, alcohols, esters, amines, nitroso compounds, organic sulfur compounds, organic halides (fluorine, chlorine, bromine, iodine), and organometallic compounds. Unless methane is of interest, it is usually excluded from analysis. The remaining VOCs may be analyzed as all other non-methane volatile organic compounds (NMVOCs) and may be reported either as parts per million by volume (ppmV) or as parts per million by carbon (ppmC). The unit ppmC is the accepted way to report ambient hydrocarbons (Boubel *et al.*, 1994).

Comprehensive analysis of atmospheric hydrocarbons is a complex process involving collection, separation, and analysis for each species or groups of species. Separation is usually performed by gas chromatography. Various column packing materials are used depending on the species required to be analyzed.

Instrumental analysis methods include flame ionization detectors (most suitable for hydrocarbons); flame photometry (used for organic sulfur compounds), electron capture detectors (suitable for halogenated organics and nitrogen compounds), and mass spectrometry (often used to assist in the qualitative identification of species).

The tunable diode laser (TDL) has been used for the direct measurement of some airborne VOCs, in particular, fornaldehyde (HCHO). This technique makes use of absorption of visible light by molecules at particular frequencies and, depending on which other species are present, may be suitable for the measurement of one species in the presence of others, thus avoiding the need for separation. The diode laser output is "tuned" to the frequency at which maximum absorption occurs for the species of interest. The attenuation of the beam is proportional to the species concentration. Sample collection can be as grab samples returned to a laboratory for analysis or automated, semi-continuous batch-analysis systems in field locations.

Differential optical absorption spectrometry (DOAS) has been used to detect HCHO in long path operation (several kilometers). The UV/visible light source is viewed by a reflecting telescope and the light passed through a grating to a photodiode array. The absorption spectrum is analyzed using a computer to give concentration values.

Air Particulates

The major characteristics required of air particulates are total mass concentration by volume (g/m^3), chemical composition, and size distribution. In most cases, the size range of interest is 0.01-10 μm (corresponding to that posing the greatest human inhalation risk). Mass concentration is generally measured by either filtration or impaction from a known air volume with pre- and post-weighing of filter or impaction surface. Size distributions are best obtained from multi-stage impactors with each stage optimized to remove particles of a particular size range. Impactors are appropriate for particle sizes above about0.1μm. Electric field devices are used to obtain the size distribution in the 0.01-1.0 μm diameter range. These work by attracting progressively larger particles as the electrical field strength is incremented. The charge collected by electrodes is proportional to the number of particles present of a particular size.

Light scattering is also used to determine a number-size distribution. The intensity of scattered light is proportional to particle size and refractive index. Instruments have been designed to dilute and proportion the sample air stream such that single particles are detected, and the scattered light is detected and analyzed by a photomultiplier tube. Thus, the number of particles by diameter range increments is reported.

The chemical composition of particulate pollutants can be determined either by elemental composition or as specific compounds or ions. Commonly used techniques include X-ray fluorescence spectroscopy and neutron activation analysis. In X-ray fluorescence spectroscopy, the particulate material is bombarded with X-rays. The inner shell electrons become excited and emit light characteristic of the element as the electrons return to their normal energy levels. This light is detected and analyzed. In neutron activation analysis, the particulate material is bombarded with neutrons to produce radioactive isotopes of the original elements. The gamma ray emissions from decay of the radioactive isotopes are analyzed, and the composition of the original sample is computed. This can be a very sensitive technique but can be expensive as it requires a neutron source and highly trained personnel (Boubel *et al.*, 1994).

References

Batchelor, G.K. (1967) An *introduction to fluid dynamics,* Cambridge University Press.

Boubel, R.W., Fox, D.L., Turner, D.B., and Stern, A.C. (1994) *Fundamentals of Air Pollution (Third Edition),* Academic Press, San Diego.

Briggs, G.A. (1975) Plume rise predictions, in: *Lectures on Air Pollution and Environmental Impact Analyses, Workshop Proceedings*, A.M.S. Boston, MA. pp. 59-111.

Businger, J.A. (1973) Turbulent transfer in the atmospheric surface layer. in D.A. Haugen (ed.), *Workshop on Micrometeorology,* Amer. Meteor. Soc., Boston, pp. 67-100.

Businger, J.A., Wyngaard, J.C., Izumi, Y., and Bradley, E.F. (1971) Flux-profile relationships in the atmospheric surface layer, *J. Atmos. Sci.,* **28**,181-189.

Caughey, S.J. (1982) Observed characteristics of the atmospheric boundary layer, in F.T.M. Van Nieuwstadt and H. van Dop (eds.), *Atmospheric Turbulence and Air Pollution Monitoring,* Riedel, pp. 107-158.

Chandrasekhar, S. (1968) *Hydrodynamic and Hydromagnetic Stability*, Oxford University Press.

Csanady, G.T. (1973) *Turbulent Diffusion in the Environment*, D. Reidel Publishing Company, Boston.

Csanady, G.T. (1967) Concentration fluctuations in turbulent diffusion, *J. Atmos.* Sci., **24**, 21-28.

Currie, I.G. (1974) *Fundamental Mechanics of Fluids*, McGraw-Hill, New York.

Deardorff, J.W., and Willis, G.E. (1984) Groundlevel concentration fluctuations from an buoyant and a non-buoyant source within a laboratory convectively mixed layer, *Atmos. Environ.*, **18**, 1297-1309.

Fleagle, R.G., and Businger, J.A. (1980) An *Introduction to Atmospheric Physics,* Academic Press, New York.

Gifford, F.A. (1976) Nuclear Safety, **17**(1), 68-86.

Irwin, J.S. (1983) Estimating plume dispersion - a comparison of several sigma schemes, *J Climate Appl. Meteor.*, **22**, 92-114.

Irwin, J.S. (1979) Estimating plume dispersion - a recommended generalized scheme, in preprints of *"Fourth Symposium on Turbulence, Diffusion and Air Pollution"*, A.M.S. Boston, MA., pp. 62-69.

Kaimal, J.C., and Finnigan, J.J. (1994) *Atmospheric Boundary Layer Flows,* Oxford University Press, Oxford, UK.

Kaimal, J.C. (1973) Turbulent spectra, length scales, and structure parameters in the stable surface layer, *Boundary Layer Meteorology,* **4**, 289-309.

Kolmogoroff, A.N. (1935) La transformation de laplace dans les espaces Linéaires, *Comptes Rendus Acad. Sci.*, 200, **21**, 1717-1718.

Lamb, R.G. (1978) A numerical simulation of dispersion from an elevated point source in the convective boundary layer, *Atmos. Environ.*, **13**, 733-741.

Lepage (1993) Wind flow in a building wake region, *RWDI report number* 93-174, RWDI, Inc. Guelph, Ontario, Canada.

Lyons, W. A., and Cole, H.S. (1973), Fumigation and trapping on the shores of Lake Michigan during stable onshore flow, *J Appl. Meteor.*, **12**, 494-510.

Maher, W.A., Cullen, P.W., and Norris, R.H. (1994) Framework for designing sampling programs, *Envir. Monitoring and Assessment,* **30**, 139-162.

Misra, P.K. (1982) Dispersion of nonbuoyant particles inside a convective boundary layer, *Atmos. Environ.*, **16**, 239-243.

Monin, A.S., Yaglom, A.M. (1973) *Statistical Fluid Mechanics,* MIT Press.

Nieuwstadt, F.T.M., and VanUlden (1978) A numerical study on the vertical dispersion of passive contaminants from a continuous source in the atmospheric surface layer, *Atm. Env.*, **12**, 2119-2124.

Nieuwstadt, F.T.M., and van-Dop (eds.) (1982) *Atmospheric Turbulence and Air Pollution Modeling*, Reidel, Dordrecht, 358 pp.

Pasquill, F. and Smith, F.B. (1983) *Atmospheric Diffusion,* Ellis Horworth Limited, Chichester, U.K.

Pasquill, F. (1976) Atmospheric dispersionparameters in Gaussian plume modeling: Part II. Possible requirements for change in the Turner workbook values: *Report EPA-600l4-760306,* US Environmental Protection Agency.

Reid, N. (1996) Field measurements, instruments and network design, *Seminar notes for a short course in atmospheric chemistry,* York University, Canada.

Slawson, P.R., and Csanady, G.T. (1971) The effect of atmospheric conditions on plume rise, *Journal of Fluid Mechanics,* **47**, 33.

Snyder, W.H and Lawson R.E. (1993) Wind-tunnel simulation of building downwash from electric-power generating stations, *US EPA internal report.*

Sorbjan, Z. (1989) *Structure of the Atmospheric Boundary Layer,* Prentice Hall, New Jersey.

Taylor, G.I. (1921) Diffusion by continuous movements, *Proc. London Math. Soc.,* (2), **20**, 196-211.

Tennekes, H. and Lumley, J.L. (1972) A *First Course in Turbulence,* MIT Press, Cambridge, Mass.

Trujillo-Ventura, A., and Ellis, J.H. (1991) Multiobjective air pollution monitoring network design, *Atmospheric Environment,* **25A**, 469-479.

Van Ulden, A.P. (1978) Simple estimates for vertical diffusion from sources near the ground, *Atmos. Environ.*, **12**, 2125-2129.

Venkatram, A., and Wyngaard, J.C. (1988) *Lectures on Air Pollution Modeling,* American Meteorological Society, Boston.

Weil, J.C. (1988) Dispersion in the Convective Boundary Layer, in A. Venkatram and J.C. Wyngaard (eds.), *Lectures on Air Pollution Modeling,* American Meteorological Society, Boston.

Weil, J.C., Sykes, R.I., and Venkatram, A. (1992) Evaluating air-quality models: Review and outlook, *J. Appl. Meteorology,* **31**, 1121-1145.

Willis, G.E., and Deardorff, J.W. (1976) A laboratory model of diffusion into the convective planetary boundary layer, *Quart. J. Roy. Meteor. Soc.*, **102**, 427-445.

CHAPTER 3

DATA NEEDS FOR SOLID WASTE MANAGEMENT

M. N. Alpaslan

Abstract. The solution to all kinds of environmental problems requires informational support to be provided by observed data. This chapter focuses on methods of data collection for solid waste management, including supplementary information on the management process itself.

3.1. Introduction

As it may be followed from the contents of this book, environment is considered herein to be made up of air, solid, and liquid (water) components. If we look back into history, we observe that the initial concern of mankind was over the supply of water to maintain life conditions. Centuries later, the problem of wastewater and wastewater treatment has emerged, followed by the present issues of air pollution and handling of solid wastes. The latter two have become significant only in the 20th century, and they still remain as unresolved problems of our age.

As noted in Chapter 1, the solution to all kinds of environmental problems requires informational support to be provided by observed data. This chapter focuses on methods of data collection for solid waste management, including supplementary information on the management process itself.

Solid wastes are all the wastes arising from human and animal activities that are normally solid and that are discarded as useless or unwanted. The generation of solid wastes may simply be described as in Figure 3.1, (Tchobanoglous, 1977). This figure indicates that solid wastes are generated at every step in the process of the conversion of raw material to goods.

The disposal of wastes leads to significant problems related to public health and ecological impacts. This fact implies the need for proper management of solid wastes.

Solid waste management involves successive processes extending from the generation of wastes to the final disposal stage. In other words, to solve the solid waste problem, functionalities indicated in Fig. 3.2 are performed, in what is usually known as "solid waste management". In practice, some or all of these steps are applied. The basic goal of solid waste management is optimization of the described functions. In this framework, data collection practices are employed for better management practices. Data collection procedures are often different for each step; yet, they constitute an essential component in the planning of management stages.

N.B. Harmancioglu et al. (eds.), Environmental Data Management, 49-60.

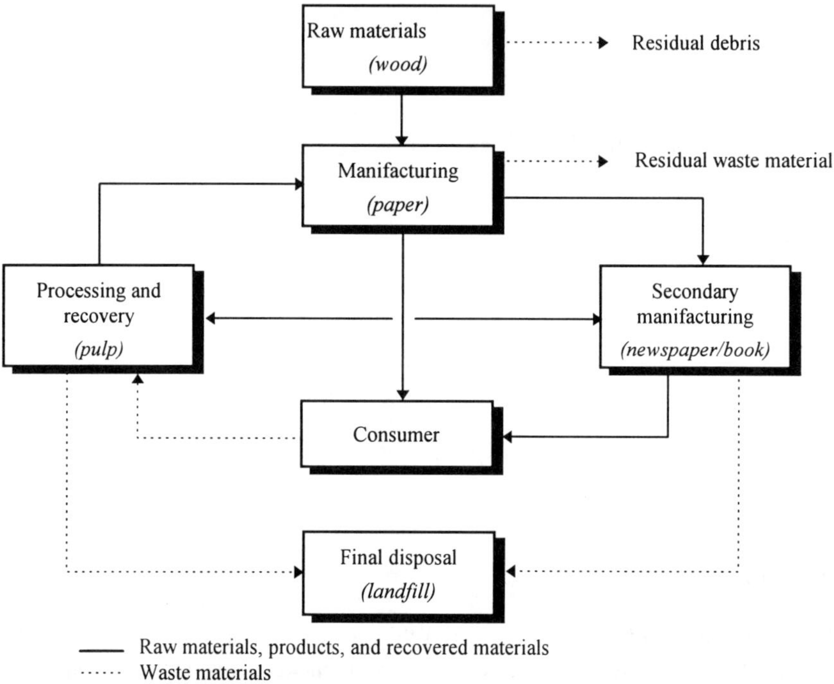

Figure 3.1. Generalized flow of materials and the generation of solid wastes in society (Tchobanoglous *et al.*, 1977)

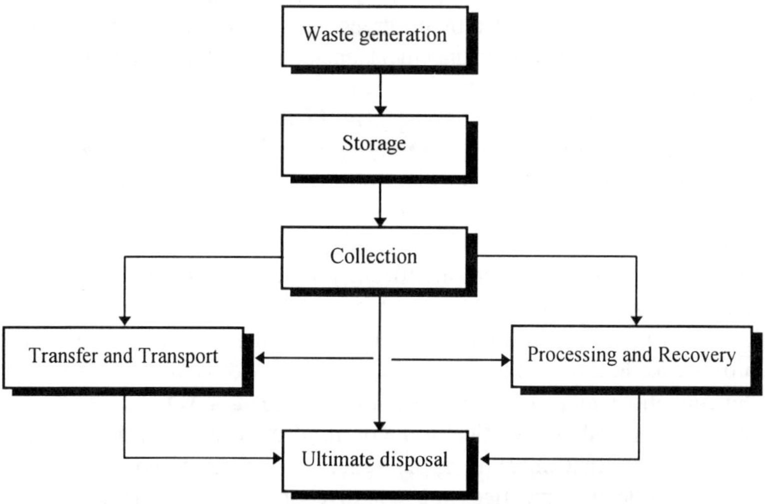

Figure 3.2. Simplified diagram showing the interrelationships of the functional elements of a solid waste management system

It should be noted here that the data collection policy for solid waste depends completely on the objectives of each management step, as is the case with water quality monitoring. However, as it will be discussed later, since objectives of solid waste management and water quality management are very different from each other, the philosophy behind solid waste data collection is also different from that underlying the collection of water quality data. For example, in water quality monitoring, one of the objectives is to assess whether the original quality of water has degraded. A possible approach to determine this situation is to apply trend analyses where long-term (historical) data are essential; whereas in solid waste management, historical data may only help for future estimations to a limited extent. Besides, solid waste quantity and quality are not functions of nature; the level of data analyses performed to interpret the two types of data (water quality and solid waste data) are different.

Another major difference between solid wastes and water is that the latter one is liquid and flows when any potential (topographic elevation difference) is provided. This means that it reflects a significant spatial scale. Furthermore, its quality varies both by natural and anthropogenic impacts. A spatial distribution is also valid in the case of solid wastes; however, solid waste accumulation at a point in space results not from a natural transport mechanism but from production at anthropogenic settlements. Unlike the case of water, there is no correlation between spatial points of solid waste accumulation.

And finally, human beings are absolutely obliged to have clean water. The degree of water contamination is a significant figure for treatment as well as for assessment activities. On the other hand, solid wastes remain where they are disposed. Reactions may only occur (leachate or biogas generation) and cause detrimental effects upon the confined environment.

Therefore, when water (or wastewater) is disposed to the environment, its spatial distribution may have an importance. On the other hand, in solid waste management, spatial distribution of data is significant only for the identification of waste sources within an area (city or town).

According to the above discussion, one can state that solid waste data collection efforts are basically problem or project-oriented. Problems occur in the form of hazards to public health, and projects comprise the planning of the major steps of solid waste management.

3.2. Need for Solid Waste Data / Objectives of Solid Waste Data Collection

As stated in the previous section, whoever decides to collect data, he or she should ask the question "why should data be collected?". In other words, "what are the objectives of solid waste data collection?". The answer to this question is more simple than that for water quality monitoring. The basic objective in solid waste data collection is to provide data for the planners, engineers, and the operators (executors) to provide the background information for solid waste management. Figure 3.2 shows the solid waste management issues of planning, engineering investments, and the operational features of management. In relation with these issues, solid waste data are collected with an engineering point of view to:

a) select the appropriate on-site storage facilities, i.e., bins, containers, ground storage, etc. without endangering public health and aesthetic conditions but assuring economic solutions;

b) collect solid waste effectively, (determine the hauling distances, types and/or capacities of collection vehicles, etc.);

c) determine the locations and capacities of transfer stations (plan the transfer of wastes from smaller collection vehicles to larger transport equipments and the subsequent transport of wastes, usually over long distances, to the disposal site);

d) determine the appropriate processing and recovery techniques, (equipments and facilities used both to improve the efficiency of the functional elements of solid waste management and to recover usable materials, conversion products or energy from solid waste);

e) and finally to design and operate the ultimate disposal (landfill area) facilities, (location of the landfill, leachate and gas collection, covering material, etc.).

The above points are the functional elements of solid waste management that transfer solid waste from cradle to grave. Among them, item (d) recently has acquired special attention. Human beings are no longer willing to discard solid wastes as useless or unwanted; at least, they want to "recover", "reuse" and "recycle" (the 3R principle) all the wastes (or goods) as much as possible, so that the amount of wastes and eventually management efforts will significantly decrease.

Essentially, data are needed on solid wastes to realize the above steps of solid wastes management, to design efficient solid waste disposal systems in technical and economic terms, and to operate such systems effectively. The types of data to be collected basically depends on objectives of solid waste management.

3.3. Types of Solid Waste Data

Type of solid waste data may be interpreted in two ways. The first one relates to the sources where solid wastes are generated, e.g., residential, commercial, agricultural, open areas, etc. Data may also relate to the nature of the solid waste such as food wastes, rubbish, ashes and residues, demolition and construction wastes, etc. However, the characteristics of solid waste data do not depend on the "type" of wastes. A better classification would be as follows:

a) *Massy characteristics of solid wastes*: This involves the amount of solid wastes (mass/time, mass/time. ca or mass/production) generated from a source (city, community, industry or any agricultural activity);

b) *Quality characteristics of solid wastes* (e.g., composition, calorific values, moisture contents, etc.).

As discussed before, the selection of the type of data basically depends on the objectives of solid waste management, which essentially are the objectives of the steps shown in Fig. 3.2. Accordingly, data requirements for each step may be specified as follows:

Step 1. Waste Generation

This step is only a description of the generation process and does not imply any engineering activity. Thus, there is no need here to focus on data requirements.

Step 2. Solid Waste Storage

Two factors are significant with respect to the types of data needed:

a) appropriate duration of storage,
b) the volume of the storage equipment, e.g., that of bins, containers, etc.

The putrecibility of solid waste must be identified to determine the first factor; therefore, data are needed on the composition of wastes. For the second point, the rate of solid waste generation is significant. In this case, the amount of solid waste generated in a time unit, or the specific solid waste weight (kg solid waste/ca. day) must be known. Essentially, the determination of the volume of storage bins depends on the duration of storage. Thus, in the design of data collection systems. one first has to determine the storage duration and then calculate the bin volume.

Step 3. Solid Waste Collection

This is the step that reflects significant data needs. Some of the required data are easy to collect; whereas some are highly difficult to obtain. In most cases, one may have to make assumptions.

Data required for this step comprise two types:

a) data that serve to determine the number and the type of collection vehicles;
b) data that are needed to identify the route of the collection vehicle and the time of collection.

In the planning process, one observes that, the above two objectives are interrelated. Yet, the number of required vehicles relates to the capacity of the vehicle and the amount of solid waste to be collected.

To determine the collection route and the time of collection, it is necessary first to select the collection system. In general, collection systems are classified into two categories:

a) hauled-container systems;
b) stationary container systems.

For both systems, a detailed map is required for the region (city) where solid wastes are to be collected. Apart from this, data are needed on capacities of containers, speeds of vehicles, length of the workday (h/d), time from garage to the first container location (h), time from the last container location to the garage (h), time required between the containers (h) and the similar.

Step 4a. Processing and Recovery

This step deals with the better handling of solid wastes. There are several means for it, and, in general, they may be classified as:

a) Mechanical volume reduction (compaction),
b) Chemical volume reduction (incineration, pyrolysis, chemical conversions, etc.),
c) Mechanical size reduction (shredding),
d) Component separation,
e) Drying and dewatering.

For all the above processes, data are needed on both the quantitative and qualitative properties of solid waste. For example, if the objective is incineration of solid waste, then data only on the amount, calorific value, moisture content and inert residue content will be necessary; whereas if the objective is to "reuse or recycle" the solid waste, then the amounts of paper, glass, metal and organics will have prime importance.

Step 4b. Transfer and Transport of Solid Wastes
This step involves:

a) transfer of solid waste from small collection vehicles to larger ones;
b) transport of solid waste to the selected destination via transport vehicles. The destination may be processing/recovery site or more often, the disposal site. Then, for this step, data needs cover:
    (a) a detailed map of the region (which also shows the roads);
    (b) amounts of solid waste generated by different parts of the city;
    (c) land use conditions to permit transfer/transport of solid waste;
    (d) the relative location of the transport point with respect to solid waste collection points.

In addition, depending on the type of transfer station (i.e., direct-hauling or storage-hauling), data must be collected or assumed, regarding the amount of daily loads, number incoming vehicles and outgoing trailers.

Step 5. Ultimate Disposal of Solid Wastes
Here, disposal refers to the landfilling stage other than dumping or its alternatives, i.e., ocean dumping, deep mine dumping, space dumping, etc.; however, the most common means of disposal is the ultimate landfilling.

There are two significant points relevant to landfills:

a) the operation time until the completion of landfill (when the landfill will reach the full stage);
b) estimation of the amount of leachate and gas production.

The first point requires data on the amount of solid waste dumped. It may equally be important to know the compressibility of solid waste and the degree of consolidation in the landfill for the cases when filling is realized under compaction.

For point (b) above, the composition and the moisture content of the solid waste must be determined. One may as well require data on precipitation that reaches the landfill site. For operational purposes, it is also important to identify the amounts of hospital wastes, hazardous wastes, bulky material, tires, etc., that are landfilled. In this way, special sites may be reserved in the landfill for these materials.

In addition, during the operation of the landfill, one of the most hazardous effects may be the contamination of groundwater. In order to observe this situation, monitoring wells (boreholes) should be drilled around the landfill area, for example, with three downstream and one upstream of the landfill. To detect the movement of contaminants from the site at an early stage, it will be necessary to place at least one hole close to the site boundary. Systematic groundwater quality analyses should be performed both during the operation period and certain years after the operation is terminated. Generally, heavy metals, trace elements and micropollutants are observed. The frequency of sampling is normally quarterly for routine checking of leachate levels and quality within the site. However, there may be specific reasons for more frequent analysis of some variables. For example, the monthly monitoring of conductivity, dissolved oxygen, and pH, using portable instruments, identifies changes in the quality of groundwater. Attention must be paid to sampling techniques if meaningful results are to be obtained. Monitoring of groundwater and surface water provides background data which will assist in the event that, once the landfill operation has commenced, there is a dispute regarding, the source of contamination in a pollution incident. Ideally, the site should be monitored for a twelve-month period to identify seasonal variations in water quality prior to disposal of any waste. Without these background data, it may be difficult or impossible to prove that an alternative source is responsible for a pollution incident.

Therefore, it can be stated that if any "a priori assumptions" are made by the planners, as to the characteristics of solid waste then the selection of the type of data will be more robust; otherwise, if there is no a priori assumption then a wide range of solid waste data types should be collected.

3.4. Sampling of Solid Wastes

3.4.1. NATURE OF SOLID WASTES

Solid wastes have a heterogenous character in mass. Therefore, it is not possible to have an ideal representative sample for a city or for an area. However, if appropriate methods are applied with care, acceptable samples can be taken and analyzed. There is a good source book for this purpose entitled "*Methods of Analysis of Sewage, Solid Wastes and Compost*" published by WHO Waste Disposal International Reference Center.

In addition, there exists no standard method in order to analyze the characteristics of solid wastes (Curi, 1997). The generalization of results derived from small samples are often doubtful since, as noted above, wastes do not have a homogeneous character. In addition, the characteristics of solid wastes change from the shape of the bin storage up to the final disposal site. Therefore, the corresponding sampling point of each datum should clearly be specified.

3.4.2. SELECTION OF SAMPLING SITES

Monitoring site selection to determine the amount and contents of solid wastes occurring as a result of anthropologic factors is a significant decision since the sample has to represent the overall population. There are generally two alternatives for this decision:

1) sampling of solid wastes at the generated site (in the city);
2) sampling of solid wastes at the disposal site: (during the unloading of collection vehicles in the landfill area).

Before going further, it should be noted here that solid waste data collection is really a rough and tough job and needs intensive labor. Furthermore, working conditions are not pleasant.

If the samples are to be collected from the source (the former site alternative), then the plan view of the city is divided into squares, and the most representative point in the square is selected as the sampling area. Here, as it can be imagined, the numbers and/or the dimensions of the squares become effective over the sensitivity and the reliability of collection practices. Another division may be made by considering rich and poor areas of the city since the data from each part will likely be different. As it is known, the value of the data are assessed by their information capacity. If the values of the data are significantly variable, then the information content of the data will be high. Considering this fact, it is very well known that both the quantity and the quality of solid wastes generated from rich and poor districts of a city are different from each other. Therefore, in order to increase the efficiency of data collection activities, the city is ranked and divided into rich to poor districts, and sampling is performed on each rank.

Besides the "rich" and the "poor" type of classification, a city may also be divided into "residential areas", "commercial areas" and "industrial areas". These major divisions can then be subdivided as, for example, "high rise apartments - low rise apartments", "houses in yards" or "shopping centers, trade centers", and the similar. Data collected from such areas will again have high information content and will represent the city properly.

In such collection practices, three sampling caps (or nylon sacks) are distributed to houses (dwellings) in order to store solid wastes for 24 or 48 hours. The first sack (or cap) is for organic wastes (good for composting), the second one for the wastes appropriate for recycling, and the third one for ash and dirt. Then, the sacks are collected, weighed, separated into components, and other desired experiments are performed.

The second sampling site for solid waste data collection is the processing or landfill areas. Here, the samples are taken when the collection vehicles empty their wastes. Then, the samples are taken from different vehicles by using shovel, ladle or similar equipments. The collected samples are mixed, grouped and then mixed, to eventually handle at least 50 kg of a sample is handled. The schematic explanation of this procedure is given in Fig. 3.3.

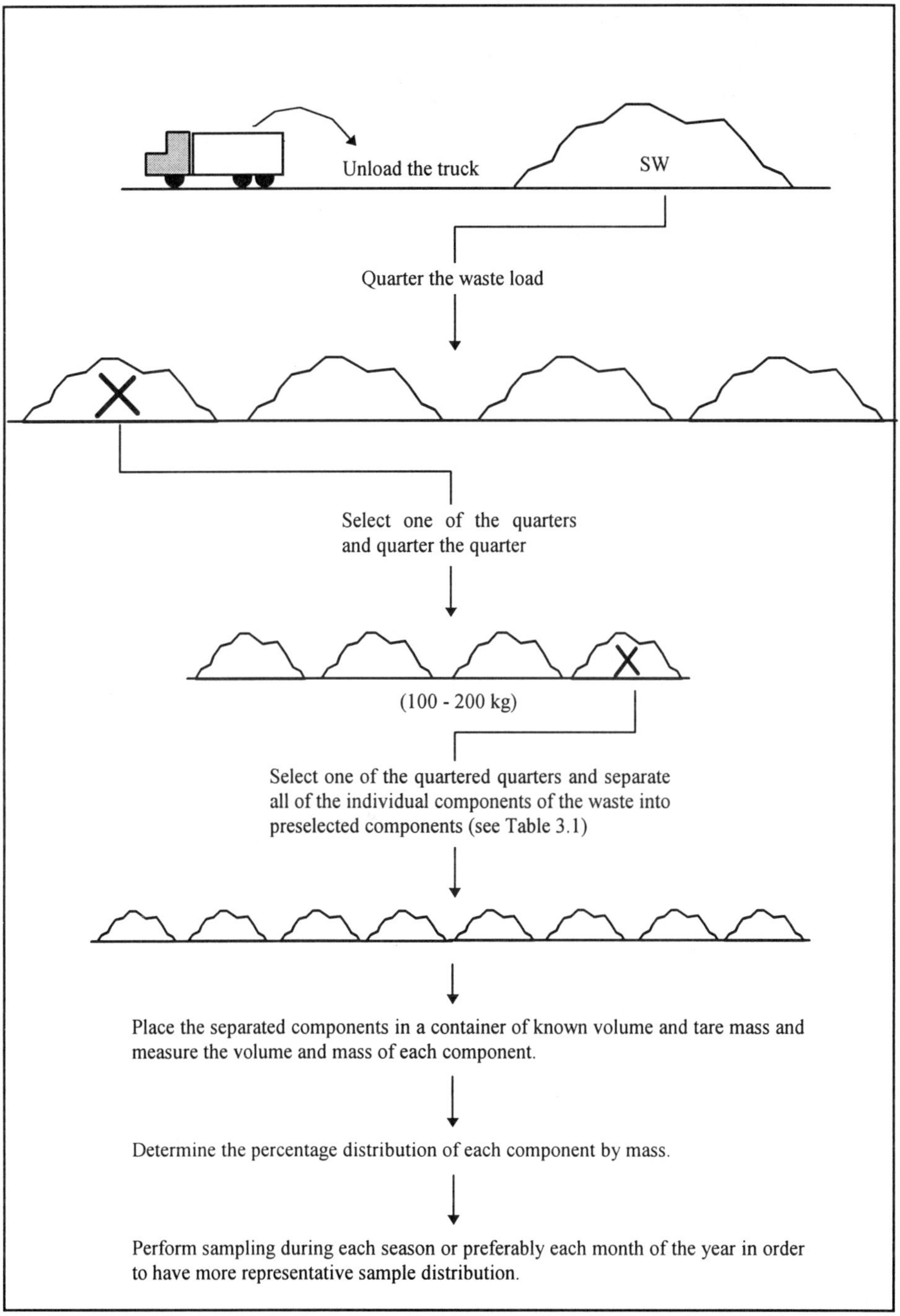

Figure 3.3. Sampling at a disposal site

For example, if residential and commercial areas are intermixed within a district, one may want to identify the amount of waste load per person (specific solid waste production kg/ca.day) from the residential area and that from the commercial area. In this case, vehicles from each area are weighed. For example, let's assume that 100 tons of load come per day. The amount of load that originates from the residential area is sampled by the 1st method as 80 tons; then, the difference, SW (in commercial areas) = SW (in landfill) - SW (in houses), gives 20 tons as the load from commercial areas.

The disadvantage of such sort of sampling is that it is very difficult to distinguish whether the wastes are coming from the poor or the rich parts, or from residential or commercial areas. By examining the routes of the trucks, this disadvantage may be lessened to a certain extent; however, this is rather difficult in practice.

Sometimes, in order to have reliable data, both collection practices (i.e., in source and in disposal site) are applied to both sites, and the relationship between the data collected from the former and the latter methods is investigated.

3.4.3. DETERMINATION OF SAMPLING FREQUENCY

Following the selection of sampling sites, the frequency of sampling should be determined. As stated before, the efficiency of monitoring is increased if observations can detect significant variations in wastes. In solid waste management, the quantity and the quality of wastes vary slightly from day to day. The magnitude of variation also slightly increases over the weekends in contrast to weekdays. Significant variations are expected with respect to seasons. Therefore, the frequency of solid waste data collection can be set on a seasonal basis. However, this does not mean that such a program will be applied 4 times per year. Since each sampling duration for each season should cover a week, the sampling times should be adjusted as "alternating week, alternating day". This means that, at the beginning week of the season, if a sampling program is performed on Monday, the next should be done on Tuesday of the following week, and so on.

This means that throughout a season, sampling duration lasts seven weeks. If a season is considered as 3 months or 12 weeks, sampling duration covers 7 weeks out of 12. Therefore, sometimes during the last 5 weeks, sampling practices are continued so that the procedure is repeated throughout the year. The alternating day alternating week approach eliminates the errors created by the high cross correlation (or time lag correlation) characteristics of solid waste data.

The determination of the number of samplings plus the amount of samples (or that of sample sites) is another important issue. The required number and the minimum amount of samples are determined by statistical analyses.

3.4.4. VARIABLES TO BE MEASURED

As stated before, the composition of solid waste is of prime importance in solid waste management. The major components may be glass, paper, organic, plastic, metal, dirt and ash (residue). The selection of the variable to be sampled depends basically on the objectives of solid waste management. On the other hand, the determination of solid waste composition requires analyses by statistical methods.

As noted in Section 3.3, properties of solid wastes that must be monitored can generally be classified as massy and quality characteristics. Massy characteristics are the specific solid waste weight (mass/capita. time; kg/ca. a) specific solid waste volume (volume/capita.time), and volumetric weight or bulk density (mass/volume; tons/m3). The most reliable indicator of solid waste properties is the specific solid waste weight. Since the volume of solid wastes is variable, specific solid waste volume does not give a reliable figure. The variability of solid waste volume is due to the compaction applied when collecting or landfilling the wastes, rather than to its natural compaction or consolidation. This variability is also reflected in the volumetric weight or bulk density. Thus, specific solid waste weight is the most significant massy parameter characterizing solid wastes.

There are several ways of determining the specific solid waste weight. One may also use the following simple procedure:

a) weigh the collection vehicle prior to emptying (A);
b) determine the number inhabitants served by that collection vehicle (B);
c) divide A into B by considering the time and the empty weight of vehicles.

For practical purposes, 1000g of solid waste/ca.day may be used.

Quality characteristics of solid wastes comprise:

a) Composition;
b) Moisture content;
c) Calorific value.

The composition of solid wastes can be classified as physical, chemical and microbiological. Otherwise, it may be described in terms of solid waste components as seen in Table 3.1 (Peavy, et.al, 1985).

TABLE 3.1. Composition of solid wastes

Component	Typical Value (%)
Food wastes	14
Paper	34
Cardboard	7
Plastics	5
Textiles	2
Rubber	0.5
Leather	0.5
Garden trimmings	12
Misc. organics	2
Wood	2
Glass	8
Tin cans	6
Nonferrous metals	1
Ferrous metals	2
Dirt, ashes, brick, etc.	4

3.5. Shortcomings of Solid Waste Data Collection

Solid waste management is still an unresolved problem particularly for developing countries, as well as for developed countries to a certain extent. Several difficulties may be noted. Among the management steps shown in Fig. 3.2, storage cannot be realized in a systematic manner; collection, transfer and transport of solid wastes are not handled in an economic way; processing techniques are often costly and inefficient; disposal is either continued as open dumps or sanitary landfills cannot be operated effectively. To handle these problems in the most efficient way, reliable and accurate data are needed. Often, such data are not available. When data from one area is used to represent another area, results are often misleading since characteristics of solid waste depend upon standards of living, degree of industrialization, gross national production, etc. Thus, the major difficulty with respect to data use on solid wastes is the lack of sufficient data and the fact that available data are of a local nature, representing only the area and the time when solid wastes are collected. For example, the water content in the municipal solid wastes of Germany during the month of August is around 20%; whereas it is 80% in Turkey (Curi, 1997).

Another limitation on solid waste data is that, often, solid waste data are presented as individual values in the form of tables with no statistical interpretation. Thus, the reliability of such figures is doubtful.

The highly variable characteristics of solid waste with regard to life standards, habits, seasons, etc., make it also pretty difficult to collect representative data.

Furthermore, collection of solid waste is often an unpleasant, messy, and difficult procedure. In most countries, this activity is performed by untrained personnel. These points also hinder the collection of reliable data, because the requirement of collaboration between the data collector and solid waste collector is not assured.

References

Curi, K. (1997) Sampling of municipal solid wastes, in: N.B. Harmancioglu, M.N. Alpaslan, S.D. Ozkul and V.P. Singh (eds.), *Integrated Approach to Environmental Data Management Systems*, Kluwer Academic Publishers, NATO ASI Series, 2. Environment, vol. 31, pp. 149-152.

Peavy, H.S., Rowe, D.R., and Tchobanoglous, G. (1985) *Environmental Engineering*, McGraw-Hill International Editions, Civil Engineering Series, 699 p.

Tchobanoglous, G., Theisen, H., and Eliassen, R. (1977) *Solid Wastes*, McGraw-Hill Book Company, New York, 592 p.

CHAPTER 4

WATER QUALITY MONITORING AND NETWORK DESIGN

N. B. Harmancioglu, S. D. Ozkul and M. N. Alpaslan

Abstract. In recent years, shortcomings of both the available data on water quality and the existing networks have led designers to focus more critically on the design procedures used. Within this respect, this chapter addresses the prevailing problems associated with water quality monitoring networks and discusses current attempts towards improvement of existing networks.

4.1. Introduction

In recent years, problems observed in available water quality data and shortcomings of current monitoring networks have led designers and researchers to focus more critically on the design procedures used. Developed countries have felt the need to assess and redesign their monitoring programs after having run their networks for more than 20 years. Developing countries are still in the process of expanding their rather newly initiated networks; yet they also find it necessary to evaluate what they have accomplished so far and how they should proceed from this point on. In both cases of the developed and the developing countries, the major problem is that there are no universally confirmed guidelines to follow in the assessment and design of water quality monitoring networks. Upon this need, significant amount of research has been initiated to evaluate current design procedures and investigate effective means of improving the efficiency of existing networks (Ward *et al.*, 1990; Chapman, 1992; Harmancioglu *et al.*, 1992; Adriaanse et al, 1995; Ward, 1996; Timmerman *et al.* 1996; Niederlander *et al.*, 1996; Dixon and Chiswell, 1996).

At present, the adequacy of collected water quality data and the performance of existing monitoring networks have been seriously evaluated for two basic reasons. First, an efficient information system is required to satisfy the needs of water quality management plans and to aid in the decision making process. Second, this system has to be realized under the constraints of limited financial resources, sampling and analysis facilities, and manpower.

Despite all efforts made on monitoring of water quality, the current status of existing networks shows that the accruing benefits are low. That is, most monitoring practices do not fulfill what is expected of monitoring. Thus, the issue still remains controversial among practitioners, decision makers, and researchers for a number of reasons. First, proper delineation of design considerations is often overlooked. That is, objectives of

N.B. Harmancioglu et al. (eds.), Environmental Data Management, 61-106.

monitoring and information expectations for each objective are not clearly identified. Second, there are difficulties in the selection of temporal and spatial sampling frequencies, the variables to be monitored, and the sampling duration. Third, benefits of monitoring cannot be defined in quantitative terms for reliable benefit/cost analyses. There are no definite criteria yet established to solve these problems.

In view of the above difficulties, water quality monitoring and network design has become one of the most significant problem areas in environmental management. Chapter 18 of Agenda 21, declared at UNCED of 1992, emphasizes that information on the hydrological cycle, including both water quantity and water quality, constitutes the basis for effective water management. Yet, it is also stated in this chapter that current monitoring systems experience a declining trend in terms of effectiveness. This situation is highly unfortunate since, globally, there is an increasing trend in our demand for water (WMO, 1994). In simple terms, the basic problem is that, despite all the investments and efforts devoted to monitoring, we still lack the information to define water quality (Ward, 1996).

Considering the current problems in water quality management, this chapter is intended to review the current status of monitoring networks and design procedures on examples from developed and developing countries. This review covers details of the network design problem with respect to monitoring objectives, sampling sites, time frequencies, sampling duration, and variables to be sampled. Current attempts at redesign of monitoring programs and recommendations for better designs are presented

4.2. Water Quality Monitoring

4.2.1. DEFINITION

Water quality monitoring comprises all sampling activities to collect and process data on water quality for the purpose of obtaining information about the physical, biological, and chemical properties of water. Besides collecting data, monitoring activities cover the subsequent procedures, such as laboratory analyses, data processing, and data analyses to produce the expected information. These procedures are essentially the basic steps of a data management system, presented in Fig. 1.1 of Chapter 1.

Water quality monitoring practices are basically designed to achieve specific purposes which lead to various types of monitoring, i.e., trend monitoring, biological monitoring, ecological monitoring, compliance monitoring, and the similar. Among these types, collection of data for purposes of assuring compliance with standards has probably been the oldest practice. In the past, these activities were carried out in a problem, project, or user-oriented framework. Recently, however, as the emphasis is shifted more to water quality management and control efforts in a larger perspective, the major concern has become the assessment of the quality of surface waters in a wide area or a river basin. In achieving this specific purpose, trend monitoring is required to evaluate both changing quality conditions and the results of control measures.

One of the developments in the late 80's with respect to the types of monitoring is that sampling for stream standard violations has gradually been replaced by effluent sampling. This is due to the inadequacies of the former in realistically detecting possible violations (Sanders, *et al.*, 1983; Warn, 1988; Alpaslan and Harmancioglu, 1990). Compliance monitoring can be most efficiently realized only by means of continuous sampling, which in most cases is costly. On the other hand, intermittent sampling poses some difficulties in detecting what is a true violation and what is not, in addition to uncertainties in pinpointing the possible violators (Sanders, *et al.*, 1983). Under these conditions, the preference goes for effluent monitoring rather than for in-stream monitoring when the concern is compliance with standards. This is also a change in favor of trend monitoring because it enables the assessment of both prevailing and/or changing water quality conditions and the effectiveness of control measures. In fact, some researchers have claimed that the basic function of monitoring is to determine long-term trends in water quality, once compliance is assured by effluent monitoring (Dandy and Moore, 1979). Effluent monitoring is not considered within the scope of this chapter because it has its own very specific aspects that are beyond the general design procedures of monitoring networks.

In some studies, two basic functions are defined for water quality monitoring: prevention and abatement (Dandy and Moore, 1979; Karpuzcu, *et al.*, 1987). The first one has the objective of maintaining the existing unpolluted or acceptable status of water quality; while the second one puts the emphasis on a control mechanism by reducing or moderating pollution conditions. Prevention foresees the enforcement of effluent standards and, thereby, requires effluent monitoring plus trend monitoring. For abatement, compliance with in-stream standards is significant, so that compliance monitoring has the highest priority among other types of monitoring.

4.2.2. COMPLEXITY OF WATER QUALITY MONITORING

Whatever the specific purpose of monitoring may be, it must first be recognized that water quality monitoring is a highly complex issue. Apart from technical features of monitoring, this complexity may be attributed to two factors:

a) uncertainties in the nature of water quality; and
b) uncertainties in delineating a specific purpose for monitoring.

Uncertainties in the nature of water quality are due to the two fundamental mechanisms underlying these processes: the natural hydrologic cycle and man-made effects, which are often referred to as the "impact of society". Both of these mechanisms, particularly the first one, are affected by the laws of chance so that water quality has to be recognized as a random process by nature (Sanders, *et al.*,1983). Monitoring activities, then, are required to reflect the stochastic nature of water quality to efficiently produce the expected information. This is why most researchers like Sanders *et al.*(1983), Cotter (1985) and Karpuzcu, *et al.* (1987) specify the term "monitoring" further to mean "statistical sampling".

Second, it is not quite easy to define a specific purpose for monitoring. The technical part of this problem is best described by Praskins (1989) as: "Today ... we face water quality problems whose sources are diffuse, impacts subtle, and solutions unproven". Besides, specification of purposes is subject to social, economic, and legal constraints, which are also subject to unexpected changes in time.

4.2.3. SIGNIFICANCE OF WATER QUALITY MONITORING

As complex as it is, water quality monitoring is also highly significant because it is our only means of being informed about water quality. Thus, monitoring constitutes the link between the actual process and our understanding, interpretation, and assessment of the highly complex phenomena. Therefore, water quality monitoring is the most crucial activity on man's side with respect to all management and control efforts.

For example, Schad (1984) claimed that "we are not really sure of the cost-effectiveness of some of the programs accomplished to date because of the lack of adequate monitoring of water quality in our streams, lakes, and estuaries", a statement which holds true even today. According to Ward (1989), our understanding of environmental processes and problems evolve quite rapidly, whereas monitoring systems develop at a slower pace, often becoming out of date with respect to recently emerging issues and purposes of water quality assessment. On the other hand, the decision making process in water quality management is highly sensitive to the reliability and accuracy of available data. Unreliable data, further, the misinterpretation of the information they convey may lead to wrong decisions. This situation is apparently worse than taking no action at all. In such a case, "the underlying data can be said to have a negative economic value" (Moss,1989).

4.2.4. WATER QUALITY MONITORING NETWORKS

Assessment of water quality conditions over a wide area (such as a river basin) with respect to time and space requires the monitoring activities to be carried out in a network. A monitoring network comprises a number of sampling sites which collect data on particular water quality variables at selected time intervals. At this point, one has to distinguish between the terms "monitoring" and "network". The former refers to the actual sampling process at a site; whereas, the latter describes a number of monitoring stations at selected sites, which operate in coordination with each other. Such a coordination is realized by the selection of appropriate sampling sites, sampling frequencies, and variables to be sampled. Therefore, monitoring a number of variables at random points with random time intervals does not constitute a network unless this coordination is established.

To be more specific, a network is a family of systematically operated monitoring stations which, as a whole, represent the water quality conditions over a wide area. The systematic (or coordinated) operation of the network is realized by the selection of three basic factors: sampling sites, sampling frequencies, and variables to be sampled. Thus, network design covers basically the determination of these factors to produce required information. Other components of monitoring, i.e., laboratory analyses, data processing and data analysis procedures, can be evaluated as the subsequent steps of the network design problem.

Developed countries have established water quality monitoring networks within the past 30 years, and developing countries are in the process of enlarging the scope of their monitoring activities into a network.

4.3. Existing Networks

4.3.1. BACKGROUND

Water quality observations date back to almost 100 years. Yet the need for systematic measurements has become eminent only recently as a result of:

a) the recognition of water quality as a hydrologic process; and
b) the increased concern over water quality and, thereby, the demand for a better understanding of the process.

Regular observations coupled with necessary laboratory analyses were then started basically with a problem or project-oriented approach to collect data as needed and where needed. These early attempts at monitoring water quality were by no means considered in connection with regular hydrologic networks. Several variables were observed at a large number of sites, but with temporal frequencies as low as four samples per site per year (Starosolszky, 1987). Later, as the need arose for more data, the frequency of sampling has been increased to at least monthly and finally to daily observations. These developments were apparently coupled with parallel advances in both sampling and laboratory analysis techniques (automatic samplers, analyzers and monitors). In the meantime, the relation between water quantity and water quality variables was emphasized, as the former has been recognized to be the major carrier of pollutants. As a consequence, the sampling procedures were modified to include simultaneous measurements of discharge and water quality variables. This led to the consideration of information transfer between water quality variables themselves (Starosolszky, 1987; Sanders *et al.*, 1983; Harmancioglu *et al.*, 1992).

Developed countries have already experienced the above-mentioned progress and developed water quality monitoring networks within the last two decades. Yet, the problem is not over; the design of water quality monitoring networks is still a current issue receiving considerable attention from researchers and decision makers. This indicates the presence of unsolved problems still remaining in the existing networks.

On the other hand, it is by no means surprising that existing water quality monitoring networks have shortcomings, if one considers that even water quantity or other hydrometerological data networks have similar problems despite their much longer historical background. There are still the questions of how many gages are needed to present a particular basin, how frequently should the variables be sampled, how long should the gages continue operating, or should they be discontinued. Yet these issues are more intensified in case of water quality variables as they are more costly and time-consuming to sample. Furthermore, the information needs of water quality variables are much more diverse than in the case of other hydrometeorological variables. For example, if one inquires about the quantity of water at a certain time and space along a river, the expected reply will be a single value to represent the discharge. However, the answer to the question "what is the quality of water?" has to include the outcomes of several variables so that one has to deal with a "vector" of variables instead of a "single" discharge variable. Sanders *et al.* (1983) point out that "several hundred variables have already been identified that may be of interest to different users in a comprehensive description of water quality processes". Thus, in general, hydrological data network design is a fairly complicated issue, and in particular for water quality, it becomes more complex due to the nature of and information needs on water quality processes.

In most developing countries water quality data collection practices have not yet evolved into what may be called a "network". These countries basically live through the beginner's problem to face, in time, the present difficulties of the developed countries. Consequently, the shortcomings of existing networks need to be considered separately for developed and developing countries.

4.3.2. NETWORKS IN DEVELOPING COUNTRIES

Developing countries are experiencing fast economic growth through industrialization, urbanization and agricultural activities. The cost is also fast growth of environmental pollution, the majority of which occurs in surface waters. Much of the progress in these countries depends on the development of their water resources; therefore water use for various purposes is particularly important. Yet, the case is often that such countries spend intensive efforts to develop their water resources on one hand, but other areas of economic growth limit the amount of usable water by pollution on the other hand. Thus, water quality control has become an urgent issues requiring rapid remedies.

The first step in efforts toward mitigation of water quality is to gather data, and this activity has started in highly polluted areas. Therefore, data collection procedures are often of a problem or project-oriented character. However, the demand on water for various purposes also requires information about the status of unpolluted waters. Therefore, most developing countries have recently started the design of water quality data networks to monitor both the polluted and the nonpolluted areas.

In contrast to developed countries, developing countries have not yet stored enough data to start questioning what they should do with the observed samples or how they should evaluate these data. Rather, they are concerned more with the selection of sampling sites, intervals, and variables to be observed. In time, these countries, as soon as they build up sufficient data banks, will face the current problems of developed countries, the major one being the lack of agreement between objectives and available data. Actually, developing countries live through the same experiences as developed countries did 10-20 years ago. However, they have the advantage of learning through the mistakes of developed countries. Consequently, they can design their networks so as to prevent the experienced shortcomings of present networks in developed countries.

On the other hand, developing countries have to deal with some other problems before catching up with the developed countries. For example, economic pressures are extremely significant in the former so that the cost-effectiveness of any network will have a basic consideration in design procedures. Further, monitoring methodologies need to be adjusted at all levels, including sampling techniques, laboratory analyses, data retrieval, storage and eventually data transmission. These procedures will require training of personnel to run a reliable system.

In developing countries, the basic problems related to selection of variables, sampling sites and frequencies appear to be solved on the basis of demand for particular data and available facilities for data collection. However, there are no clearly defined objective criteria to be followed as guidelines in the development of the network. Therefore, it is fairly difficult to assess the significance of any new development (e.g., addition of new monitoring sites, changing of sampling frequencies, addition or exclusion of variables to be observed) in the network. As soon as the current monitoring practices fill in sufficiently rich data banks, the shortcomings of earlier decisions, if any, will likely become more apparent because all data may not be found useful. This actually means that, if the network develops without the required guidelines, it may turn out to be a "data-rich, information-poor" system (Ward *et al.*, 1986, Harmancioglu *et al.*, 1992; Harmancioglu and Singh, 1991).

4.3.3. EXAMPLES OF NETWORKS IN DEVELOPING COUNTRIES

Turkey

Turkey, as a developing country, is presently experiencing the progress described in Sections 4.3.1 and 4.3.2 to develop a nation-wide water quality monitoring network. There are two major monitoring agencies in Turkey: State Hydraulic Works (DSI) and Electrical Works Authority (EIE). The monitoring practices of the two agencies extend to all Turkish river basins and thus are developing into what may be called a “nation-wide network”.

DSI started sampling in 1979 at 65 sites, most of which were located in polluted areas. By 1982, these activities were intensified to cover 182 sampling sites along with an increase in the number and type of variables observed. The program for the year 1986 enlarged the system to 489 sites which included not only polluted areas but also surveys of unpolluted streams. In 1988, the number of sampling points increased to 679 to reach more than 1000 sites as of 1995 (Harmancioglu *et al.*, 1994 a). The given figures indicate a fairly rapid development of a network; yet, the basic questions of where, when and what to observe still remain unsettled. In selection of sites, the basic considerations are the locations of polluting sources, easiness of access to sampling sites, representative capacity of sites, presence of water quantity gaging stations, and availability of required facilities (laboratories, personnel, equipment, etc.). The sampling frequencies happen to be a more significant problem with respect to utilization of available data. The measurements are basically realized on a monthly basis with several gaps and missing values. Available data records are also pretty short (the longest being 7 to 8 years). In the selection of sampling frequencies, time periods are considered when significant variations in water quality are expected. These periods cover low flow time points during warm and dry seasons. Next, the problem of what variables to observe is simplified by specifying two groups. The first one includes variables that are to be monitored at every site; whereas the second group covers more specific variables depending on water use and sources of pollution at particular sites.

EIE has started water quality sampling in early 70's. Currently, at 79 of 285 streamgaging stations, water quality is also observed to assess the quality of rivers for purposes of irrigation and assuring safety of hydraulic structures. Like DSI, EIE also performs monthly observations in a rather uniform basis.

It is observed that both DSI and EIE are expanding their networks in terms of both sampling sites and variables sampled. However, this expansion is realized without clearly defined objectives and guidelines. In essence, neither of the two agencies has questioned the performance of their networks until recent times, since their activities have always been subsidized by the government. Currently however, the government itself has started to foresee assurance of productivity in financed sectors. Thus, the monitoring agencies are now questioning the performance of the existing networks with respect to both efficiency and cost-effectiveness. To this end, the first step to be taken must be the assessment of the current monitoring practices in view of network efficiency and cost-effectiveness. The result of such an evaluation should then lead to a redesign process to assure an optimal network. Harmancioglu *et al.* (1994 a) have initiated the assessment process in Gediz and Sakarya basins. They have disclosed that networks in Turkish rivers are far from being systematic and further that monitoring objectives and information expectations are not clearly defined (Alkan *et al.*, 1995).

India

In India, there are two projects that cover water quality monitoring activities across the country (Naidu and Khan, 1987). The first one is realized by the Central Pollution Control Board in cooperation with the GEMS (Global Environmental Monitoring System) program. This project has increased the number of sampling stations from 33 in 1970's to 51 in 1986. The second project is called Minars (Monitoring of Indian National Aquatic Resources); and with its contribution, the entire network in India comprises about 170 sampling locations. At these stations, 23 water quality variables are analyzed on a monthly basis to determine the present status of water quality and the possible future trends in rivers. However, it is claimed that the number of stations is not sufficient to achieve these two basic purposes of monitoring.

Indonesia

In recognition of the need for a systematic water quality monitoring network, Indonesia has put significant efforts into developing a national network. Yet, difficulties are experienced particularly in selecting the basic network features, i.e., sampling sites, sampling frequencies, and variables to be sampled (Harmancioglu *et al.*, 1994 a).

Taiwan

Water quality monitoring in Taiwan is carried out under two major programs:

a) coastal monitoring;
b) river monitoring.

Significant levels of water pollution are experienced in the country since all types of wastewaters (domestic, industrial, agricultural, etc.) are discharged to surface waters without any treatment. Thus, weekly observations of particular quality variables have been considered necessary. Apart from this program, three major rivers are monitored for physical and chemical water quality variables on a monthly basis (Harmancioglu *et al.*, 1994 a).

4.3.4. NETWORKS IN DEVELOPED COUNTRIES

Several agencies in developed countries have established data networks to assess the quality of their surface waters. In the United States, the U.S. Geological Survey (USGS) and the U.S. Environmental Protection Agency (EPA) are the two institutions that have developed nationwide networks of fixed water quality stations on the major rivers of the United States. Apart from these two major networks, many states run fixed station water quality data collection networks (Lettenmaier, 1988; Harmancioglu *et al.*, 1994 a). Similar institutions in other countries routinely collect water quality data at fixed stations like, for example, the Canadian Department of Environment (CDOE) does in Canada. Australia has developed networks to monitor and control water quality in streams and storages. For example, the existing network in Queensland dates back to the 1960's and currently involves 400 sampling points. However, due to various inadequacies observed, the Water Resources Commission of Queensland has recently started to redesign the network to meet future needs (McNeil *et al.*, 1989). In European countries, similar activities are observed; however, monitoring practices on international rivers are of particular interest. Along these rivers, such as the Rhine, monitoring is realized in a river-based manner with the contribution of riparian countries. Such a practice is intensified especially when significant levels of pollution are observed. Equally important in developed countries are specific surveys carried out for a particular period of time. Often, these monitoring practices are problem or project-oriented activities applied in polluted areas to measure the levels of particular effluents. For a more detailed review of current monitoring practices and recent trends in developed countries, one may refer to the extensive research report prepared by Harmancioglu *et al.* (1994 a).

One of the major problems in developed countries is the lack of coordination between monitoring agencies with respect to purposes of monitoring and activities involved in monitoring. Consequently, an overall perspective of the total monitoring system can hardly be preserved neither to evaluate the existing system nor to add new objectives and activities. In the United States, the local, state, and federal governments have intensively emphasized the legal aspect of water quality managements in the recent years, so that new objectives and methodologies for monitoring have developed. As a result, the evaluation of the total system becomes much more complicated since the new developments often lead to more sophisticated monitoring procedures. Furthermore, if each monitoring agency subscribes to a different perspective of goals and practices, this would eventually mean a proliferation of monitoring activities.

A natural consequence of the above described situation is to have too many data that one doesn't know what to do with. In fact, this appears to be the major problem in developing countries. Ward *et al.* (1986) express it as the "data-rich but information-poor syndrome in water quality monitoring". In early practices of water quality monitoring, every measurement was significant so that one could say "the more data the better" (Langbein, 1979). At those times, the problem was to conceive what available data showed about prevailing water quality conditions. Presently, the situation is reversed as new objectives have developed in water quality management. The question nowadays is whether the available data convey information relevant to a certain objective. The failure of existing networks appears at this point. Monitoring activities have indeed become sophisticated with new methods and technologies. However, when it comes to utilizing collected data, no matter how numerous they may be, one often finds that available samples fail to meet specific data requirements foreseen for the solution of a certain problem. In this case, one is inclined to describe the current monitoring practices as being unsatisfactory. Yet the basic problem is often the failure to define prior to sampling what is expected from collecting data, rather than the failure of available data themselves (Harmancioglu *et al.*, 1992 and 1994 a).

This means that developed countries have already fed their data banks and now have come to the point of asking how they should utilize these data banks. Furthermore, when they attempt to use them, they find that available samples do not meet their information requirements. This situation may even lead to consideration of station discontinuance if, for example, a monitoring station has plenty but information-poor samples.

It appears from the above that the basic problem in developed countries is the discrepancy between information expected from a monitoring network and the information produced by that network. That is, developed countries suffer from "data-rich but information-poor" networks. In view of the prevailing shortcomings, most developed countries have started assessment programs to evaluate the performance of existing networks. Within this framework, they have also begun to critically review their design methodologies and network assessment procedures. A significant output of these developments is the initiation of the redesign process, where the basic purpose of a monitoring network is considered to be the assessment of water quality trends on a basin-wide or even country-wide basis (Harmancioglu *et al.*, 1994 a).

4.3.5. EXAMPLES OF NETWORKS IN DEVELOPED COUNTRIES

USA

In the USA, water quality is monitored by several agencies at federal, state, regional and local levels. Among these are the USGS (United States Geological Survey), USEPA (United States Environmental Protection Agency), NOAA (National Oceanic and Atmospheric Administration), U.S. Fish and Wildlife Service, Soil Conservation Service, U.S. Department of Agriculture Forest Service, and others. Monitoring practices of state agencies differ from each other as water management strategies and regulations are often specific to each state. On the other hand, USGS and USEPA have developed nation-wide networks.

As of 1984, USGS had 4610 stations for monitoring lake and river water quality. Continuous monitoring has been applied at 784 of these stations although types of variables monitored differ. 2906 stations monitored river water quality in a systematic framework with long-term programs. A wide range of variables are monitored including inorganic and organic constituents, trace elements, nutrients, pesticides and radioactive constituents. The sampling frequencies vary from daily to yearly. As of 1990, USGS has been cooperating with about 1000 federal, state and regional agencies to monitor river water quantity and quality at 49000 sites.

In 1973, the NASQAN (National Stream Quality Accounting Network) was initiated solely for water quality monitoring by including 50 of the above-mentioned 2906 stations. The number of stations reached 516 in 1978. The basic objectives of this network have been to provide informational basis for water quality management in the country, to determine the spatial variability of surface water quality across the continent, and to assess long-term trends in water quality. The specific feature of the NASQAN network is that it realizes a uniform monitoring practice across the country by observing the same variables at all stations with the same frequencies, the same sampling and analysis procedures. Such a practice permits comparisons among stations and regions. The sampling frequency at NASQAN stations vary from continuous to daily and monthly observations. On the other hand, recent assessments of the network have reflected certain deficiencies, e.g., incompatibility between information produced by the network and that required by data analysis and decision making procedures. These deficiencies hindered the evaluation of the effects of various network modifications upon monitoring objectives (Moss, 1989; Harmancioglu *et al.*, 1994 a). Thus, USGS has initiated the redesign of the network by adding to its objectives, the requirements of consequent data analyses (particularly trend analyses).

Similar to NASQAN, USEPA runs the NWQSS (National Water Quality Surveillance System). This network included 200 stations with monthly sampling for the years between 1970-1981. Apart from NWQSS, EPA contributes to significant amount of monitoring activities at state levels. The basic objective of EPA in monitoring water quality is to produce information for regulatory management, i.e., to assess compliance with state and federal standards.

Apart from USGS and EPA, several states run fixed water quality monitoring stations. Eventually, this practice resulted in several agencies monitoring the same river. Often, data from different agencies cannot be merged as they are incompatible in terms of sampling frequencies, variables monitored, sampling durations, units used and data reliability.

USGS and EPA have also developed national data banks called WATSTORE (National Water Data Storage and Retrieval System-USGS) and STORET (Storage and Retrieval System of EPA). These two data banks comprise water quantity and quality data for both surface and ground waters. The data are made available to users in the form of tables, graphics and statistical analyses.

In 1976, USGS has developed a more comprehensive information system called NAWDEX (National Water Data Exchange). WATSTORE, STORET, and large numbers of other data banks have been linked to this system via computer networking. The data bank WATDOC (Water Resources Document) of CDOE (Canadian Department of Environment) in Canada has also been interconnected to this system

As it is followed from the above, the USA has developed and expanded its monitoring efforts within the last 20-30 years. From time to time, monitoring agencies have felt the need to assess the performance of their networks so that they have started assessment programs at state and federal levels. Among these is the NAWQA (National Water Quality Assessment Program) which was initiated by USGS in 1986 as a pilot program in 7 states. The program foresees: the evaluation of surface water quality across the country to produce relevant information for water management; assessment of spatial and temporal trends in water quality; determination of monitoring needs by evaluating the performance of existing networks; and, redesign and modification of existing networks in respect of specified information needs (Harmancioglu *et al.*, 1994 a).

Canada

In Canada, DOE (Department of Environment) is responsible for water quality management through its IWD (Inland Waters Directorate) of WQB (Water Quality Branch). WQB was initiated in 1970 to develop the scientific/technical basis for water quality management, and this basis foresaw the monitoring of major rivers in the country (Kwiatowski, 1986). In 1982, agreement was reached between states and the federal government to cooperate in monitoring activities.

WQB considers major river basins as the basic monitoring units. Among its objectives are:

a) development of an informational basis for water quality management;
b) identification of trends in water quality;
c) assessment of consequences of management decisions;
d) assessment of consequences of water quality control efforts;
e) development of an informational basis for revision of regulations.

The first and the third objectives are served by fixed station networks and the others by specific survey stations. Thus, the first group stations constitute the Index Station Network, and the second group makes up the Recurrent River Basin Networks which are established on a basin scale. An ecosystem approach is adopted for selection of variables to be monitored in the two types of networks (Dafoe *et al.*, 1989).

WQB has also started the redesign of a National Reference Network, where the objectives of monitoring were reevaluated, and network features redesigned as a result of basin studies based on advanced tools of modeling, GIS, and the similar.

In the meantime, WQB has also developed a national data bank known as NAQUADAT (Canada's National Water Quality Data File).

European Countries

Similar monitoring activities and redesign procedures are observed in the European countries. A special feature of the European practice relates to the monitoring of transboundary rivers such as the Rhine and the Danube (Harmancioglu *et al.*, 1994 a). In these cases, riparian countries apply river-based monitoring systems to control river sections exposed to intensive water pollution. For example, the Monitoring, Laboratory and Information Management Sub-Group (MLIM-SG) in the Environmental Programme for the Danube River Basin (EPDRB) is responsible to harmonize the water quality monitoring in the Danube catchment. In 1996, implementation of Phase 1 of the Trans-National Monitoring Network (TNMN) is continued in 11 countries. Each country is represented by a National Reference Laboratory (NRL) (Literathy, 1997).

International Programmes

Both the developing and the developed countries contribute to a WHO and UNEP supported project called "the Global Environmental Monitoring System" or GEMS, which investigates air and water quality among other environmental issues, as part of the UN's Earthwatch program. The GEMS/Water project, with the contribution of about 60 countries, supports the collection of data on more than 50 representative variables of water quality through a network of about 350 sampling locations covering lakes, rivers, and groundwater. The project has been in effect for more than ten years to provide information on the global water quality conditions (Ongley, 1994).

Furthermore, a World Hydrological Cycle Observing System (WHYCOS) is proposed to facilitate access to global data and to support Hydrological Services in need. A world-wide network of about 1000 stations is planned for the largest rivers, together with associated databases and products to meet the needs of users. The concept of WHYCOS was initiated by WMO (World Meteorological Organization) and the World Bank in 1993 (Paulson, 1994; WMO, 1994).

4.3.6. FURTHER SHORTCOMINGS OF EXISTING NETWORKS

Within the major problem of coordination between available data and objectives, others of a more specific nature may be cited. These difficulties are related to such questions as what to measure, where, when, and how long. In fact, these are the issues that cause the failure of available samples to meet data requirements.

First, the selection of water quality variables to be observed is a complicated issue since there are several variables to choose from. Different approaches are used to handle this problem. In some cases, the chemical, physical and biological parameters of water quality that need to be observed are determined on the basis of various water uses (e.g., domestic, industrial, agricultural or multipurpose uses). Sometimes, levels of monitoring efforts are defined for a network to include different variables at each level. These levels may be surveillance, intensive control or project oriented programs, respectively, in order of priority (UNESCO-WMO, 1972; Chapman, 1992; Harmancioglu *et al.*, 1992). Another approach, more of a statistical character, is to investigate relationships between regularly observed water quality variables and those with small number of sporadic observations to reduce the number of variables to be observed. Sanders *et al.* (1983) suggest ranking of water quantity and quality variables among which information may be transferred. In this ranking, water quantity appears as the basic variable followed by "associated quality variables of aggregated effects" (often regularly observed) and then by "quality variables that produce aggregated effects" (often unobserved or observed sporadically). If information transfer between the first and second group of variables is possible, then the required number of variables to be observed may be reduced as long as there is no doubt as to the reliability of information transfer (Harmancioglu *et al.*, 1992; Harmancioglu and Singh, 1991).

The next problem is the selection of temporal frequencies with which to observe quality variables. The major limitation of water quality data is that they often have short records. However, what's worse is that there are gaps and missing data in most available series (Lettenmaier, 1988). Although some quality variables are regularly monitored, most of them are sampled sporadically for laboratory analyses. In this case, samples cover only a relatively short period of observations with many missing values. The situation is more serious when the variables are observed at highly unequal time intervals. The result is difficulty in the evaluation of available data for a reliable assessment of water quality conditions.

Another problem of prime importance is the selection of observation sites. This is also a controversial issue like the selection of sampling frequencies, although it has received the least attention. Early considerations on this matter led to problem-oriented selection procedures to detect the origin and levels of pollution at particular sites. Later, as new objectives of monitoring developed, several sites had to be observed. The basic problem with multisite monitoring is the realization of representative sampling. This means to select the sampling points in such a way that the river reach investigated is best represented by these sites. If this approach can be realized, then the variability of water quality along the reach may be assessed and further, information transfer among sites may be effectively carried out. However, most of the existing networks reflect shortcomings related to representative sampling so that the issue is still investigated to improve the network designs.

The question of how long a station should be run is again a controversial issue. Station continuance is related basically to objectives of monitoring and information expectations from observed data. There are no definitive criteria yet established to decide whether monitoring should be continued or terminated at a particular site.

Other difficulties related to utilization of water quality data are concerned with their reliability and accuracy. Water quality processes are strongly subject to nonhomogeneities created by man while similar effects also occur naturally. Furthermore, some water quality variables can be easily monitored, yet some others require complex laboratory analyses. Errors in laboratory experimental analyses plus changes either in monitoring or laboratory practices may often lead to inconsistencies (systematic errors). Another problem is censored data which occur when some concentrations are below detection limits, and cannot be described numerically by laboratory practices. All these limiting factors eventually make the utilization of water quality data difficult. Furthermore, the reliability of the output information is poor. Chapman (1992) summarizes data limitations as the following:

a) *missing values*: these may occur due to equipment breakdowns, lost samples, contaminated samples, poor weather, and employee illness; they may be random or systematic;
b) *sampling frequencies that change over the period of record*: this limitation often occurs when monitoring agencies are faced with budget restrictions; shifting water quality problems or a new crisis can also cause this change;
c) *multiple observations within one sampling period*: a common reason for this to occur in a water quality data record is when QA/QC results are stored in the same computer record as the original water quality observation;
d) *uncertainty in the measurement procedures*: this uncertainty is due to random analytical errors; it varies with calibration of the measuring equipment;

e) *censored data*: this problem becomes more complicated when the detection limit changes over the period of record; multiple censoring levels occur when different analytical techniques are used over the period of record, or when different laboratory protocols are used, or when data from different laboratories are analyzed as one data set;

f) small sample sizes;

g) *outliers*: these may be due to erroneous measurements or extreme events; it is difficult to differentiate between the two.

Recognition of data limitations during the design phase may help to minimize them; however, they are often recognized during the analysis of data.

The major problems associated with available water quality data are their incompleteness, inadequacy, and inhomogeneity. It is stated by a report by WMO (1994) that much emphasis in water quality monitoring (physical, chemical, and biological) has been put on sampling frequency and laboratory analyses, while the assessment and interpretation of available data have not developed at the same rate. The report further points out that "water quality data are very heterogenous in nature: they may be numerical, orderial, or descriptive; they may be distributed in time and space; they may reside in different locations; they may exist in a variety of formats ... Considered individually, these data sets are of limited use for water management and assessment of ecosystem health". It is indicated further that successful environmental management requires an integrated management of "mixed-mode" data sets (WMO, 1994).

The above-mentioned shortcomings of existing networks in developed and the developing countries may be summarized as:

a) lack of coordination between various agencies running different networks;

b) lack of agreement between collected data and water quality management objectives, resulting in "data-rich, information-poor" monitoring practices;

c) problems related to:

    1. selection of variables to be observed,
    2. selection of sampling techniques,
    3. selection of sampling sites, and
    4. how long monitoring of certain variables at certain sites should be continued; and

d) lack of reliable and accurate data (messy data).

4.4. Current Methods in the Design of Water Quality Monitoring Networks

4.4.1. REVIEW OF THE GENERAL APPROACH

As discussed in section 4.1, problems observed in available data and shortcomings of current networks have led researchers to focus more critically on the design methodologies used. In addition, recent advances in sampling and analysis techniques for water quality have also led to expansion of networks and thus to a growth in economic features of monitoring. Accordingly, researchers have started to question both the efficiency and the cost-effectiveness of existing networks with regard to design methodologies used.

The first data collection procedures for water quantity foresaw the gaging of major streams at potential sites for water resources developments. The approach in initiating water quality observations has been practically similar, namely to collect data at potential sites for pollution problems. Thus, as described earlier in section 4.3, the early water quality monitoring practices were often restricted to what may be called "problem areas", covering limited periods of time and limited number of variables to be observed (Harmancioglu *et al.*, 1992). Recently, however, water quality-related problems have intensified so that the information expectations to assess the quality of surface waters have also increased. The result has been an expansion of monitoring activities to include more observational sites and larger number of variables to be sampled at smaller time intervals. These efforts have indeed produced plenty of data; yet they have also led to the "data-rich information-poor" networks as information expectations have not always been met.

The above considerations have eventually led to the realization that a more systematic approach to monitoring is required. Following up on this need, monitoring agencies and researchers have proposed and used various network design procedures either to set up a network or to evaluate and revise an existing one.

Current methods of water quality monitoring network design basically cover two steps: first, the description of design considerations, and second, the actual design process itself. Researchers emphasize the proper delineation of design considerations as an essential step before attempting the technical design of the network. This step is to provide answers to the questions of why we monitor and what information we expect from sampling water quality. In other words, objectives of monitoring and information expectations for each objective must be specified first. Various objectives or goals for monitoring have been proposed up to date by different researchers, i.e., assessment of trends, delineation of water quality characteristics for water use, assessment of compliance, evaluation of water quality control measures, etc. (Whitfield, 1988; Ward and Loftis, 1986; Tirsch and Male, 1984; Sanders *et al.*, 1983; Langbein, 1979). In practice, the definition of objectives is not an easy task since it requires the consideration of several factors, including social, legal, economic, political, administrative and operational aspects of monitoring goals and practices. Therefore, the delineation of design considerations, inevitably includes assumptions and subjective views of the designers and decision-makers no matter how objectively the problem is approached. In this case, design considerations are often presented as general guidelines, rather than fixed rules to be pursued in the second step of actual design process (Sanders *et al.*, 1983; Harmancioglu *et al.*, 1992).

The technical design of monitoring networks relates to the determination of:

a) sampling sites,

b) sampling frequencies,

c) variables to be sampled, and

d) the period or duration of sampling.

It is only at this actual design phase that fixed rules or methods are proposed. Current literature provides considerable amount of research carried out so far on the above-mentioned four aspects of the design problem. One may refer to Sanders *et al.* (1983), Tirsch and Male (1984), Whitfield (1988), Ward *et al.* (1989) or to WMO (1994) for a rather thorough survey of research results and practices on the establishment of sampling strategies with respect to these factors.

Basically, designers and researchers recognize water quality monitoring as a statistical procedure and address the design problem by means of statistical methods. Ward and Loftis (1986) stress that information expectations from a monitoring system must be defined in statistical terms and that these "expectations are to be in line with the monitoring system's statistical ability to produce the expected information". This implies that one can infer on the types of data needed to perform the statistical methods which, in turn, will eventually lead to the expected information. Then, the selection of sampling strategies (sampling sites, variables, frequencies, and duration) can be realized by starting off with such a statistical approach (Ward and Loftis, 1986; Sanders *et al.*, 1983). Statistical analyses based on regression theory as well as decision theory and optimization techniques are used to select the spatial and temporal design features of a network. Although none of these methods are widely accepted, they serve at least to assess the effectiveness of design decisions and the efficiency of an existing network. The problem is much more difficult in case of variable selection as there are no methods established yet for defining objective selection criteria (Harmancioglu and Alpaslan, 1992; Harmancioglu *et al.*, 1992; Ozkul *et al.*, 1996; Harmancioglu *et al.*, 1996).

Moss (1989) has emphasized that network design should be realized with a combined approach based on hydrology, optimization techniques, decision theory and data analysis methods. In particular, he states that networks should produce data that permit the application of statistical data analysis techniques. Since such considerations are not taken into account in current design methodologies, it is often very difficult to assess the information conveyed by current networks. Developed countries have realized this deficiency and started to redesign their networks in respect of new considerations such as those proposed by Moss (1989) and other researchers (Harmancioglu *et al.*, 1994 a).

Monitoring networks are expected to reveal three basic statistical characteristics of water quality: means, extremes and changing water quality conditions (or trends). Designers point out that a network which is highly intense with respect to time and space is required to detect extremes with confidence (Karpuzcu *et al.*, 1987; Sanders *et al.*, 1983). However, such a design on a routine basis is pretty costly, so that networks which reliably detect means and trends are more preferred (Ward *et al.*, 1979). Yet, there are also researchers who argue that modern information technology and electronic engineering provide the means of revealing more variability in the behavior of water quality processes, including the extremes (Beck and Finney, 1987). The latest redesign procedures started by developed countries consider trend assessment as the basic objective of a basin-wide or country-wide monitoring network. In this respect, compliance monitoring is to be realized via frequent or continuous effluent monitoring. With these two types of monitoring activities (i.e., trend and compliance), it is then possible to statistically assess the mean values of water quality variables.

4.4.2. SITE SELECTION

Allocation of sampling sites is the initial and the most crucial step of the network design process. It conveys all the difficulties and the complicated aspects of the design problem; furthermore, it can not be dissociated from other three design criteria, i.e., selection of variables, temporal frequencies and sampling durations.

The most reasonable approach to allocation of sampling sites seems to be the selection of locations so as to comply with the objectives of monitoring. However, the issue is not as easy as expressed; it is subject to assumptions and subjective considerations of designers, because first, it is difficult to state precisely the objectives and second, each monitoring objective entails with it diverse constraints. The result is that there are no fixed rules or standard methods to be pursued in selecting sampling sites. Some approaches do provide a scientific basis for allocation; however, the designer's judgment is equally important. Thus, the problem is as much an art as it is a science (Harmancioglu and Alpaslan, 1994).

The early practices of water quality sampling started at sites of easy access or often at streamflow gaging points without any systematic approach to selection of sampling locations. The number of these sites have increased in time to include stations "at points of interest" such as those located at upstream and downstream of highly industrialized or highly populated areas, areas with point pollution sources, or areas of intensive land use (Tirsch and Male, 1984). Such nonsystematic approaches in the selection of sampling sites are still valid, especially in developing countries where monitoring efforts have not yet evolved into a network. The basic criteria used in such practices are the locations of polluting sources, easiness of access to sampling sites, representative capacity of the sites, presence of streamflow gaging stations, and the availability of required facilities (laboratory, personnel, equipment, etc.) (Harmancioglu, *et al.*, 1992 and 1994 a).

Later, methodologies were proposed to select both the locations and the numbers of sampling stations. Some of these methods used drainage area or flow characteristics. Others were more sophisticated, such as those proposed by Scheidegger (1965) and Sharp (1971), the former using stream order numbers and the latter foreseeing a hierarchical order in establishing sampling stations. Sharp's approach is intended to locate possible sources of pollutants by analyzing a trade-off between sample source uncertainty and sampling intensity. Later, Sanders and Clarkson followed up on Sharp's procedure (Sanders *et al.*, 1983).

Spatial design of water quality networks is also attempted by regression techniques. Tirsch and Male (1984) propose a multivariate linear regression model where the corrected regression coefficient of determination between sampling stations is considered as a measure of monitoring precision. The monitoring precision changes with the addition or deletion of some number and location of stations within a basin. Whitlatch (1989) examines the spatial adequacy of NASQAN (USGS) water quality data by testing the differences between two sample means as a direct method and then by regression analyses between water quality variables and basin characteristics.

Some researchers stress the use of optimization techniques in selection of both sampling sites and sampling frequencies (Reinelt *et al.*1988; Palmer and MacKenzie, 1985; MacKenzie *et al.*, 1987; Dandy and Moore, 1979). In such design procedures, two requirements are expected to be fulfilled by the network: cost-effectiveness and statistical power. The latter is often investigated by analysis of variance (ANOVA) techniques, and optimization methods are used to maximize the statistical power of the network while minimizing the costs (Harmancioglu, *et al.*, 1992).

Sanders *et al.* (1983) consider the problem of selecting sampling sites at two levels: macrolocation and microlocation. Macrolocation is handled by any one of the above methods, whereas microlocation relates to representative sampling at a point and requires an analysis of complete mixing within a river reach. Statistical methods (e.g., regression analyses, two-way analysis of variance) are proposed for microlocation purposes. Sanders *et al.* (1983) claim that, in practice, microlocation and representative sampling with respect to station location are not sufficiently evaluated by monitoring agencies.

On the other hand, macrolocation encompasses the identification of sampling reaches in a river basin when the intent is to allocate monitoring sites along the entire basin. The method proposed by Sanders *et al.* (1983) is originally based on Horton's (1945) stream ordering procedure to describe a stream network. Horton assigns each unbranched small tributary the order of one, a stream made up of only first order tributaries the order of two, and so on. Later, Sharp (1970) used Horton's approach to measure the uncertainty involved in locating the source of pollutants observed at the outlet of a network. Then, Sanders *et al.* (1983) followed Sharp's procedure by selecting sampling sites on the basis of the number of contributing tributaries. Next, they modified the same method by considering the pollutant discharges as external tributaries.

Macrolocation can be performed by three approaches (Sanders *et al.*, 1983):

a) allocation by the number of contributing tributaries;

b) allocation by the number of pollutant discharges;

c) allocation by measures of BOD loadings.

These approaches, although each may produce a rather different system of stations, work pretty well in initiating a network when no data or very limited amounts of data are available. It must be noted that, by applying these methods, one may roughly specify the appropriate sampling sites. To pinpoint the locations more precisely, microlocation and representative sampling considerations will have to be followed. As a case study for the above three approaches, allocation of sampling sites is performed in the Gediz River in Turkey (Alpaslan and Harmancioglu, 1990; Harmancioglu *et al.*, 1992; Harmancioglu *et al.*, 1994 a and b). The results of these investigations have shown that allocation by one of the above approaches divides the basin into equal subbasins with respect to the number of tributaries or discharges. A comparison between the existing network in the basin and that delineated by using these approaches discloses that the two do not coincide. The reason for this discrepancy is that the existing network is established on the basis of particular project requirements so that it does not reflect the quality conditions within the entire basin. As a result of these investigations, it is concluded that Sanders' method (Sanders *et al.*, 1983) may be effectively used to allocate sampling sites by considering all the polluting sources or discharges within the basin.

Once macrolocation is realized by the above-mentioned approaches, the network may be revised and modified by statistical approaches such as the classical correlation and regression theory. Within this context, Harmancioglu and Alpaslan (1992a) have proposed the use of the entropy theory to decide upon the required numbers and locations of stations. According to this theory, decisions may be made to reduce the number of stations where information is redundant or to increase sampling sites at regions where additional information is required. This methodology is later applied to site selection problems in Gediz and Sakarya River basins in Turkey (Harmancioglu *et al.*, 1994a; Ozkul *et al.*, 1995).

Ozkul (1996) applied the entropy principle of Information Theory to assess spatial frequencies of water quality observations along the Mississippi River in Louisiana, USA, for basin segment 07. The methodology she used resulted in a spatial orientation of sampling stations where the redundant information among these stations was minimized by an appropriate choice of the number and locations of monitoring stations.

4.4.3. SELECTION OF TEMPORAL FREQUENCIES

Since temporal frequencies significantly affect sampling costs, selection of temporal design criteria has received the highest attention from designers and researchers. Quimpo and Yang (1970) address this problem as: "On the one hand, by sampling too often, the information obtained is redundant and thus expensive, and on the other hand, sampling too infrequently bypasses some information necessitating an extended period of observation". The selection of sampling frequencies is significant, then, not only in terms of the cost-effectiveness of a monitoring system, but also in terms of information that may be extracted from available data (or data utility).

The early practice in determining temporal frequencies has been pretty random just as it was in the case of locating sampling sites. Observations were made when the time, budget, and routine capabilities of the monitoring system were available. Experience also showed that there could be more reasonable times to sample or that data need to be collected at "times of interest" like periods of low flow. Later, temporal frequencies were selected on the basis of river basin characteristics and river flow variability. This approach was relatively more systematic yet still did not provide a quantitative basis for evaluating information expectations (Sanders *et al.*, 1983).

In practice, the case is often that cost considerations and professional judgment call for constant frequencies to be applied at all stations as the most convenient procedure. Recently, however researchers propose the use of sound statistical methods as quantitative criteria in selection of sampling intervals. The basic consideration underlying this approach is the recognition of monitoring as a statistical process and the evaluation of information expectations by means of statistical measures. Sanders *et al.* (1983) summarize some of the statistical methods in selection of frequencies as determination of statistical properties of water quality series (e.g., cyclic frequencies, autocorrelations), ratios of maximum flows, determination of confidence intervals of mean values, evaluation of sampling errors and their variance, or the determination of required numbers of data for testing statistical hypotheses (Harmancioglu *et al.*, 1992).

The use of most of these techniques is shown by data analyses, often including regression techniques and standard error criteria in estimation of parameters that relate to the information conveyed by data (Hipel, 1988; Hirsch, 1988; Lettenmaier, 1988;

Tirsch and Male, 1984; Ward and Loftis, 1986). For example, Gupta (1982) uses the Modified Langbein Method, which is essentially based on regression techniques, to maximize information gain via optimum data lengths determined for primary and secondary gaging stations.

Other approaches include trend analysis techniques, Bayes decision theory and extended applications of optimization techniques (Harmancioglu *et al.*, 1992, 1994 a; Alpaslan *et al.*, 1992; Liebetrau, 1979; Lettenmaier, 1979; Mueller, 1989; Richards, 1989; Lachance *et al.*, 1989; Loftis *et al.*, 1991).

Whitfield (1988) claims that different sampling frequencies are to be selected for different goals of monitoring. He identifies five objectives as:

a) assessment of trends;
b) compliance of standards;
c) estimation of mass transport;
d) assessment of environmental impact; and
e) general surveillance.

For each of these goals of monitoring, Whitfield stresses the selection of different sampling frequencies to maximize the information gain via sampling. The idea here is again that information expectations for each objective are different. In a more recent study, Valiela and Whitfield (1989) propose monitoring strategies to determine compliance with water quality objectives. They analyze fixed frequency sampling at frequent intervals versus exceedance-driven sampling for data that are seasonal and significantly autocorrelated.

Tirsch and Male (1984) address temporal design of networks by a similar method they use for spatial design. This time, monitoring precision as described by the corrected regression coefficient of determination is expressed as a function of sampling frequency.

Sanders and Adrian (1978) proposed a statistical method to select temporal sampling intervals when the objective is to determine the true mean value of a water quality variable. The method is based on the expected half-width of the confidence interval of the mean value. Although this approach was intended for water quality variables, Sanders and Adrian (1978) applied it to streamflow data due to lack of sufficient water quality data and found it to be a reliable method (Sanders *et al.*, 1983; Sanders, 1988).

Lettenmaier (1976) proposed another approach to determine optimum sampling intervals. His method is based on the parametric trend test where the required sampling frequency is one that corresponds to a specified power of the trend test. This approach was later used by Schilperoort *et al.* (1982) in an optimization framework to select optimum sampling intervals when the objective of the monitoring network is to determine trends in water quality.

The above two methods are employed by Tokgoz (1992) and Harmancioglu and Tokgoz (1995) to assess sampling frequencies in case of the water quality network in Porsuk river basin. The results have shown that Sanders' method is not applicable to water quality time series that have a short duration of observation and large numbers of missing values. This is because the underlying assumptions of the method cannot be satisfied with such messy water quality data. On the other hand, Lettenmaier's method is found to be more suitable and to better adjust to deficiencies in observed series.

If the temporal frequencies are to be assessed on the basis information expectations, the entropy concept, as defined in Information Theory, may also be used to evaluate the trade-off between information loss due to the increased time intervals and the reduced cost of frequent sampling (Harmancioglu and Alpaslan, 1992; Harmancioglu *et al.*, 1994 a; Ozkul, 1996). Assessment of temporal frequencies by entropy measures is based on the minimization of redundant information among successive measurements. The method was applied to water quality observations in the Gediz and Sakarya basins and was found to produce effective results in assessing sampling intervals although, in some cases, entropy computations were hindered by the messy character of water quality data. Ozkul (1996) applied the same methodology to the regular water quality observations of the Mississippi river and obtained satisfactory results.

4.4.4. SELECTION OF COMBINED SPACE/TIME FREQUENCIES

Some design procedures combine both the spatial and temporal design criteria to evaluate space-time trade-offs. The approach in such combined design programs is to compensate for lack of information with respect to one dimension by increasing the intensity of efforts in the other dimension (Harmancioglu and Alpaslan, 1992).

Statistical analyses based on the regression techniques as well as decision theory may be employed to solve the multidimensional design problem. There are several studies carried out in this area, most of which investigate networks for the other hydrometeorological variables, but which may, in principle, also be valid for water quality variables, too (Moss, 1979 a and b; Moss, 1976; Moss and Karlinger, 1974; Dawdy, 1979; Tasker and Moss, 1979). Tirsch and Male (1984) have combined spatial and temporal design by multivariate linear regression and again used the corrected coefficient of determination as a measure of monitoring precision. They incorporated cost and benefit considerations into their design by using Bayesian analysis. In fact, this is the only study up-to-date that investigates combined space/time frequencies in case of water quality.

Another group of techniques that combine both spatial and temporal design covers optimization methods. Some researchers like Schilperoort *et al.* (1982) emphasize the need for the optimization of monitoring networks to achieve cost-effective designs while fulfilling the objectives of monitoring. Such a procedure enables the evaluation of space-time trade-offs in design.

Harmancioglu *et al.* (1994 a) have attempted the selection of combined space/time frequencies via entropy method. They tried to apply the method to water quality observations in the Porsuk basin; however, the presence of significant numbers of missing values and the short duration of observations did not permit a reliable analysis to test the applicability of the method.

Ozkul (1996) investigated space/time dimensions of the water quality monitoring network in the Mississippi River basin using the entropy principle. She derived curves of redundant information with respect to both the number of stations and the sampling frequencies. Figure 4.1 shows the results of this application for Cl^-, where redundant information (transinformation) increases with an increase in the number of sampling locations and decreases with a decrease in temporal sampling frequencies. Here, for a constant level of transinformation, a number of space/time alternatives exist such that one may evaluate:

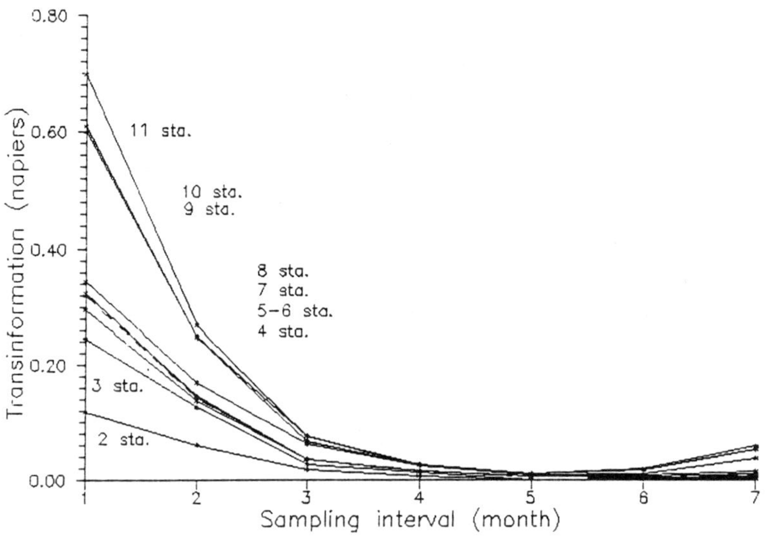

Figure 4.1. Changes in redundant information for different space/time frequencies in case of Cl^-

a) whether to increase the number of stations and decrease the frequency; or
b) decrease the number of stations and increase the temporal frequency.

The final decision to select among alternatives depends on evaluation of cost reduction with respect to decreases in space or time frequencies.

4.4.5. SELECTION OF VARIABLES

Selection of variables to be sampled depends basically on the objectives and economics of monitoring. It is a highly complicated issue since there are several variables to choose from in representing surface water quality. Some of the selection procedures stress water uses as the major criterion to be pursued; some define levels of monitoring efforts (e.g., surveillance, intensive control, or project-oriented programs) with different groups of variables included at each level (UNESCO-WHO, 1972).

There are also studies which apply quantitative statistical techniques in selection of variables to be sampled. These techniques are basically regression-type methods to investigate the relationship between water quantity and water quality variables, or between water quality variables themselves. If significant correlations are detected, then the number of variables to be observed is reduced since some can be estimated by the assumed regression models. Yevjevich and Harmancioglu (1985) and Harmancioglu and Yevjevich (1986) investigate the transfer information between daily observed water quality variables for the purpose of determining those variables that need to be sampled continuously and those that can be estimated via their correlation with other variables. Similar analyses were carried out by Harmancioglu *et al.* (1987) on monthly observed data of a highly polluted river basin. Entropy-based measures were also used in these studies to evaluate the goodness of information transfer by regression. The results of these studies have basically revealed that the association between most of the water quality variables is insignificant.

Some studies use multivariate statistical methods, such as the principal component analysis (Karpuzcu *et al.*, 1987), to reduce the number of variables to be observed. It is claimed in these studies that such methods give better estimates of the most representative water quality variables than those obtained by conventional correlation analyses.

In an earlier study, Huthmann (1979) simulated the values for a water quality variable at a downstream point along a river reach using data collected at upstream points. He also addressed the problem of reducing the number of sampling stations and temporal frequencies by developing a multiple input-single output system model for the river.

Another study (Chapman, 1992) considers three groups of variables to be sampled:

a) base variables to be monitored at every station;
b) variables that need to be monitored with respect to water use;
c) variables that need to be monitored with respect to impact assessment.

The second and third groups are further divided into industrial and nonindustrial water-use and impact assessment variables. Harmancioglu et al (1994 a) have further developed this approach to define:

a) variables that need to be sampled at every station in a basin-wide network;
b) variables that need to be sampled specifically at each station.

This methodology considers basin characteristics and local features at each station to determine the variables to be monitored. Next, all variables are ordered with respect to their significance. Finally, the list is screened once more by regression to reduce the number of variables if strong correlations exist among them.

4.4.6. SAMPLING DURATION

The question how long sampling should be continued is basically treated together with the problem of temporal design. Therefore, much of the cited work above for selection of sampling frequencies refer also to the duration of sampling. Yet sufficient amount of research effort has not been devoted particularly to this aspect of the design problem. Among the few available, one may refer to Hirsch (1988) who compares long-term continuous sampling versus rotational sampling where data are collected in bursts of small periods. Hirsch claims that more research is needed in this area because his results, although very interesting, could not be generalized for practical network design before the approach is tested on other sets of data.

The more extensive problem of station discontinuance is a controversial issue even for other hydrometric data networks. There are no definite criteria yet established to decide whether monitoring should be continued or terminated at a particular site, although there are some studies carried out for streamflow gaging stations (Wood, 1979).

Harmancioglu and Alpaslan (1992) and Harmancioglu (1994) have used the entropy principle to investigate the problem of station discontinuance. The entropy method may be used to assess the status of an existing station with respect to information gathering. To solve the problem in the space domain, the spatial orientation of stations within a network may be evaluated for redundancy of information so that a particular site that repeats the information provided by other stations can be discontinued. The problem is similar in the time domain. A monitoring site is again evaluated for redundancy of information, this time with respect to temporal frequency and the duration of observations (Harmançioglu, 1994).

4.5. Shortcomings of Current Design Methodologies

As discussed in section 4.4, there are still problems in the design of water quality monitoring networks at both stages, i.e., statement of objectives and the actual technical design. At the current state of matters, there are no definitely prescribed and widely accepted standard procedures to solve the above problems.

Deficiencies related to current design procedures are primarily associated with an imprecise definition of information and value of data, transfer of information in space and time, and cost-effectiveness. The major difficulty associated with these current design methods is related to the lack of a precise definition for "information". They either do not give a precise definition of how information is measured, or they try to express it indirectly in terms of other statistical parameters like standard error or variance. One important consequence of failure to define information can possible be the interchangeable use of the terms "data" and "information". Although current methods stress the distinction between the two, a direct link between them has not yet been established (Harmancioglu *et al.*, 1992; Harmancioglu *et al.*, 1994 a).

Another difficulty with current design methods is how to define the value of data. In every design procedure, the ultimate goal is an "optimal" network. "Optimality" means that the network must meet the objectives of the data gathering at minimum cost. While costs are relatively easy to assess, the major difficulty arises in the evaluation of benefits because such benefits are essentially a function of the value of data collected. The value of data lies in their ability to fulfill information expectations. However, how this fulfillment might be assessed in quantifiable terms still remain unsolved. As in the case of information, the value of data has been described indirectly (Dawdy, 1979; Moss, 1976), often by Bayesian decision theory (Tirsch and Male, 1984).

Another criticism of the current design methods relates to how the techniques are used in spatial and temporal design. The majority of current techniques are based on classical correlation and regression theory, which basically constitutes a means of transferring information in space and time. The use of regression theory in transfer of information has some justification. However, regression approaches transfer information on the basis of certain assumptions regarding the distributions of variables and the form of the transfer function such as linearity and nonlinearity. Thus, how much information is transferred by regression under specified assumptions has to be evaluated with respect to the amount of information that is actually transferable. One may refer to Harmancioglu *et al.* (1986) for the definition and comparison of the terms "transferred information" and "transferable information".

To summarize the above discussions, one may state that the existing methods of water quality network design are deficient because of the following specific difficulties:

a) a precise definition of "information" contained in the data and how it is measured is not given;
b) the value of data is not precisely defined, and consequently, existing networks are not "optimal" either in terms of the information contained in these data or in terms of the cost of getting the data;
c) the method of information transfer in space and time is restrictive;
d) cost-effectiveness is not emphasized in certain aspects of monitoring;
e) the flexibility of the network in responding to new monitoring objectives and conditions is not measured and not generally considered in the evaluation of existing or proposed networks (Harmancioglu et al., 1992; Harmancioglu and Alpaslan, 1992).

Within this context, a methodology based on the entropy theory can be used for design of efficient, cost-effective, and flexible water quality monitoring networks to alleviate many of the above shortcomings of the existing network design methods (Harmancioglu *et al.*, 1992; Harmancioglu *et al.*, 1994 a and b).

4.6. Needs for Better Designs

4.6.1. INFORMATION NEEDS

A report prepared by WMO (1994) emphasizes that we need to resolve three questions for better water management:

a) What minimum physical, chemical, biological, and socioeconomic information is required to plan and manage water resources?
b) What minimum data are needed to produce the required information?
c) How do we efficiently produce the required information from data?

The last question essentially relates to methods used to transfer data into information. The first two questions, however, impose significant requirements on the design of monitoring systems.

As described in the previous sections, design of water quality networks is a highly complicated issue because it requires the consideration of numerous diverse factors. Consequently, it seems impossible to prescribe just one design procedure that will satisfy everyone's needs in all areas. However, the current status in both the developed and developing countries shows that the shortcomings of existing networks stem from some common factors. In this case, one may conclude that the basic principles of network design can be derived and agreed upon by all designers. Then, within the fundamental framework, adjustments may be made to account for local or site-specific factors.

Two fundamental factors may be recognized as the basic needs for better designs. The first one is the delineation of monitoring objectives to settle the "demand" part of the problem. The second one is a "reply" to the first factor to answer the basic question of how monitoring should be realized to meet the demand. In this way, the network design problem is viewed as one of matching the "demand" and the "reply". Then all other aspects of design considerations (i.e., what variables to observe, when, where and how long) fit into appropriate places within this basic framework.

The "demand" part of the problem incorporates the objectives of monitoring. These objectives comprise the particular data requirements of various water quality management practices. The basic areas where data on water quality are needed range from the assessment of trends in water quality to general surveillance of quality conditions over an area. Then, types and characteristics of data required for each objective may be specified.

The "reply" part of the problem will cover the actual design of the network to meet the data requirements of a priori determined objectives of monitoring. Then, the basic questions of what variables to observe, where, when and how long can be solved directly to satisfy the requirements of each objective.

Furthermore, in matching the "demand" and the "reply", there is a need for a common reference level with respect to which the two factors should be evaluated. The success of the design depends on how objectively this evaluation is done. This implies that the common reference level should be concretely defined in tangible terms. Otherwise, a subjective definition of demands, replies and the reference level should be subject to diverse personal views and assumptions which hardly converge to a common point of agreement. The result is a network still marked with the shortcomings of the existing ones.

Under these considerations, one may define the above-mentioned reference level on the basis of "information". This is because "demand" implies the need for some kind of information expected from data. The "reply" side of the problem has to provide the expected information by monitoring. Then, both the "demand" and the "reply" can be defined in terms of information so that the problem turns out to be the determination of:

a) information needs prior to sampling; and
b) information conveyed by data which is the result of sampling.

If these two factors can be made to match each other well, then the design of the network can be justified.

Then the problem is to define "information" intangible terms. At this point, a distinction has to be made between the two terms "data" and "information". The term "data" means a series of numerical figures which constitute our means of communication with nature. On the other hand, what these data tell us or what they communicate with us is "information". In this respect, it is possible that data tell us all we need to know about what occurs in nature (full information), or they may tell us some but not all about nature (partial information), or they may tell us nothing at all (no information). This means that availability of data is not a sufficient condition unless they have utility, and the term "information" describes this utility or usefulness of data.

There are basically two requirements expected from the design of monitoring: efficiency and cost-effectiveness. The efficiency of a network is considered here as a function of the information it provides for various objectives. If the design of the network is realized such that this information is maximized, then the requirement of efficiency will be satisfied. Cost-effectiveness, as the other requirement in design, needs to be investigated on the basis of costs of monitoring versus information gain through monitoring. A sound quantitative measure is needed to incorporate these two factors into an optimum solution. The approach here is to maximize the amount of information or utility of data while minimizing the accruing costs.

4.6.2. GENERAL GUIDELINES FOR IMPROVEMENT

The current trend in water quality monitoring network design shows two significant developments:

a) emphasis on delineation of design considerations before the actual design is attempted; design considerations cover the definitions of objectives of monitoring, information expected for each objective, and various types of constraints (e.g., social, legal, economic, political, administrative, and operational) to set up the framework for technical design;

b) emphasis on defining information expectations in statistical terms and use of statistical methods in the actual design phase to assure that the information expectations are met by the specified sampling programs.

The general guidelines to be followed for improvement must also cover these two steps, i.e., delineation of design considerations and technical design of the network, to achieve an efficient and cost-effective network design. The crucial point here is the specification of the objectives of monitoring to define data requirements or information needs required by each objective. Having also evaluated various types of constraints, the technical design procedures will then be set up so as to match with the information expected from monitoring. This means that sampling locations, frequencies, variables, and sampling durations will be specified to produce the expected information. Such an approach covers both the "demand" (objectives of monitoring) and the "reply" (monitoring practices) part of the problem. Efficiency of the network, or its informativeness, can be realized by matching these two parts. Economic considerations will also have to be incorporated into the basic technical framework to ensure a cost-effective design. In this case, costs of monitoring have to be evaluated with respect to information gain via monitoring. The issue then is an optimization problem to maximize the amount of information while minimizing the accruing costs. What one basically needs in the technical design is a sound quantifiable measure of information because, after all, a monitoring system is essentially an information system. Such a measure has to be a statistical one if one follows the argument that monitoring is statistical sampling. To this end, one may, for example, refer to the entropy-based measures to quantify information as these measures describe the utility or the usefulness of data.

It is important to recognize here that no one monitoring system can meet every information need as there may be several local or site-specific factors to cover. However, once one handles the broad objectives monitoring and the common problems of current design procedures, adjustments may be made to account for site-specific objectives.

The above mentioned guidelines are to be followed in establishing a basic framework in the development of new networks. In addition, the design must also be flexible enough to provide the following:

a) evaluation of the existing networks with respect to new objectives of monitoring (needs for data) and the efficiency of the existing system (capacity and informativeness of available data to meet the needs);
b) evaluation of new objectives or new practices of monitoring when they are to be incorporated into either the existing networks or the newly designed systems (for purposes of detecting changes in the efficiency or cost-effectiveness of the network with respect to changes in objectives and/or practices of monitoring).

This last point is particularly significant because the delineation of objectives, information expectations, and constraints is subject to a change in time, so that the design and the operation of water quality monitoring networks have an essentially dynamic character.

4.6.3. PROPOSED APPROACHES TO IMPROVEMENT

As the need is recognized for improvement of current water quality monitoring networks, researchers and designers have come with proposals on how this improvement can be realized. A few of these recommendations will be cited here.

Chapman (1992) addresses the basic problems in water quality monitoring and recommends steps to be taken for each problem. These problems and proposed solutions are listed in Table 4.1.

Regarding the design of water quality monitoring networks, M. Moss has reported, in Harmancioglu *et al.* (1997), the major conclusions derived at the NATO Workshop on Integrated Approach to Environmental Data Management Systems. The following points outline these conclusions:

a) *Environmental data networks can benefit from integrated approaches to their design.* There are both philosophical and pragmatic reasons for the integration of environmental data networks across various environmental phenomena.

   The philosophical basis for this conclusion is that environmental processes are interdependent in nature. Thus, if one wants to understand any particular aspect of the environment, the data describing the web of processes whose interactions influence that aspect must be studied to attain adequate understanding.

TABLE 4.1. Some possible sources of errors in the water quality assessment process with special reference to chemical methods

Assessment step	Operation	Possible source of error	Appropriate actions
definition of objectives	Statement	– lack of specific objective	Clearly specific and state objective
Conceptual understanding	Forces and interactions	– lack of understanding or conceptualizing	Field work, investigation, training
Monitoring design	Site selection	– station not representative (e.g., poor mixing in rivers)	Preliminary surveys
	Frequency determination	– sample not representative (e.g., unexpected cycles or variations between samples)	
Field operations	Sampling	– sample contamination (micropollutant monitoring)	Decontamination of sampling equipment, containers, preservatives
	Filtration	– contamination or loss	Running field banks
	Field measurement	– uncalibrated operations (pH, conduct., temperature)	Field calibrations Replicate sampling
		– inadequate understanding of hydrological regime	Hydrological survey
Sample shipments to laboratory	Sample conservation and identification	– error in chemical conservation	Field spiking
		– lack of cooling	
		– error in biological conservation	Appropriate field pretreatment
		– error and loss of label	
		– break of container	Field operator training
Laboratory	Preconcentration	– contamination or loss	Decontamination of laboratory equipment and facilities
	Analysis	– contamination	Quality control of laboratory air, equipment and distilled water
		– lack of sensitivity	
		– lack of calibration	Quality assurance tests (analysis of control sample; analysis of standards)
		– error in data report	Check internal consistency of data (e.g., with adjacent sample, ionic balance, etc.)
Computer facility	Data entry and retrieval	– error in data handling	Checks by data interpretation team
Interpretation	Data interpretation	– lack of basic knowledge	Appropriate training of scientists
		– ignorance of appropriate statistical methods	
		– omission in data report	
Publication	Data publication	– lack of communication and dissemination of results to authorities, the public, scientists, etc.	Setting of goals and training to meet the need of decision makers

From a pragmatic point of view, integration of environmental data networks makes sense because the interdependencies of the environmental processes permit information transfer among the processes. Thus, synergy and cost-effectiveness can result from integrated data networks.

b) *Design of data networks should be based on the purposes for which the data are to be collected.* There are many purposes for the collection of environmental data, and thus many network design tools are required. However, multipurpose networks are difficult to design rationally, so an approach that permits interactive designs of single purpose networks is the most feasible means of performing integrated design.

c) *A taxonomy of environmental data network purposes is useful in developing a strategy for integrated network design.* The use of the following taxonomy for the classification of network design purposes could highlight commonalities among network design technologies that would facilitate their use under a more robust set of situations:

   I. Decision-support networks;

   II. Academic-curiosity networks;

   III. Contingency networks.

   A decision-support network has an explicit purpose that results in specific types of data being collected in specific locations, at specific times, with the use of specific data collection technologies to provide the most cost-effective information for decision making.

   An academic-curiosity network ideally is designed to test a hypothesis about one or more environmental phenomena. The demands for specific locations of data collection, the frequencies of data collection, and the data collection technologies employed for such networks usually are not as rigid as for decision-support networks.

   Contingency networks are designed to ameliorate the impacts of not being able to forecast perfectly the future demands for environmental information. Such networks serve as insurance against unanticipated information needs, and they have designs that are less sensitive to location, frequency, and technology than do networks with other purposes.

d) *Basic understanding of environmental phenomena is the starting point for the design of environmental data networks.* Knowledge of the phenomena of interest is required to select an appropriate suite of network design tools. The choice of the actual tool or tools to be used for the design should be based on any existing data from the region of interest.

e) *Feedback from data collected in the initial network permits more complete description of the environmental phenomena and the subsequent use of more complex approaches to redesign the network.* Knowledge and information gained from an environmental data network can be used for improvement of the network.

f) *Network design is but one link in an integrated environmental data management chain, and it must be harmonized with the constraints and opportunities provided by the complementary links.* The design of data networks should not be performed in isolation from the technologies that will be used to convert the data to environmental information.

g) *There currently is a paucity of robust technologies for the design of environmental data networks, and technology transfer for the existing technologies is not being carried out satisfactorily on an international scale.* Because of the great interest in the environment that exists today, there is a large investment internationally in the collection of environmental data. With the lack of adequate network design support, many of the data collection programs probably are not being conducted in a cost-effective manner.

The following points are recommended for the design of monitoring networks:

a) Environmental data networks should be designed and operated in an integrated manner to take advantage of the international synergies that exist among environmental phenomena.

b) Environmental data networks should be resigned periodically to incorporate the new knowledge that is contained in the added data.

c) The development of more robust technologies for the design of environmental data networks should be supported by international environmental agencies.

d) New vehicles for the transfer of the technologies of data network analysis and design should be sought and implemented as they are demonstrated to be effective.

4.7. Redesign of Existing Networks

4.7.1. A CURRENT APPROACH

As discussed in section 4.3, the majority of developed countries have already started to redesign their networks. However, generally accepted guidelines do not exist on how the redesign process should be addressed.

A multilateral project has been initiated by research teams from six countries (i.e., USA, Canada, Italy, Turkey, Hungary, and Russia), including the authors, to focus on the development of rules for network assessment and redesign. The purpose here is to identify the basic guidelines of network design which may be followed by both the developed countries in their redesign process and the developing countries in their efforts for initiating and expanding their monitoring practice. The methodology investigated has the potential for application to design and assessment of other types of networks, including air pollution, streamgaging, rainfall, and soil moisture networks.

The current studies carried out within the above-mentioned multilateral project foresee the development of a methodology comprising technical and economic guidelines for design of efficient, cost-effective, and flexible water quality monitoring networks. This process involves an investigation into water quality time series, the water quality processes themselves, the relationships between water quality variables, and the value of the information contained in the various water quality parameters. The project has been initiated in view of the following basic objectives:

a) to develop a methodology for evaluation of the existing networks with respect to the efficiency of the existing system (capacity and informativeness of available data to meet the needs) and to the new objectives of monitoring (needs for data);
b) to develop a methodology for establishment of a basic framework in the development of new networks with monitoring agencies designed to match data needs and objectives of monitoring and meet economic constraints.
c) to develop a methodology for evaluation of new objectives or new practices of monitoring when they are to be incorporated into either the existing networks or the newly designed systems (for purposes of detecting changes in the efficiency or cost-effectiveness of the network with respect to changes in objectives and/or practices of monitoring).
d) to establish measures of flexibility in the information gathering efficiency of existing or planned networks in order to identify the ease by which the networks can be modified for collection of new data and ascertain the cost of achieving that flexibility.
e) to test and evaluate the methodologies on field data.

The basic approach in the current study is that monitoring of water quality should be evaluated within an integrated data management system. The process of data collection involves a number of activities that include not only the design of monitoring systems, but also physical sampling, data processing, data storage, data analysis and dissemination of information. Although the current state of technology has produced sophisticated means of handling each activity, there are still problems encountered in production of information from such a system. These problems are mainly due to lack of coherence between:

a) each step of data management;
b) different disciplines involved in each activity;
c) monitoring agencies involved; and
d) different countries in view of international exchange of information for the solution og global environmental problems.

Thus, a major consideration is the establishment of harmonization among different steps of data management, different disciplines and agencies involved, and different countries so as to ascertain availability and comparability of data. This issue requires standardization not only in monitoring principles but also in the other stages of data management (Harmancioglu, 1997).

The objectives of the above-mentioned study are accomplished through the following research tasks:

1) delineation of monitoring objectives;
2) determination of temporal sampling frequencies;
3) determination of spatial sampling frequencies;
4) selection of water quality variables for sampling;
5) selection of sampling duration;
6) evaluation of existing networks;
7) matching data needs and objectives of monitoring;
8) incorporation of new objectives into network design;
9) assembly of data;
10) verification of the methodology.

The first task covers the description of monitoring objectives and data needs; whereas, the others relate to the actual technical design phase. Tasks (1), (6), (7) and (8) set the basis for redesign of existing networks where two types of analyses are also carried out: (a) a spatial analysis of physiography and ecology which produce a system of zones of similarity; (b) analysis of the temporal information structure of a need to focus on smaller scales of watershed to produce transferable data and data which is of ecological significance (Whitfield and Clark, 1997). The above steps are preceded by an extensive review of the current monitoring practices in developed and developing countries (e.g., objectives of monitoring, shortcomings of existing networks, assessment of network efficiency and cost-effectiveness, projections into the future). This review helps to identify the basic problems and needs so that eventually an integrated approach to the network design problem can be developed. Such an approach is particularly significant for sampling on transboundary rivers and for monitoring to detect global environmental change.

The current tasks realized for the redesign process, within the framework shown in Fig. 4.2, are summarized in the following as the set of basic questions and relevant rules to be stated in the form of universal guidelines. Such questions and rules comprise the following steps:

1) identification of information needs and setting of realistic goals;
2) investigation of driving and modifying forces;
3) identification of sources of noise;
4) selection of proper sampling methods;
5) statistical analysis of data;
6) selecting and deciding on monitoring strategies as they relate to data quality; and
7) setting of operational rules.

Essentially, there is a feedback in the redesign process such that, once a set of decisions are made, they are to be assessed for the current network following the same steps denoted in Fig. 4.2. These steps are briefly described in the following.

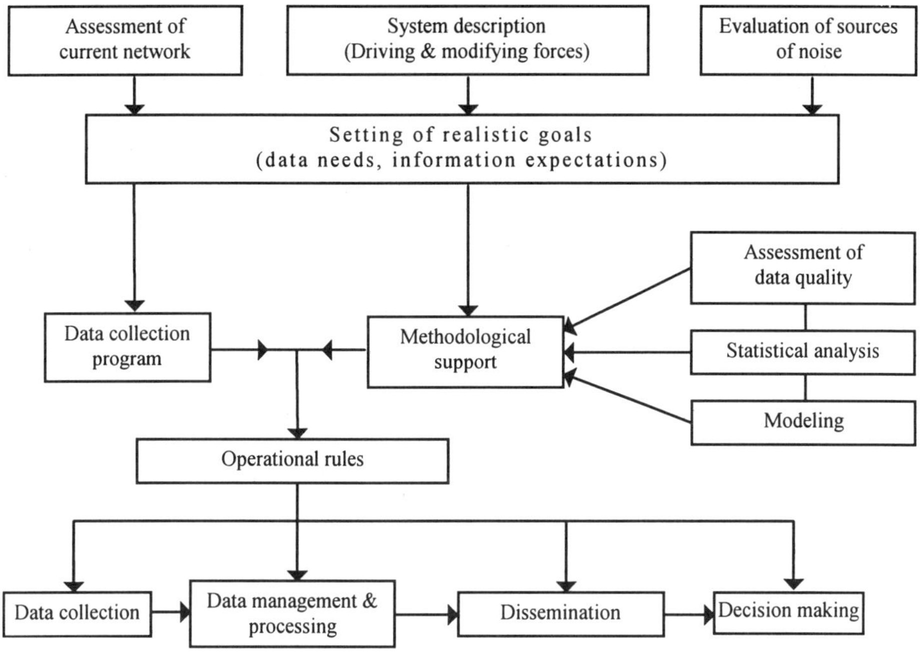

Figure 4.2. Proposed set of activities for network redesign

4.7.2. RULES PROPOSED FOR THE REDESIGN PROCESS

Settings of Realistic Goals

One of the most significant problems associated with current networks is the lack of a precise and proper definition of monitoring objectives. Since objectives delineate the eventual information expected from the network, setting of goals is the most crucial step of the redesign process (Harmancioglu *et al.*, 1992).

Literature provides a wide range of objectives specified for water quality monitoring objectives specified in water quality monitoring. These definitions relate to different perspectives, e.g., social, academic, economic, etc. Essentially, water quality observations have started for the same reasons that led to water quantity monitoring; yet the interesting point is that it is only now that designers are trying to define objectives.

A review of various definitions given in literature has led to specifications of three major objectives for water quality monitoring:

a) to assess water quality for water use and impacts;
b) to meet ecological demands; and
c) to carry out scientific research.

In this regard, the key objective of a monitoring network should relate to sustainable exploitation of water resources.

The two major conclusions related within the current study are:

a) that one should not collect any data that doesn't have an objective; and
b) that objectives should be clearly stated as specific goals for monitoring.

It is also considered essential that the following questions be addressed when initiating an assessment and redesign process of an existing network:

a) What do we already know about the design of the existing network? What are the problems? The solution to these questions can be derived by proper assessment of the existing network.
b) What is the "information" expected from the network? In this case, a detailed inventory of the basin has to be established, including such factors as the climate, population, industry, hydrology, water and land-use, pollutants, and the similar.
c) To elaborate further on question (b), one has to ask whether the information produced by the current network is sufficient. If not, we need to define what type of information is required.
d) The next question relates to the specification of the extent of the network with regard to both temporal and spatial scales. Provided that this extent is defined, designers have to consider technical and economic constraints to see if such a network can be afforded.
e) How important is the network in comparison with the existing situation or with other competitive networks?
f) Whom should the network address (e.g., agencies, cities, industries, society, public users, etc.)? It is important to identify here to whom the network will be of interest.
g) Apart from the general objective of the network, specific goals should be stated by working out the above questions in more detail.

These questions should be periodically reviewed to see if they are still valid after the network is redesigned. The goals should be rechecked to ascertain that collected data meet the specified needs.

Identification of Basic Forces and Functions in the Physical System
Whitfield (1997) proposes an "ecosystem approach" which he defines as "a geographically comprehensive approach to environmental planning and management which recognizes the interrelated nature of environmental media, and that humans are a key component of ecological systems". He further attempts to link data collection to ecosystems by assessing the forces acting on the ecosystem. Whitfield's (1997) approach is also adopted as one of the key features of the multilateral project mentioned above. The basic idea here is to define how the ecosystem functions and to identify the forces which dominate the input/output balance in the system (Fig. 4.3). Such forces may be natural driving forces and modifying forces, the latter referring to impacts by man. The effects of these forces and their interactions are to be determined with respect to the relevant time scales or periods (i.e., short-term or long-term) and spatial scales (i.e., global, regional, or watershed scale) (Whitfield and Wade, 1992).

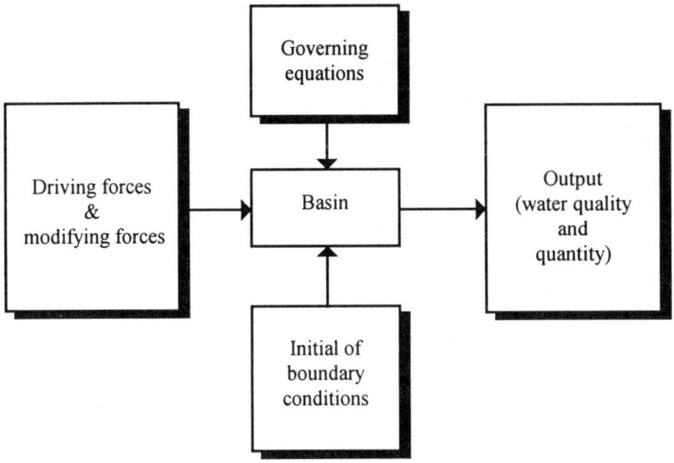

Figure 4.3. Identification of the physical system

Identification of Sources of Noise

Data are collected to attain information about the ecosystem and the way it functions under basic forces and their interactions. Data are essentially signals from the ecosystem; however, they do not represent perfect information about the natural system due to various sources of noise. When assessing the information content of data, sources of noise must be accounted for as they lead to blurring of information. "Noise" refers to a number of uncertainties which stem from monitoring practices. Such uncertainties may be due to:

a) mistaken assumptions and bias in the conceptual description of the ecological system as well as in the evaluation of data representativeness;
b) detectability of true signals (detection limits);
c) failure to accomplish representative sampling;
d) failure to select the proper methods in measurement;
e) various interferences that occur during sampling and laboratory analyses;
f) failure to look at the right place for the right material (e.g., water, air, biota, bottom sediments, etc.);
g) lack of quality assurance at various stages of monitoring;
h) lack of consistency with respect to sampling methods and sampling sites;
i) changes in sampling programs with respect to changing objectives or funding;
j) errors in sampling;
k) changes in sampling and analytical techniques (e.g., changes in methods, equipment, or detectability);
l) lack of completeness in information production due to missing data.

The above sources of noise should be assessed when trying to extract the information contained in available data. Basically, these sources indicate three major areas where uncertainties may prevail:

a) conceptual understanding of basic processes;
b) available data; and
c) statistical noise.

To minimize noise, the major recommendations formulated within the ongoing project foresees that:

a) all procedures, standards, and information expectations should be documented for traceability;
b) personnel involved in the monitoring program should be skilled and trained;
c) each organization and agency involved in different steps of data management should check for the quality of their inputs and outputs;
d) there is a need for standards and standardized procedures in each step of data management;
e) risks in the monitoring system should be identified.

Selection of Proper Sampling Methods

There are several options regarding sampling and analytical methods (Literathy, 1997). These options relate to:

a) *variables to be observed*: variables may be abiotic in the form of physical or chemical constituents (i.e., those that pertain to water and sediment); or, they may be biotic (e.g., biotic communities, biodiversity, and pollutants in tissue);
b) *field observations*: field measurements may be realized by in-situ sensors, visual methods (e.g., floating materials, foam, oil, color, etc.) organoleptic means (odor), photologs (aerial photos, remote sensing, etc.), or anecdotal evidences;
c) *sample collection from water, sediment, and biota*: sample collection is subject to spatial alternatives (e.g., sites, points, or sediment cores), temporal alternatives (frequencies), and alternative methods (e.g., manual or automatic);
d) *sample transport and storage*: selection may be made among pretreatment alternatives and freezing;
e) *analytical methods*: analysis may be made on a group of compounds or on specific substances. Here, performance characteristics of the applied methods (e.g., range of application, detectivity, etc.) and analytical quality control (intralab and interlab) are significant.

The redesign process should evaluate the above alternative options, which are significant for both the design and the operation of a network.

Statistical Analysis of Data

This is the stage when data are transformed into information by data analyses realized under certain assumptions as in Fig. 4.4 (Harmancioglu and Alpaslan, 1992). It is essential here to test and assess:

a) validity of assumptions as they set the basis for data interpretation;
b) data analysis features so as to check for the presence of errors in data and for the completeness, homogeneity, and representativeness of data;
c) reliability of the data analysis performed;
d) information content of data.

The basic requirement here is to define "information" in specific, preferably quantitative terms.

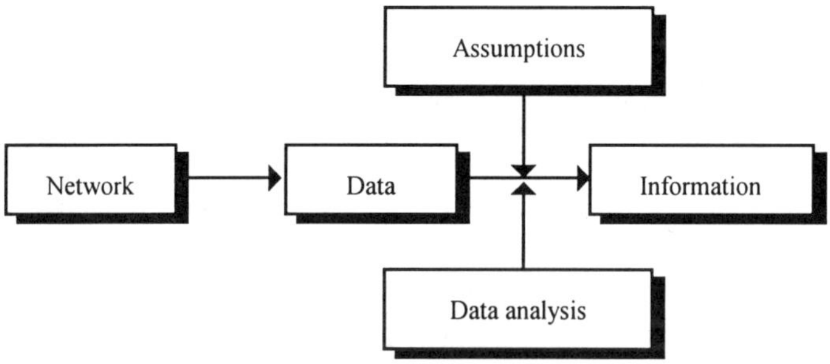

Figure 4.4. Basic components in transfer of data into information

Selecting and Deciding on Monitoring Strategies as They Relate to Data Quality

This is the stage where monitoring strategies are finalized on the basis of:

a) assessment of the quality of information (also of data) required;
b) data quality standards and classification;
c) impact of modeling on information content of data; and
d) evaluating the above in view of the goals specified for the network.

The quality of information required (and hence that of data) should be evaluated with respect to spatial scales (e.g., network density per unit area or length), temporal scale (i.e., frequency), reliability of sampling, and flexibility of data collection procedures. Next, data to be collected should be classified on the basis of quality standards in view of information utilization (e.g., for forecasting or prediction). For example, if information is needed on ecological balance for purposes of prediction, a certain quality of data (e.g., class A) has to be ascertained; whereas a lower standard may be satisfactory for forecasting purposes. A third component may be added to the above assessment, and that is the impact of modeling on the required quality of data. For instance, selection of distributed models may lead to increased data and information requirements as compared to lumped models. An important issue here is to assess how much a good model can save on monitoring efforts.

Setting of Operational Rules
This stage involves setting of rules for data management, processing, and dissemination for the eventual decision making process. Operational rules have three components: informational, technological, and institutional.

The informational component comprises the development of databases. Databases include metadata together with observed, derived, and modeled data to constitute a knowledge base. Forms of data presentation within the database have to be specified as one of space-oriented, cartographic (numerical), or textual forms. Next, standard operation procedures (SOPs) and standards for input data stream and output informational products must be selected.

The technological component of operational rules relate to the selection of means for:

a) Database Management System (DBMS) applications;
b) processing of data;
c) expert system applications for decision making; and
d) hardware development.

In the organizational component, the basic issue is the dissemination of data for relevant users. Two problems must be solved at this stage: data exchange policies must be specified among involved agencies and institutions; and specifications of system operation should be delineated on local, regional, and state levels.

4.7.3. CONCLUSION

At the current stage of the multilateral project, the above rules are stated in a draft form to be eventually finalized as guidelines for the redesign process. Three test cases are specified for testing of rules: the Danube river basin in Hungary to represent the case of transboundary rivers, the Gediz river basin in Turkey, and a watershed in Italy comprising lakes for drinking water supply. Specific details of the rules will be elaborated by assessing the results of the case studies. As each case has different characteristics with respect to technical, economic and institutional factors, it is expected that these results will allow to develop a comprehensive and universal set of guidelines for the network redesign problem.

Acknowledgment

The research leading to this chapter has been supported by the NATO Linkage Grant ENVIR.LG.950779. This support is gratefully acknowledged.

References

Adriaanse, M.J., van de Kraats, J., Stoks, P.G., and Ward, R.C. (1995) Conclusions monitoring tailor made, in: *Proceedings, Monitoring Tailor-Made*, An International Workshop on Monitoring and Assessment in Water Management, Sept. 20-23, pp. 345-347. Beekbergen, The Netherlands.

Alkan, A., Ozkul, S., Alpaslan, N., and Harmancioglu; N. (1995) Developments of water quality monitoring networks in Turkey and other countries (in Turkish), *Turkish Chamber of Civil Engineers, 13th Technical Congress, Proceedings,* pp. 559-572.

Alpaslan, N. and Harmancioglu, N.B. (1990) *Water Quality Monitoring-Site Selection*, Stuttgart, Seminar Umweltschutz, Sept. 1990, pp.185-205.

Alpaslan, N., Harmancioglu, N.B., and Singh, V.P. (1992) The role of the entropy concept in design and evaluation of water quality monitoring networks, in V.P. Singh and M. Fiorentino (eds.), *Entropy and Energy Dissipation in Water Resources*, Kluwer Academic Publishers, Water Science and Technology Library, Dordecht (pp. 261-282.

Beck, M.B. and Finney, B.A. (1987) Operational water quality management: Problem context and evaluation of a Model for river quality, *Water Resources Research* **23(11)**, 2030-2042.

Chapman, D. (ed.) (1992) *Water Quality Assessments*, (published on behalf of UNESCO, WMO and UNEP), Chapman & Hall, London.

Cotter, A.J.R. (1985) Water quality surveys: a statistical method based on determinism, quantiles and the binomial distribution, *Water Research* **19(9)**, 1179-1189.

Dafoe, T.J., Watt, E.R., and Stevens, R. (1986) Water quality monitoring branch activities: a dynamic approach to evolving issues, in: R.C. Ward, J.C. Loftis, and G.B. McBride (eds.), *Proceedings, International Symposium on the Design of Water Quality Information Systems,* Fort Collins, CSU Information Series no. 61, pp. 47-58.

Dandy, G.C. and Moore, S.F. (1979) Water quality sampling programs in rivers, *J. of Env. Eng. Div., ASCE* **105(EE4)**, 695-712.

Dawdy, D.R. (1979) The worth of hydrologic data, *Water Resources Research* **15**, 1726-1732.

Dixon, W. and Chiswell, B. (1996) Review of aquatic monitoring program design, *Water Research* **30(9)**, 1935-1948.

Gupta, V.L. (1982) Hydrologic data network design by modified Langbein Method, in: V.P. Singh (ed.), *Modeling Components of Hydrologic Cycle*, Proceedings of the International Symposium on Rainfall Modeling, May 1981, Water Resources Publications, pp. 51-70.

Harmancioglu, N.B. (1997) The need for integrated approaches to environmental data management, in: N.B. Harmancioglu, M.N. Alpaslan, S.D. Ozkul and V.P. Singh (eds.), *Integrated Approach to Environmental Data Management Systems*, Kluwer Academic Publishers, NATO ASI Series, 2. Environment, vol. 31, pp. 3-14.

Harmancioglu, N. B. (1994) An entropy based approach to station discontinuance, in: (K. W. Hipel et al. (eds.), *Stochastic and Statistical Methods in Hydrology and Environmental Engineering, Vol. 10/3 (Time Series Analysis in Hydrology and Environmental Engineering),* Kluwer, Water Science and Technology Library, pp. 163-176.

Harmancioglu, N. and Tokgoz, S. (1995) Selection of sampling frequencies in water quality monitoring network design (in Turkish), *Journal of Water Pollution Control* **5(1)**,. 9-20.

Harmancioglu, N.B. and Alpaslan, N. (1994) Basic approaches to design of water quality monitoring networks, *Elsevier, Water Science and Technology* **30(10)**, 49-56.

Harmancioglu, N.B. and Alpaslan, N. (1992) Water quality monitoring network design: a problem of multi-objective decision making, *Water Resources Bulletin* **28(1)**, 179-192.

Harmancioglu, N.B., and Singh, V.P. (1991) An information-based approach to monitoring and evaluation of water quality data in advances in water resources technology, in: G. Tsakiris (ed.), *ECOWARM, Proceedings of the European Conference on Advances in Water Resources Technology,* (Abstract: Water Resources Management, n.4/4, Dec. 1990), A.A. Balkema Publishers, Athens, pp. 377-386.

Harmancioglu, N.B. and Yevjevich, V. (1986) *Transfer of Information Among Water Quality Variables of the Potomac River, Phase III: Transferable and Transferred Information,* Report to D.C. Water Resources Research Center of the University of the District of Columbia, Washington, D.C., June 1986, 81p.

Harmancioglu, N.B., Alpaslan, M.N., and Ozkul, S.D. (1997) Conclusions and recommendations, in: N.B. Harmancioglu, M.N. Alpaslan, S.D. Ozkul and V.P. Singh (eds.), *Integrated Approach to Environmental Data Management Systems,* Kluwer Academic Publishers, NATO ASI Series, 2. Environment, vol. 31, pp. 423-434.

Harmancioglu, N. B., Alpaslan, N., and Singh, V. P. (1994) Design of a basin-wide water quality monitoring network in Turkey, in: G. Tsakiris and M.A. Santos (eds.), *Advances in Water Resources Technology and Management,* A.A. Balkema, Rotterdam, pp. 29-36.

Harmancioglu, N.B., Singh, V.P., and Alpaslan, N. (1992) Design of water quality monitoring networks, in: R.N. Chowdhury (ed.), *Geomechanics and Water Engineering in Environmental Management,* A.A. Balkema Publishers, Rotterdam, ch. 8, pp. 267-296.

Harmancioglu, N.B., Ozer, A., and Alpaslan, N. (1987) Procurement of water quality information (in Turkish), *IX. Technical Congress of Civil Engineering, Proceedings, the Turkish Society of Civil Engineers, v. II,* pp. 113-129.

Harmancioglu, N.B., Alkan, A., Alpaslan, N., and Singh, V.P. (1996) Entropy-based approaches to assessment of monitoring networks, in: K.S. Tickle, I.C. Goulter, C. Xu, S.A. Wasimi, and F. Bouchart (eds.), *Stochastic Hydraulic'96, Proceedings of the Seventh IAHR International Symposium, Mackay, Queensland, Australia,* A.A. Balkema Publishers, pp. 183-190.

Harmancioglu, N., Alpaslan, N., Alkan, A., Ozkul, S., Mazlum, S., and Fistikoglu, O. (1994 a) *Design and Evaluation of Water Quality Monitoring Networks for Environmental Management* (in Turkish), Report prepared for the research project granted by TUBITAK, Scientific and Technical Council of Turkey, Project Code: DEBAG-23, Izmir, 514 p.

Hipel, K.W. (1988) Nonparametric approaches to environmental impact assessment, *Water Resources Bulletin, AWRA,* **24(3)**, 487-492.

Hirsch, R.M. (1988) Statistical methods and sampling design for estimating step trends in surface-water quality, *Water Resources Bulletin, AWRA*, **24(3)**, 493-503.

Horton, R.E. (1945) Erosional development of streams, *Geological Society Am. Bull.*, **56**, 281-283.

Huthmann, G. (1979) Modeling of water quality systems by multiple frequency response analysis, in: H.J. Morel-Seytoux (ed.), *Surface and Subsurface Hydrology*, Proceedings of the Forth Collins Third International Hydrology Symposium on Theoretical and Applied Hydrology, July 27-29, 1977, Water Resources Publications, pp. 662-681.

Karpuzcu, M., Senes, S. and Akkoyunlu, A. (1987) Design of monitoring systems for water quality by principal component analysis and a case study, *Proceedings, Int. Symp. on Environmental Management: Environment'87*, pp. 673-690.

Kwiatkowski, R.E. (1986) The importance of design quality control to a national monitoring program, in: A.H. El-Shaarawi and R.E. Kwiatkowski (eds.), *Statistical Aspects of Water Quality Monitoring,* Elsevier, Proceedings of the workshop held at Canada Centre Inland Waters, October 1985, pp. 79-98.

Lachance, M., Bobee, B., and Haemmerli, J. (1989) Methodology for the planning and operation of a water quality network with temporal and spatial objectives: application to acid lakes in Quebec, in: R.C. Ward, J.C. Loftis and G.B. McBride (eds.), *Proceedings, International Symposium on the Design of Water Quality Information Systems*, Fort Collins, CSU Information Series no. 61, pp. 145-162.

Langbein, W.B. (1979) Overview of conference on hydrologic data networks, *Water Resources Research* **15(6)**, 1867-1871.

Lettenmaier, D.P. (1988) Multivariate nonparametric tests for trend in water quality, *Water Resources Bulletin, AWRA* **24(3)**, 505-512.

Lettenmaier, D.P. (1979) Dimensionality problems in water quality network design, *Water Resources Research* **15**, 1692-1700.

Lettenmaier, D.P. (1976) Detection of trends in water quality data from records with dependent observations, *Water Resources Research* **12**, 1037-1046.

Liebetrau, A.M. (1979) Water quality sampling: some statistical considerations, *Water Resources Research* **15**, 1717-1725.

Literathy, P. (1997) Transboundary water pollution monitoring: data validation and interpretation, in: N.B. Harmancioglu, M.N. Alpaslan, S.D. Ozkul and V.P. Singh (eds.), *Integrated Approach to Environmental Data Management Systems*, Kluwer Academic Publishers, NATO ASI Series, 2. Environment, vol. 31, pp. 199-214.

Loftis, J.C., McBride, G.B., and Ellis, J.C. (1991) Considerations of scale in water quality monitoring and data analysis, *AWRA, Water Resources Bulletin* **27(2)**, 255-264.

MacKenzie, M., Palmer, R.N. and Millard, S.T. (1987) Analysis of statistical monitoring network design, *J. of Water Resources Planning and Management* **113(5)**, 599-615.

McNeil, V.H., McNeil, A.G. and Poplawski, W.A. (1989) Development of water quality monitoring system in Queensland, in: R.C. Ward, J.C. Loftis and G.B. McBride (eds.), *Proceedings, International Symposium on the Design of Water Quality Information Systems*, Fort Collins, CSU Information Series No. 61, pp. 73-86.

Moss, M.E. (1989) Water quality data in the information age, in: R.C. Ward, J.C. Loftis, and G.B. McBride (eds.), *Proceedings, International Symposium on the Design of Water Quality Information Systems*, Fort Collins, CSU Information Series No. 61, pp. 8-15.

Moss, M.E. (1979a) Some basic considerations in the design of hydrologic data networks, *Water Resources Research* **15(6)**, 1673-1676.

Moss, M.E. (1979b) Space, time and the third dimension (model error), *Water Resources Research* **15(6)**, 1797-1800.

Moss, M.E. (1976) Decision theory and its application to network design, in: *Hydrological Network Design and Information Transfer*, World Meteorological Organization, WMO, no. 433, Geneva, Switzerland.

Moss, M.E. and Karlinger, M.R. (1974) Surface water network design by regression analysis simulation, *Water Resources Research* **10(3)**, 425-433.

Mueller, D.K. (1989) Use of box plots and trend analyses to evaluate sampling frequency at water quality monitoring sites, in: R.C. Ward, J.C. Loftis, and G.B. McBride (eds.), *Proceedings, International Symposium on the Design of Water Quality Information Systems*, Fort Collins, CSU Information Series No. 61, pp. 88-104.

Naidu, B.R. and Khan, K.R. (1987) Water quality monitoring data analysis-case studies on rivers Sabarmati, Godavari and Mahi, *Proceedings, Int. Symp. on Environmental Management: Environment'87*, pp. 993-1012.

Niederlander, H.A.G., Dogterom, J., Buijs, P.H.L., Hupkes, R., and Adriaanse, M. (1996) *UN/ECE Task Force on Monitoring & Assessment, Working Programme 1994/1995, Volume:5: State of the Art on Monitoring and Assessment of Rivers,* RIZA report: 95.068.

Ongley, E.D. (1994) Global water quality information needs-GEMS/WATER, in: *Advances in Water Quality Monitoring - Report of a WMO Regional Workshop (Vienna, 7-11 March 1994)*, World Meteorological Organization, Technical Reports in Hydrology and Water Resources, No. 42, WMO/TD-NO 612, Geneva, Switzerland, pp. 32-40.

Ozkul, S. D. (1996) *Space/Time Design of Water Quality Monitoring Networks by the Entropy Method*, Ph. D. Thesis on Civil Engineering, Dokuz Eylul University, Graduate School of Natural and Applied Sciences, Izmir, 196 p., (Advisor: Prof. Dr. N. B. Harmancioglu).

Ozkul, S., Fistikoglu, O., Harmancioglu, N.B., and Singh, V.P. (1996) Statistical evaluation of monitoring networks in space/time dimensions, in: K.S. Tickle, I.C. Goulter, C. Xu, S.A. Wasimi, and F. Bouchart (eds.), *Stochastic Hydraulic'96, Proceedings of the Seventh IAHR International Symposium, Mackay, Queensland, Australia,* A.A. Balkema Publishers, pp. 357-364.

Ozkul, S., Alkan, A., Harmancioglu, N., and Alpaslan, N. (1995) Evaluation of sampling frequencies in the design of water quality monitoring networks, in: *Proceedings, Advances in Civil Engineering, Second Technical Congress,* September 18-20, 1995, Bogazici University, Istanbul, pp. 302-312.

Palmer, R.N. and MacKenzie, M. (1985) Optimization of water quality monitoring networks, *J. of Water Resources Planning and Management* **111(4)**, 478-493.

Paulson, R.W. (1994) Observing the hydrological cycle in the western hemisphere via GOES-a first step towards WHYCOS?, in: *Advances in Water Quality Monitoring - Report of a WMO Regional Workshop (Vienna, 7-11 March 1994)*, World Meteorological Organization, Technical Reports in Hydrology and Water Resources, No. 42, WMO/TD-NO 612, Geneva, Switzerland, pp. 302-312.

Praskins, W. (1989) Monitoring to improve decision making in EPA and state surface water quality problems, in: R.C. Ward, J.C. Loftis, and G.B. McBride (eds.), *Proceedings, International Symposium on the Design of Water Quality Information Systems*, Fort Collins, CSU Information Series No. 61, 54-58.

Quimpo, R.G. and Yang, J. (1970) Sampling considerations in stream discharge and temperature measurements, *Water Resources Research* **6(16)**, 1771-1774.

Reinelt, L.E., Horner, R.R., and Mar, B.W. (1988) Nonpoint source pollution monitoring program design, *J. of Water Resources Planning and Management* **114(3)**, 335-352.

Richards, R.P. (1989) Determination of sampling frequency for pollutant load estimation using flow information only, in: R.C. Ward, J.C. Loftis, and G.B. McBride (eds.), *Proceedings, International Symposium on the Design of Water Quality Information Systems*, Fort Collins, CSU Information Series no. 61, pp. 136-144.

Sanders T.G. (1988) *Water quality monitoring networks in water and wastewater systems analysis,* D. Stephenson (ed.), Elsevier Science Publishers, The Netherlands, B. V., pp. 204-216.

Sanders, T.G., and Adrian, D.D. (1978) Sampling frequency for river quality monitoring, *Water Resources Research* **14**, 569-576.

Sanders, T.G., Ward, R.C., Loftis, J.C., Steele, T.D., Adrian, D.D., and Yevjevich, V. (1983) *Design of Networks for Monitoring Water Quality*, Water Resources Publications, Littleton, Colorado, 328p.

Schad, T.M. (1984) Introduction, in: T.M. Schad (ed.), *Options for Reaching Water Quality Goals*, Symposium Proceedings, 20th Annual Conference of AWRA, Washington, D.C., 1.

Scheidegger, A.E. (1965) *The Algebra of Stream Order Number*, U.S. Geological Survey, Prof. Paper 525-B, B187-B189.

Schilperoort, T., Groot, S., Wetering, B.G.M., and Dijkman, F. (1982) *Optimization of the Sampling Frequency of Water Quality Monitoring Networks*, "Waterloopkundig" Laboratium Delft, Hydraulics Lab, Delft, the Netherlands.

Sharp, W.E. (1971) A topologically optimum water - sampling plan for rivers and streams, *Water Resources Research* **7(6)**, 1641-1646.

Sharp, W.E. (1970) Stream order as a measure of sample uncertainty, *Water Resources Research* **6(3)**, 919-926.

Starosolsky, O. (ed.) (1987) *Applied Surface Hydrology*, Water Resources Publications, Littleton, Colorado, pp. 175-380.

Tasker, G.D. and Moss, E.M. (1979) Analysis of Arizona flood data network for regional information, *Water Resources Research* **15(6)**, 1791-1796.

Timmerman, J.G., Gardner, M.J., and Ravenscraft, J.E. (1996) *UN/ECE Task Force on Monitoring and Assessment, Working Programme 1994/1995, Volume:4: Quality Assurance*,. RIZA report: 95.067.

Tirsch, F.S. and Male, J.W. (1984) River basin water quality monitoring network design: options for reaching water quality goals, in: T.M. Schad (ed.), *Proceedings of Twentieth Annual Conference of American Water Resources Associations*, AWRA Publications, pp. 149-156.

Tokgoz, S. (1992) *Temporal Design of Water Quality Monitoring Networks*, Master of Science thesis in Civil Engineering, Dokuz Eylul University, Graduate School of Natural and Applied Sciences, Izmir.

UNESCO-WMO (1972) *Hydrologic Information Systems: Studies and Reports in Hydrology*, (G.W. Whetstone, and J.J. Grigoriev (ed.)), prepared by the Panel on SAPHYDATA, no.14, 74p.

Valiela, D. and Whitfield, P.H. (1989) Monitoring strategies to determine compliance with water quality objectives, *Water Resources Bulletin, AWRA* **25**, 63-69.

Ward, R.C. (1996) Water quality monitoring: where's the beef?, *Water Resources Bulletin* **32(4)**, 673-680.

Ward, R.C. (1989) Water quality monitoring - a systems approach to design, in: R.C. Ward, J.C. Loftis, and G.B. McBride (eds.), *Proceedings, International Symposium on the Design of Water Quality Information Systems*, Fort Collins, CSU Information Series No. 61, 37-46.

Ward, R.C. and Loftis, J.C. (1986) Establishing statistical design criteria for water quality monitoring systems: Review and synthesis, *Water Resources Bulletin, AWRA* **22(5)**, 759-767.

Ward, R.C., J.C. Loftis and G.B. McBride (1990) *Design of Water Quality Monitoring Systems*, Van Nostrand Reinhold, New York.

Ward, R.C., Loftis, J.C., and McBride, G.B. (eds.) (1989) *Proceedings International Symposium on the design of Water Quality Information Systems,* Information Series No. 61, Colorado Water Resources Research Institute, 472 p.

Ward, R.C.; Loftis, J.C.; and McBride, G.B. (1986) The data-rich but information-poor syndrome in water quality monitoring, *Environmental Management* **10**, 291-297.

Ward, R.C., Loftis, J.C., Nielsen, K.S., and Anderson, R.D. (1979) Statistical evaluation of sampling frequencies in monitoring networks, *J. of WPCF* **51(9)**, 2292-2300.

Warn, A.E. (1988) Auditing the quality of effluent discharges, in: *Workshop on Statistical Methods for the Assessment of Point Source Pollution*, 12-14 September, Canada Centre for Inland Waters, Burlington, Ontario, Canada.

Whitfield, P.H. (1997) Designing and redesigning environmental monitoring programs from an ecosystem perspective, in: N.B. Harmancioglu, M.N. Alpaslan, S.D. Ozkul and V.P. Singh (eds.), *Integrated Approach to Environmental Data Management Systems*, Kluwer Academic Publishers, NATO ASI Series, 2. Environment, vol. 31, pp.107-116.

Whitfield, P.H. (1988) Goals and data collection designs for water quality monitoring, *Water Resources Bulletin, AWRA* **24(4)**, 775-780.

Whitfield, P. and Clark, M. (1997) Driving forces, *Water Quality International* **March/April 1997**, 20-21.

Whitfield. P.H and Wade, N. (1992) Monitoring transient water quality events electronically, *Water Resources Bulletin* **28(4)**, 703-711.

Whitlatch, E.E. (1989) Spatial adequacy of NASQAN water quality data in Ohio river basin, *J. of Env. Eng.* **115(1)**, 173-191.

WMO (1994) *Advances in Water Quality Monitoring - Report of a WMO Regional Workshop (Vienna, 7-11 March 1994)*, World Meteorological Organization, Technical Reports in Hydrology and Water Resources, No. 42, WMO/TD-NO 612, Geneva, Switzerland, 332 p.

Wood, E.F. (1979) A statistical approach to station discontinuance, *Water Resources Research* **15(6)**, 1859-1866.

Yevjevich, V. and Harmancioglu, N.B. (1985) *Modeling Water Quality Variables of Potomac River at the Entrance to its Estuary, Phase II (Correlation of Water Quality Variables within the Framework of Structural Analysis)*, Report to D.C. Water Resources Research Center of the University of the District of Columbia, Washington, D.C., Sept. 1985, 59p.

CHAPTER 5

RISK IN WATER QUALITY MONITORING

I. Goulter and A. Kusmulyono

Abstract. Any consideration of risk in water quality monitoring must be placed in a framework of the objectives defined for monitoring. Actual determination of risk is based upon the likelihood (probability) of events, e.g., missing a violation of an important water quality standard, and the consequences of those events. Risk in water quality monitoring is associated not only with different aspects of data collection but also with data processing and analysis to provide the required information. The following sections discuss the consideration of risk and reliability with respect to objectives of monitoring. Within this framework, various components of risk are assessed as they relate to data collection, processing, and analysis.

5.1. Introduction

As in all other fields, consideration of risk and actual assessment of risk in water quality monitoring must be framed in terms of both the consequences of events and the likelihood or probability of those events occurring. It is also important to recognize that risk in water quality monitoring does not arise solely in the data collection phase. Risk in data collection is associated with whether the correct parameters are being measured, whether these parameters are being measured at sufficient locations, whether the measurements are being taken at sufficiently frequent intervals, and whether the measurements themselves are accurate. Another aspect of risk arises where water quality data are collected, or environmental conditions monitored, and where they are subsequently subjected to further analysis for such purposes as identification of potential trends in conditions, or to provide information on a parameter or condition which cannot be monitored directly. This additional contribution to risk arises from questions as to how accurately, or perhaps, more correctly, with what certainty, this processing or analysis generates the required information. Reliability is also related to risk in that it is a quantitative measure of how frequently and accurately the monitoring provides the desired information.

The following sections will review the objectives of water quality monitoring and discuss the consideration of risk and reliability within that framework.

N.B. Harmancioglu et al. (eds.), Environmental Data Management, 107-122.

5.2. Objectives of Water Quality Monitoring

The initial focus of water quality monitoring was on surface water quality. (It is interesting to note that although water quality monitoring directed at pollution control began in the United States in the 1930's, Federal laws on water pollution control were not passed in that country until the 1940's and 1950's). In the 1970's, the objective of water quality monitoring in the United States was clearly directed at water quality management, primarily to control discharges to streams by creating a discharge permit system (Ward *et al.*, 1990).

In Australia, the early objectives of water quality monitoring were developed in conjunction with the establishment of the Australian Water Resources Council in 1962. These very general objectives related to the need for comprehensive and continuous assessment of Australia's water resources and to the extension of measurement to aid in the planning of future development (Department of National Development, Australia, 1965). Similar general objectives of preserving, restoring and enhancing the quality of water were specified for water quality monitoring through the Clean Waters Act enacted in 1973. The specific tasks of the Australian Water Resources Council in the context of these general objectives included the setting of standards, surveillance of water discharges and providing advice and information to interested organizations and individuals. While these objectives are for a particular national situation, they are examples of the general framework of objectives for water quality monitoring and are not dissimilar to the generic expectations for water quality monitoring in other countries.

While the early emphasis for water quality monitoring was on surface water, the growing awareness of the critical role of groundwater, together with an increasing concern about the quality of that resource, has over the last two decades, led to the placing of a very high priority on monitoring of groundwater quality. However, the same generic principles for monitoring of surface water quality in terms of the objectives and the relationship between objectives and risk also hold for monitoring of groundwater quality.

In relating these general guidelines to specific objectives, it is important to note that the underlying principle of water quality monitoring. is to extract information, i.e., to determine the condition of a water body from collected data which are normally obtained from sampling. Information derived through the data collection and subsequent analysis can then be related to the specific objectives of water quality monitoring.

These specific objectives of water quality monitoring can generally be classified into the following three categories (Ward and Loftis, 1989):

- compliance with specified standards;
- identification of trends;
- detection of extreme values.

Duckstein *et al.* (1976) and Whitfield (1988) also highlighted an additional major objective, namely, observation of the impacts on water quality, resulting, for example, from introduction of water quality treatment measures or from projects and activities having the potential to cause a deterioration in water quality. However, this objective is arguably a component of the major objectives given above.

Risk in the context of these three over-arching objectives can arise in a number of ways. The data and subsequent analysis may identify compliance or noncompliance with a standard when noncompliance or compliance respectively is in fact the case. Similarly, the trend identified through the data collection and analysis may be incorrect. With respect to the detection of extreme values, data collection must be sufficiently discriminating, i.e., with sufficient frequency o pick up the extreme and possibly rare values when they occur.

It should be recognized that each of the examples or scenarios cited above represent the outcomes or consequences which must be incorporated in the consideration of risk. These "consequences" should, in turn, be related to the environmental, social and economic costs incurred by society as a result of decisions made on the basis o£ the false or incorrect information. The magnitude of these "costs" is likely to change from one location or situation to another. This chapter will not address the costs as such but will rather concentrate on the consequences in their untransformed states, along with the issues related to determination of the probabilities of events.

5.3. Components of Risk in Water Quality Monitoring

As indicated in the introduction, risk in water quality monitoring arises from two distinct sources:

i) the completeness and accuracy of the collected raw data, and
ii) the extent to which the model used to process or analyze those raw data provides a true representation of the system.

The process of data collection itself has four distinct aspects of risk, namely:

i) the extent to which the set of parameters being measured is the correct grouping;
ii) the extent to which the parameters to be measured are collected at a sufficient number of locations to adequately characterize spatial conditions across the area of interest;
iii) whether the parameters are measured, i.e., sampled, at a sufficiently high frequency to capture the temporal variation of the parameter of interest, and
iv) whether the data are measured with sufficient accuracy and precision to provide the required level of information.

The selection of which variables to sample is very much dependent on the objectives of monitoring (Ward *et al.*, 1990). When the objective is to monitor conditions and observe long-term trends, 'indicator' type of measurements are generally sufficient. If the objective is for special investigation of a particular water quality problem, the parameters of concern should obviously be those related to the problem. When the objective of monitoring is to examine compliance, for example, to a regulatory water quality standard, the parameters chosen for sampling must, at minimum, include all those listed in the standard.

For the situation where there is some choice of parameters to monitor, Sanders *et al.* (1983) suggest the specification of categories of water quality variables as a means of assisting in data collection planning and data analysis. Their study also suggested that the parameters to be monitored be specified before the actual design of the water quality monitoring network so that the monitoring program can be undertaken in a systematic fashion. This suggestion arises from the existence of both the natural or man-made variations that can occur in every water quality variable and the inability, due amongst other things to cost considerations, to consider, all variables and their variations simultaneously. In addition, there is also the need to monitor the complete range of those parameters which. are considered important.

Heidtke and Armstrong (1979) proposed a probabilistic sampling model for water quality management. The objective of their approach was to design an optimum sampling policy for detection of stream standard violation. Lettenmaier *et al.* (1984) proposed a method based on dynamic programming for systematic consolidation of stations in an existing water quality monitoring network. The method was applied to the Municipality of Metropolitan Seattle stream and river quality monitoring network and was shown to result in a significant reduction in the number of monitoring stations with relatively little loss of information. Recognizing that increased information provides a basis for reducing risk, the small loss of information arising from the reduction in a number of monitoring stations must have, by implication, also caused only a slight increase in the exposure to risk.

Palmer and MacKenzie (1985) developed a procedure to select designs which maximize the statistical power of a monitoring network for a specified budget, or alternatively, which minimize the cost of a network for a specified statistical power requirement. The procedure was applied to aquatic monitoring in relation to a New England coastal power plant and showed that numerous solutions may exist for a specified power (level of information and risk) and cost. Pinter and Sornlyody (1987) subsequently suggested a methodology for optimizing the operation of a regional monitoring network used for water quality management. The particular objective of their study was to provide a reliable estimate of the annual nutrient load of a lake. The model itself minimizes the costs of operation of the annual monitoring system under obvious logical (physical) constraints on the expected accuracy of the monitoring program, This emphasis on reliability and accuracy is an important recognition, if only implicit, of the risk involved in water quality monitoring should important values of parameters be missed and necessary action therefore not initiated.

More recently, Harmancioglu and Alpaslan (1992) reexamined the problem of multiobjective decision making in water quality monitoring network design and proposed a statistical procedure to evaluate the efficiency and cost-effectiveness of a network. The method was demonstrated by application to the Porsuk River in Turkey.

This application indicated that their method has the capability of assessing the efficiency and cost-effectiveness of a network quantitatively. As such, the method provides a rational basis for a reduction in either the frequency of sampling or the number of sampling stations. Reduction in either the number of sampling stations or the frequency of sampling has a direct impact on the risk of not characterizing the spatial or temporal variation in water quality conditions adequately. It is also interesting to note the context of that study that 'rational' decisions as to reduce either the frequency of sampling or the number of sampling stations imply an explicit rational recognition of the costs of information and thereby of the increased exposure to risk occurring as a result of reducing the amount of information available to the decision maker.

Research on optimization of water quality monitoring networks has focused primarily on obtaining the optimal configuration of stations (Lettenmaier *et al.*, 1984) and optimum sampling frequencies (Heidtke and Armstrong, 1979; Dunnette, 1980). Relatively little research has been directed at the problem of gain in information and reliability, and hence of reduction in risk, that can generally be obtained by establishment of a new station or by an increase in sampling frequency. The same is true for the converse problem of loss of information about water quality and the associated increased exposure to risk at locations where water quality monitoring stations have been discontinued or when sampling frequencies have been reduced.

One of the existing methods for estimating values at a location where a station has been discontinued by interpolation of data from surrounding stations is the Kriging method (Gambolati and Volpi, 1979). While particularly useful for monitoring purposes in large continuous water bodies such as lakes or aquifers, the method, however, is not applicable for the river network monitoring, since the physical characteristics of the problem are different. The Kriging method is applicable to spatial predication of parameters where the nature of the parameter is changing continuously from one point to another, e.g., water quality values or water levels at various locations across a lake or piezometric levels or water quality values within an aquifer. A river network, however, is not a continuum; rather it has a skeletal structure with major discontinuities, i.e., a value at one point may not be related to a value from other points which are separated by land masses or which may be situated in different drainage basins.

5.4. Risk in Data Collection

Recent work by Kusmulyono and Goulter (1994 and 1995) has proposed a new method, based on entropy theory, for predicting water quality of discontinued water quality monitoring stations on river networks. The model was shown to perform better, i.e., to produce more accurate predictions more frequently than a traditional regression based technique. As such, the method represents a means of improving predictions of water quality across a river network, thereby improving the level of information and reducing risk.

In relation to the first requirement of collecting the correct set of data, the risk arises from a situation where an important parameter is not being measured in the collection process. The risk in this case relates directly to the consequences of requiring information on a particular water quality parameter and not having that information. As noted previously, the 'costs' of this situation will almost always be site or scenario specific. The probability of the event, i.e., the probability of information not being collected for a particular water quality parameter and that parameter subsequently being identified as important for some purpose, is also likely to be site or scenario specific. The water quality parameters which are important to collect will be identified during the design of the monitoring program. A decision in the design stage not to collect data on a particular parameter might be based upon an unforeseeable need for information on the parameter or on budgeting restriction which limits the number of parameters to be collected and evaluated. In this later case, evaluation of risk is actually undertaken in that a trade-off between the cost of monitoring that parameter versus the estimated costs of not collecting data is made in relation to other parameters of interest. As such, the probability of needing those data in the future and the consequence of not having data is evaluated, at least implicitly if not explicitly.

The probability of needing such uncollected data, whether they are anticipated but cannot be collected due to budgeting restrictions or whether they are completely unforeseen, is obviously very difficult to estimate, often because the developments that would require such information may not even be conceived of at the time of design of the water distribution. However, to ignore any possibility of needing data, and hence information, on a particular water quality parameter reflects a very short-term design perspective and would likely lead to monitoring programs of short-term value with potential for increased long-term risks and costs.

Such an argument leads naturally to a view that data should be collected on as many parameters as possible, thereby minimizing the risk of omitting parameters from the design program that later turn out to be important. There is, however, an important consideration in applying the concept of collecting data on as many parameters as possible. Large sets of data, often collected with considerable effort, do not always satisfy identified, let alone potential, objectives for water quality monitoring (Ward *et al.*, 1986). Unless the sampling program is appropriately matched to the specified (existing and anticipated) objectives for water quality monitoring, the actual data collected may be inappropriate, or at best, only partially useful. Ward *et al.*, (1986) described this situation as "data-rich information-poor", thereby reflecting the earlier statement that the objective of water quality monitoring and data collection is information rather than data per se. This same issue was revisited recently in a work by James and Gorelick (1994) in their paper "When enough is enough: The worth of monitoring data in aquifer remediation design". That study was addressed primarily at the question of the cost of taking a sample versus the reduction in cost of aquifer remediation, which the information contained in that sample can realize. While not directly linked to the question of risk, the principle of reducing remediation cost through cost-effective sampling is not dissimilar to the reduction in risk through cost-effective sampling. The key element in this discussion is that the reduction in the likelihood of not collecting appropriate or sufficient data, thereby reducing the risk involved in data collection, will not be solved by collection of more data unless such data add to the information base by which the risk in relation to the objectives can be reduced.

In terms of the second requirement to collect water quality data at sufficient locations to adequately characterize the spatial variation, the risk is related to each of the three major objectives of water quality monitoring discussed earlier. Depending on the heterogeneity of the particular water body or system (e.g., an aquifer, a lake, or a river network) being monitored, it is quite possible that compliance with the water quality standard for a particular water quality parameter at one location cannot, or should not, be construed as indicating compliance of the same water quality parameter at other locations. Similarly, trends at one location should not be construed to reflect trends at other locations. More importantly, unidentified extreme values at unmonitored locations may constitute not just noncompliance with standards but also highly undesirable conditions which may not be evident until significant damages (consequences) have already occurred.

The probabilities of each of these scenarios occurring is again very difficult to calculate, requiring both an understanding of the physical processes in the system and potential developments and activities which might change those physical processes and/or modify the water quality locally. As in the case of the choice of variables to monitor, the difficulty of estimating these probabilities does not remove the need to consider the possibilities of such events occurring. Similarly, an unfocussed gathering of data at many locations could lead to the "data-rich information-poor" condition discussed earlier. Recent work by Alpaslan and Harmancioglu (1990), Alpaslan *et al.* (1992), Harmancioglu (1981), Harmancioglu and Alpaslan (1992) Harmancioglu and Baran (1989), Harmancioglu and Yevjevich (1987), Harmancioglu and Singh (1991), and Harmancioglu *et al.* (1985) has focused on the variation in *information content* of sampling with variation in monitoring (sampling) frequency and variation in the number and locations of monitoring stations. These studies are therefore important steps towards the development of data collection programs which avoid the data-rich information-poor syndrome. Furthermore, since information is directly related to the ability, to determine risk more accurately, and at times, to take actions to actually reduce risk, these studies also have a direct bearing on the consideration of risk in the design of water quality monitoring programs.

The issue of monitoring frequency is not unlike the issue of where to monitor. Both are related to the need to capture or characterize water quality conditions in the system. In the case of monitoring frequency, the objective is characterization of the temporal variation in water quality rather than the spatial variation in water quality sought through sampling at different locations. Ideally, monitoring should be performed continuously. However, it is generally impossible to monitor all, or even a few, of the water quality parameters of interest on a continuous basis. Sampling is therefore the only viable method for monitoring many water quality values. Sanders and Adrian (1978) expressed the importance of sampling as follows:

> *"Sampling frequency is a very important consideration in the design of a water quality monitoring network. A large portion of the cost of operating a monitoring network is related directly to the frequency of sampling. In addition, the reliability and utility of water quality data derived from a monitoring network are likewise related to the frequency of sampling."*

As in the other cases, the consequences and associated costs of not identifying important or critical conditions as a result of sampling, as opposed to continuous monitoring, will tend to be site and scenario specific. however, in context to the previous requirements, a considerable amount of work based on classical statistical sampling theory has been undertaken on estimating the probability of a sampling program giving false readings.

The sampling frequency issue has been identified by Casey *et al.* (1983) as one of the most complex aspects of water quality network design and operation. From the perspective of monitoring objectives, the selection of sampling frequency becomes further complicated when a number of the objectives listed previously have been identified for a particular monitoring program, a situation that is not uncommon. To detect extreme values, for example, frequent sampling must be conducted. On the other hand, for detection of long term trends occurring over long periods of time, less frequent sampling would be more appropriate. From a statistical viewpoint, design of the sampling program, and in particular the frequency of sampling, can be considered on the basis of sampling theory, i.e., in terms of simple random sampling, systematic sampling, and stratified random sampling (Bruton, 1982).

Sanders and Adrian (1978) introduced a measure for determining frequency based on the confidence interval of the mean of the random component. Ward *et al.* (1979) also proposed a statistically based method to evaluate sampling frequencies in monitoring networks. The underlying principle of this second procedure was to obtain uniform or 'equal' information from all stations in the network. This goal was achieved by allocating the number of samples to be collected at a station as a function of the variation in the value of the water quality constituent of interest at that station, relative to the total variation for that constituent in the network. Note that this goal relates in part to the need, identified earlier in relation to the various contributors to risk, to collect samples at sufficient locations and at sufficiently frequent intervals to characterize the spatial and temporal variations of the parameter. A further complication in this process is that the variation of the constituent across the network may, not be an adequate reflection of actual conditions in the system. Hence, making decisions against an incomplete picture increases the risk of coming to an inappropriate or incorrect conclusion.

Dunnette (1980), on the other hand, proposed the use of a water quality index to optimize the sampling frequency. This water quality index implicitly aggregates the relative importance of the various parameters to be sampled through a weighting process. Dunnette (1980) asserted that the technique was therefore considered a good method for determining sampling frequency. However, a major difficulty with this approach is that an aggregate water quality index may disguise a significant variation of a single parameter within a mass of 'average' values of other parameters. Used in isolation, the technique has therefore the potential not to be sufficiently discriminating in its interpretation of water quality.

Loftis and Ward (1981) subsequently presented another point of view for the determination of sampling frequency. They suggested that three factors have to be considered in determining sampling frequency, namely:

i) random changes due to storms, rainfall, etc.,
ii) seasonal changes in temperature, rainfall, etc., and
iii) serial correlation from sample to sample.

From the analysis based on the width of the confidence interval around the sample mean and considering those three factors, three general regions of frequencies were identified:

i) a region where serial correlation is dominant (the sampling frequency is greater than 30 times per year);
ii) a region where the effects of seasonal variation and serial correlation tend to cancel each other out (the sampling frequency lies between 10 to 30 per year); and,
iii) a region where seasonal variation is dominant (the sampling frequency can be less than 10 per year).

While statistical analysis has an important role in determining the sampling frequency, and thereby in the estimation of the probabilistic component used to define risk, it should be noted that, in many cases, it is the cost constraints that play the dominant role in the selection of sampling frequency (Sanders and Adrian, 1978). Therefore, a compromise between statistical considerations and budgetary limitations is generally needed in the final determination of the sampling frequency and the associated risk exposure.

The risk associated with the precision and accuracy, or perhaps more correctly the lack of precision and accuracy of measurement and analysis, has two aspects. Accuracy of measurement is related to the need for monitoring to provide reliable information on the status of the system and, on that basis, on whether standards are actually being complied with. Risk in this case is associated with the likelihood of the data being sufficiently inaccurate so that conditions in the system as "described" by those data are erroneously represented. This outcome represents the risk of not taking appropriate actions when they are needed or the inverse case of taking action when it is not in fact necessary.

Precision in water quality monitoring, on the other hand, relates to the need to discriminate between different values measured fur a particular water quality parameter. (Any consideration of precision is of course based on an assumption that the measurements are accurate!) The need for precision may arise when small absolute changes in the value of a parameter are significant, e.g., for sharp threshold levels related to toxic conditions. Risk in these circumstances is, however, similar in nature to the risk associated with accuracy of measurements in that a false conclusion may be drawn from values which are not sufficiently precise to provide the necessary discrimination.

Accuracy and precision of measurement are related to the capabilities of the measurement and analysis equipment. As such, they relate to both the "mechanical" sampling process and the subsequent analyses. Manufacturers of the relevant equipment should be able to provide specifications on both the accuracy and the precision of their equipment. These specifications can then be used to determine the probability of false readings in the light of the requirements (objectives) of monitoring.

5.5. Risk in Data Processing and Analysis

The previous discussion focused on the risk associated with various aspects of the physical collection of raw data. However, as noted earlier, data must often be processed and analyzed before they can be used for the purposes for which they are collected. An examination of the three primary objectives for water quality monitoring discussed earlier shows that each objective requires some statistical processing of the data to obtain the required information. The risk involved in this processing is related to the extent to which the models used to analyze the data are appropriate, both in terms of any assumptions or simplifications necessary for their use and their ability to extract the desired information from the data.

Consider each of the risks in data processing and analysis in relation to the three major objectives of water quality monitoring. Determination of whether a violation of a water quality standard has occurred is a relatively complex task. One reason for this complexity is that, while the standard can be defined as a fixed value or a range of value(s), the actual water quality values are generally determined on the basis of samples which are representative, rather than absolute, statements of the actual population. Any decision based on the values derived from the sampling must therefore recognize the statistical variability of both the population at large and of the values derived from the sampling process. This statistical variability and, in the absence of continuing monitoring, the use of samples to define conditions leads to the risk of drawing erroneous conclusions. An example of the recognition of this problem is the work of Herricks *et al.* (1985), who in their examination of the assessment of compliance with water quality standards, noted that regulatory authorities involved in water quality management must recognize the stochastic nature of data collected or reported to them.

The problem of ascertaining from a set of data whether water quality is complying with a specified standard has received some considerable attention in the literature. Schaeffer *et al.* (1980) and Herricks *et al.* (1985) have examined how to determine when violations of standards occur. Berthouex and Hau (1991) have reported on how to judge compliance faced with highly censored samples, while Valiela and Whitfield

(1989) examined the issue of designing monitoring strategies to determine compliance.

A number of approaches have been proposed to address this problem of variability in sample values. Schaeffer *et al.* (1980) developed a simple graphical method which takes into account process variability, average performance, and type of sample collected (grab ox composite) when determining violations. A method to estimate the probability that a single grab sample will violate a given stream standard was subsequently, proposed by Loftis and Ward (1981). Loftis and Ward (1981) also proposed a procedure for determining the expected number of violations in a given number of samples from the cumulative density functions of the water quality random variables themselves.

Berthouex and Hau (1991) proposed a simple rule for judging compliance of effluent against effluent standards in the case when the effluent limit is set at a level below the limit of detection, i.e., outside the level of precision for the sampling and analysis method employed. Their rule was based on the binomial probability distribution and recognized the possibility of compliance even if a proportion of the samples were found to violate the effluent limit. In this latter case, effluent values observed to violate the limit are taken as a signal that initiates additional, more intensive monitoring. If the additional monitoring fails to detect further noncompliance within the time frame specified for the additional monitoring, the effluent is assumed to be still meeting the standards. The decision to initiate additional monitoring in this case is in fact a direct recognition of the risk of a false 'reading' and conclusion. Additional monitoring reduces the risk by providing further, more comprehensive information on the system being monitored.

The work by Valiela and Whitfield (1989) examined the suitability of two existing sampling strategies designed to test effluent discharge for compliance against standards established in relation to particular water quality objectives. The following results were derived from their analysis.

> *"For objectives based on long-term mean requirements, fixed frequency sampling at frequent intervals is most advantageous regardless of the underlying distribution of the data. For objectives that are based on maximum allowable concentrations, effective sampling strategies increase the likelihood o£ detecting noncompliance. If data are highly autocorrelated or sharply seasonal in distribution, an exceedence-driven sampling strategy is more effective and efficient for detecting violations than fixed frequency sampling. However, data generated by exceedence-driven sampling provide biased estimates of mean and standard deviation."*

These assertions are clearly directed at obtaining the best information for the monitoring objective. However, they also address, by association, the need to minimize, in a cost-effective fashion, the risk associated with drawing conclusions in relation to those objectives.

Despite the importance of monitoring for determining compliance with a given standard, thereby assisting water quality control authorities in enforcing rules for maintenance of the quality of water, monitoring also serves a number of other purposes, such as, assessing the existing conditions in the water body and detection of any short- and long-term trends that might be occurring in those conditions. Detection of such trends, particularly those involving deterioration in water quality, is one of the objectives in water quality monitoring noted earlier. Appropriate monitoring and appropriate trend analysis assists in ensuring that conditions of decreasing water quality can be detected as early as possible; and any measures which might be implemented to maintain water quality to relevant standards are put in place within an appropriate time scale.

The importance of detecting trends in water quality is indicated by the amount of research published on this particular subject. A number of papers by Hirsch *et al.* (1982), Hirsch and Slack (1984), Belle and Hughes (1984) and Lettenmaier *et al.* (1991) discuss techniques of trend analysis fox water quality data. These papers noted that, due to such factors as nonnormal distributions, seasonality, flow relatedness, missing values, values below the limits of detection, and serial correlation, the analysis of water quality time series is complex. A range of methods to address that complexity were then proposed. The nonparametric test, known as the seasonal Kendall test, was demonstrated by Hirsch *et al.* (1982) as being applicable for data with seasonality, missing values, or values reported as 'less than'. It can be seen that these factors, particularly the problem of missing values, are critical contributions to increase the risk associated with obtaining incomplete and possibly incorrect interpretation of reality. Hirsch and Slack (1984) subsequently modified this method to incorporate problems associated with data with serial dependence. Two types of nonparametric tests for trends called intrablock methods and aligned rank methods were also introduced by Belle and Hughes (1984) about this time.

The choice of statistical techniques used to identify trends should be based on the combination of accuracy of prediction and the confidence, as defined, for example, by specified confidence limits that can be placed on that accuracy. Hirsch (1988) discusses statistical methods and sampling designs to estimate step trends in surface water quality. That paper focused on the identification of a robust estimator appropriate for the data characteristics expected in water quality time series. Eight different existing estimators were evaluated to identify the robust estimator. The effectiveness of various existing sampling strategies were examined using Monte Carlo simulation coupled with an application of the estimator. Selection of appropriate methods for trend tests has been further examined by Hirsch *et al.* (1991). Based on their experience with a wide variety of trend detection methods, Hirsch *et al.* (1991) also provided guidance on the selection of existing statistical techniques for trend detection. The following issues were examined by Hirsch *et al.* (1991) in developing their guideline; step trend versus monotonic trend, parametric versus nonparametric methods, concentration versus flux, effect of discharge, seasonal variability, and censored data.

It is important to note that, in addition to the issues discussed above in relation to data analysis, the scale on which a conclusion will be derived should not be overlooked. It has been demonstrated by Loftis *et al.* (1991) that results of analysis of water quality data, e.g., detection of a trend, can vary significantly if different time scales are chosen. It is possible that a trend may be detected if only a short time period is of interest. However, when a longer time period is considered, the existence of the trend suspected on the basis of a short term may no longer be evident, In conjunction with its effects on the detection of trends, the choice of the time scale will also affect the estimated value of the mean and will have relevance as to the assumption of the independence of the data in the sample, i.e., data points collected over a short time scale with high frequency are likely to be more dependent than the same number of data points collected over a longer time scale. The importance of the frequency of sampling on monitoring objectives and the risks associated with erroneous conclusions is clearly evident.

Interestingly, particularly in relation to the earlier discussion on the need to have sufficient monitoring stations to characterize conditions in the system adequately, Loftis *et al.* (1991) also noted the importance of the spatial scale in estimating the condition of a water body at a particular location.

5.6. Summary

Any consideration of risk in water quality monitoring must be placed in a framework of the objectives defined for the monitoring. Objectives of water quality monitoring can generally be placed into one of the three following categories:

- compliance with specified standards;
- identification of trends; and
- detection of extreme values.

Actual determination of risk is based upon the likelihood (probability) of events, e.g., missing a violation of an important water quality standard, and the consequences of that event. The consequences of an event are site or scenario specific and will therefore vary. Determination of the probability of events is a far more generic process, involving the design of the monitoring network and monitoring program, sampling theory, the capabilities of the instruments used to measure the water quality parameters of interest, and the models used to subsequently process and analyze the data.

The components of the process of water quality monitoring that determine the likelihood of obtaining in appropriate information are:

i) the extent to which the set of parameters being measured is the correct grouping;
ii) the extent to which the parameters chosen to be measured are collected at a sufficient number of locations to adequately characterize spatial conditions across the area of interest;

iii) whether the parameters are measured, i.e., sampled, at a sufficiently high frequency to capture the temporal variation of the parameter of interest;

iv) whether the data are measured with sufficient accuracy and precision to provide the required level of information; and

v) the extent to which the models used to process or analyze those raw data provide a true representation of the system.

Determination of the probability of specified events requires evaluation of the contribution of each of these factors to those events, using appropriate techniques as outlined in the literature and then combing those probabilities in some national and coherent fashion.

An important feature of the risk involved in water quality monitoring, e.g., the event of not being aware of noncompliance of water quality standards and therefore not taking appropriate actions, is that it is often difficult to quantify and therefore has the potential not to be considered in the design of new (or reconsideration of existing) water quality monitoring networks and programs. However, neglecting to give explicit consideration to such risk increases the potential for monitoring programs which are inexpensive to operate but ineffective in achieving their objectives and prone to short-term perspectives and long-term management 'costs'.

References

Alpaslan, N. and Harmancioglu, N.B. (1990) Water quality monitoring: site selection, Stuttgart, *Seminar Umweltschultz*, pp.185-205.

Alpaslan, N., Harmancioglu, N.B., and Singh, V.P. (1992) The role of the entropy concept in design and evaluation of water quality monitoring networks, in V.J. Singh and M. Fiorentino (eds.), *Entropy and Energy Dissipation in Water Resources*, Kluwer Academic Publishers, pp. 283- 302.

Belle, G. and Hughes, J.P. (1984) Nonparametric tests for trend in water quality, *Water Resources Research*, **20**(1), pp.l27-136.

Berthouex, P.M. and Hau, I. (1991) A simple rule for judging compliance using highly censored samples, *Journal of Water Pollution Control Federation*, **63** (6), pp. 880-886.

Bruton, G. (1982) in B.T. Hart (ed.), *Water Quality Sampling and Data Analysis*, Melbourne, Water Studies Centre, Chrisholm Institute of Technology and Australian Society of Limnology, 158 p.

Casey, D., Nemetz, P.N. and Uyeno, D.H. (1983) Sampling frequency for water quality monitoring: measures of fffectiveness, *Water Resources Research*, **19**(5), pp.1107-1110.

Department of National Development (1965) *Atlas of Australian Resources*, Department of National Development, Resources Information and Development Branch, Canberra, Australia.

Duckstein, L., Kisiel, C.C., and Beckman, M. (1976) Water quality control under uncertainty: optimal stopping rules for sampling, *Journal of Hydrology*, **29**, pp. 393-406.

Dunnette, D.A. (1980) Sampling frequency optimization using a water quality index, *Journal of the Water Pollution Control Federation*, **52**(11), pp. 2807-2811.

Gambolati, G. and Volpi, G. (1979) A conceptual deterministic analysis of the Kriging Technique in hydrology, *Water Resources Research*, **15(3)**, 625-629.

Harmancioglu, N.B. (1981) Measuring the information content of hydrological processes by the entropy concept, Centennial of Ataturk's Birth, *Journal of the Civil Engineering, Faculty of Ege University*, pp. 13-38.

Harmancioglu, N.B. and Alpaslan, N. (1992) Water quality monitoring network design: a problem of multi-objective decision making, *Water Resources Bulletin*, **28(3)**, 179-192.

Harmancioglu, N.B. and Singh, V.P. (1991) An information-based approach to monitoring and evaluation of water quality data, in G. Tsakiris (ed.), *Advances in Water Resources Technology*, pp. 377-386.

Harmancioglu, N.B. and Baran, T. (1989) Effects of recharge systems on hydrological information transfer along rivers, IAHS, *Proc. Of the Third Scientific Assembly-New Directions for Surface Water Monitoring*, IAHS Publ.181, pp. 223-233.

Harmancioglu, N.B. and Yevjevich, V. (1987) Transfer of hydraulic information among river points, *Journal of Hydrology*, **91**, 103-118.

Harmancioglu, N.B., Yevjevich, V. and Obeysekera, J.T.B. (1985) Measures of information transfer between variables, *Proc. of Fourth International Hydrology Symposium*, published by Hsieh Wen Shen, Engineering Research Centre, Colorado State University, Fort Collins, Colorado 80523, U.S.A., pp. 481-499.

Heidtke, T.M. and Armstrong, J.M. (1979) Probabilistic sampling model for water quality management, *Journal of the Water Pollution Control Federation*, **51(12)**, 2916-2827.

Herricks, E.E., Schaeffer, D.J. and Kapsner, J.C. (1985) Complying with NPDES permit limits: when is a violation a violation?, *Journal of the Water Pollution Control Federation*, **57(2)**, 109- 115.

Hirsch, R.M. (1988) Statistical methods and sampling design for estimating step trends in surface-water quality, *Water Resources Bulletin*, **24(3)**, 493-503.

Hirsch, R.M. and Slack, J.R. (1984) A nonparametric trend test for seasonal data with serial dependence, *Water Resources Research*, **20(6)**, 727-732.

Hirsch, R.M., Alexander, R.B., and Smith, R.,A. (1991) Selection of methods for the detection and estimation of trends in water quality, *Water Resources Research*; **27(5)**, 803-813.

Hirsch, R.M., Slack J.R., and Smith R.A. (1982) Techniques of trend analysis for monthly water quality data, *Water Resources Research*,**18(1)**, 107-121.

James, B.R. and Gorelick, S.M. (1994) When enough is enough: the worth of monitoring data in aquifer remediation design, *Water Resources Research*, **30(12)**, 3499-3513.

Kusmulyono, A. and Goulter, I. (1995) Computational aspects in use of entropy theory in predicting water quality levels at discontinued stations, to be published in Stochastic Hydrology and Hydraulics.

Kusmulyono, A. and Goulter, I. (1994) Entropy principles in the prediction of water quality values at discontinued monitoring stations, *Stochastic Hydrology and Hydraulics*, **8 (4)**, 301- 317.

Lettenmaier, D.P., Anderson, D.E. and Brenner, R.N. (1984) Consolidation of a stream quality monitoring network, *Water Resources Bulletin*, **20(4)**, 473-481.

Lettenmaier, D.P., Hooper, E.R., Wagoner, C., Faris, K.B. (1991) Trends in stream quality in the continental United States,1978-1987, *Water Resources Research*, **27(3)**, 327-339.

Loftis, J.C. and Ward, R.C. (1981) Evaluating stream standard violations using a water quality data base, *Water Resources Bulletin*,**17(6)**, 1071-1078.

Loftis, J.C., McBride, G.B., and Ellis, J.C. (1991) Consideration of scale in water quality monitoring and data analysis, *Water Resources Bulletin*, **27(2)**, 255-264.

Palmer, R.N., and MacKenzie, M.C. (1985) Optimization of water quality monitoring networks, *Journal of Water Resources Planning and Management*, **111(4)**, 478-493.

Pinter, J. and Sornlyody, L. (1987) Optimization of regional water quality monitoring strategies, *Water Science and Technology*, **19**, 721-727.

Sanders, T.G. and Adrian D.D. (1978) Sampling frequency for river quality monitoring, *Water Resources Research*, **14(4)**, 569-576.

Sanders, T.G., Ward, R.C., Loftis, J.C., Steel, T.D., Adrian D.D., and Yevjevich, V (1983) *Design of Networks for Monitoring Water Quality*, Water Resources Publication, Colorado.

Schaeffer, D.J., Jarnardan, K.G., Kerster, H.W., and Shekar, M.S. (1980) Graphical effluent quality control for compliance monitoring: what is a violation?, *Environmental Management*, **4(3)**, 241-245.

Valiela, D. and Whitfield, P.H. (1989) Monitoring strategies to determine compliance with water quality objectives, *Water Resources Bulletin*, **25(1)**, 63-69.

Ward, R.C. and Loftis, J.C. (1989) Monitoring systems for water quality, *Critical Reviews in Environmental Control*, **19(2)**, 101-118.

Ward, R.C., Loftis, J.C., and McBride, G.B. (1990) *Design of Water Quality Monitoring Systems*, Van Nostand Reinhold, New York, 231 p.

Ward, R.C., Loftis, J.C., and McBride, G.B. (1986) The "data-rich but information-poor" syndrome in water quality monitoring, *Environmental Management*,**10(3)**, 291-297.

Ward, R.C., Loftis, J.C., Nielson, K.S., and Anderson, R.D. (1979) Statistical evaluation of sampling frequencies in monitoring networks, *Journal of Water Pollution Control Federation*, **51 (9)**, 2292-300.

Whitfield, P.H. (1988) Goals and data collection designs for water quality monitoring, *Water Resources Bulletin*, **24(4)**, 775-780.

CHAPTER 6

ENVIRONMENTAL DATA MANAGEMENT: STORAGE, HANDLING AND RETRIEVAL

A. E. Hindrichs

Abstract. Environmental data management for any agency can be a daunting task given the large number of sample sites generally needed, and the wide variety of parameters for which analysis may be required. As a result, it is imperative that a computerized system be developed which facilitates rapid storage; efficient, accurate handling; and easy retrieval of a large volume of data. Environmental data management systems currently used by a variety of State and Federal agencies are briefly described to provide an overview of such systems. One system, developed by the Louisiana Department of Environmental Quality (LDEQ), Office of Water Resources, is described in detail. The LDEQ system provides for both regularly occurring water quality monitoring data, as well as special project data as required. While this system was designed specifically for water quality data, it may be easily adapted by other agencies which utilize a wide variety of environmental data.

6.1. Introduction

6.1.1. OVERVIEW OF ENVIRONMENTAL DATA MANAGEMENT REQUIREMENTS

Environmental data managers are faced with an immense volume of data which must be accurately entered, maintained and retrieved. This volume arises both from the number of sample sites and from the number of parameters for which analyses may be required. For the Louisiana Department of Environmental Quality, Office of Water Resources (LDEQ, OWR), this amounts to over 200 Water Quality Network (WQN) monitoring sites and over 30 standard water quality parameters for each site. These WQN sites are monitored monthly or bimonthly, resulting in over 110 sites being tested each month. Separate water samples are also collected and analyzed for total and fecal coliforms at each site. As a result, one can see that thousands of data points are generated each month simply from routine monitoring. In addition to routine monitoring, agencies are often responsible for special projects such as priority pollutant scans on water, sediment or tissue samples; monitoring for selected contaminants in samples from designated problem areas; statewide fish tissue screening scans; industry enforcement issues and many other activities.

Other types of agencies may be responsible for streamflow data, erosion rates, wildlife and fisheries population dynamics, hunting or fishing statistics, water body and

N.B. Harmancioglu et al. (eds.), Environmental Data Management, 123-140.

land use descriptions, hazardous waste information, and industrial/municipal discharges. This list is in no way complete and is always expanding as environmental issues develop. While no one agency will be responsible for all these areas of concern, each agency still requires effective computerized data management to handle their specific area. In addition, it is often necessary to integrate two or more different databases in order to improve analysis and report capabilities. For example, in order to provide the most effective use of water quality data, it is helpful to combine data on streamflow and water quality parameters. This allows calculation of loading values to better determine the effect of contaminants on streams.

In order to be the most effective and user friendly, data management systems should preferably contain a series of menus which will guide the new or experienced user through such everyday operations as entry, verification, and recovery. This will allow for ease of training in data entry and recovery, in addition to promoting more frequent use as databases are developed. Second, to allow more advanced users to request customized reports, data management systems should contain some form of programmable reporting capabilities. Third, importing and exporting of data files should be available so that data can be transferred to other computer packages, such as WordPerfect, SAS, QuatroPro or other data management systems. This will permit more detailed statistical analysis or data manipulation and allow transfer of data to other agencies or institutions. Finally, any system must be flexible enough to allow for changes in standard entry, verification or reporting as monitoring requirements change or improvements are developed.

6.1.2. ENVIRONMENTAL DATA MANAGEMENT SYSTEMS CURRENTLY IN USE

All environmental management agencies have an obvious need for efficient and easily maintained data management systems. These systems are likely to have evolved over a number of years as environmental concerns expanded. This expansion brought with it a wealth of new data collected in support of environmental assessment and regulatory decisions. Because each area of the United States and each field of environmental management has its own unique interests and requirements, data management systems have developed independently to meet the needs of each agency or organization. However, as was noted in section 6.1.1, it is also necessary to maintain capabilities for integrating data from other agencies and systems. While these individual systems were developed independently, data transfer is usually possible through the use of dedicated links, modems or transfer of files contained on diskettes. Brief descriptions of a few environmental data management systems currently in use in the United States, followed by a more detailed discussion of the LDEQ, OWR water quality data management system are presented below.

United States Environmental Protection Agency

The United States Environmental Protection Agency's (EPA) Office of Water Information Systems (OWIS) maintains 20 key information systems (Table 6.1) which support program functions in 6 water resource areas (U.S. EPA, 1992). These areas include coastal and marine waters, rivers and streams, lakes, wetlands, groundwater and drinking water. In addition to the 20 key systems, EPA's OWIS also maintains nearly

100 additional information systems and provides information on 35 water-related systems developed by other EPA program offices, other Federal agencies, and special interest groups (U.S. EPA, 1992). Two of these systems, STORET (STOrage and RETrieval of U.S. waterways parametric data) and the Water Body System (WBS), will be described in more detail as examples of what is available from OWIS.

TABLE 6.1. Key information systems provided by United States Environmental Protection Agency, Office of Water Information Systems (U.S. EPA, 1992)

City and County Files	Needs Survey
Drinking Water Regulatory Impact Analysis	Ocean Data Evaluation System
Drinking Water Supply File	Permit Compliance System
Effluent Guidelines Studies	Reach File
Environmental Monitoring Methods Index	STORET-Biological System
Federal Reporting Data System	STORET-Daily Flow System
Underground Injection Control Program	STORET-Fish Kill
Grants Information/Control System	STORET-Water Quality System
Hazardous Waste Injection Well Data Base	Gage and Dam Files
Industrial Facilities Discharge File	Water Body System

STORET is subdivided into Biological (BIOS), Water Quality (WQS), and Daily Flow (DFS) systems (U.S. EPA, 1992). All 3 subdivisions of STORET are maintained by the Office of Wetlands, Oceans and Watersheds. As of 1992, approximately 800 organizations, reporting on over 735,000 sampling stations, contributed over 180 million observations on approximately 12,000 water quality parameters. Entries submitted to STORET must be verified using approved quality assurance, quality control (QA/QC) procedures and are further verified before final entry by invalid range and missing mandatory field checks built into the system. All of the STORET subsystems can be linked in order to share data. STORET is maintained at the EPA National Computer Center (NCC) at Research Triangle Park, North Carolina, using an IBM ES-9000 computer, and can be utilized by anyone with access to the NCC. All agencies which contribute to the database bear primary responsibility for its accuracy, and information may only be changed by the agency which made the contribution (U.S. EPA, 1992). State agencies like LDEQ may maintain gateways to the EPA system through SNA mainframe hardware and software (Joni DeVilbiss, 1993, personal communication).

STORET-BIOS contains information on the distribution, abundance and physical condition of aquatic organisms in waters contiguous to the United States (U.S. EPA, 1992). It also contains descriptions of organism habitats and analytical tools used for data analyses. Data include station locations identified by agency, state and county; EPA ecoregion; station ID; EPA basin code; latitude and longitude; narrative description, and United States Geological Survey (USGS) hydrologic unit code. Sampling events are identified by date. Sampling gear, meteorological and water quality conditions, as well as habitat descriptions, are included. The primary information gathered is on observed biota and must include taxonomic identities and organism counts. In addition to the STORET subsystem links, BIOS can be linked to the Permit Compliance System and a taxonomic nomenclature file maintained by the National Oceanographic and Atmospheric Administration (NOAA).

STORET-WQS is the primary component of STORET and is maintained in cooperation with the Office of Information Resources Management (U.S. EPA, 1992). This is the main repository of chemical and physical water quality information for water bodies within and contiguous to the United States. Information is organized by sites which are described in a manner similar to that used by STORET-BIOS. As of 1992, over 730,000 stations were included in the system. Date, collection method and location are provided along with results for all parameters for which analysis is done. Information is provided by states, EPA and other Federal Agencies, contractors, universities, and individuals. In particular, data from the USGS WATSTORE (WATer STOrage and REtrieval system) are transferred to STORET-BIOS periodically. Like BIOS, WQS can be linked to the Permit Compliance System.

STORET-DFS is used to maintain daily observations of streamflow and water quality parameters (U.S. EPA, 1992). Data are collected at USGS gaging stations and essentially duplicate information in the USGS Daily Values File. However, STORET-WQS provides a simplified means of linking non-USGS water databases. Flow data and water quality information on over 29,500 gaging sites are maintained in STORET-WQS. Water quality measurements include temperature, dissolved oxygen, chloride, conductivity, pH, and suspended sediment.

The WBS, also developed by EPA, is a mainframe and personal computer (PC) based program provided to state agencies to assist in preparation of reports required by section 305(b) of the Clean Water Act (U.S. EPA, 1992 and 1993). WBS is not used to store or analyze raw monitoring data; rather, it is used to store water body identifications, water quality status, sources and causes of impairment and summary assessment information. Based on this information, a series of pre-programmed reports can be generated for use in state 305(b) water quality inventory reports or for other water quality assessment needs. In addition, customized reports can be prepared by the user. Agencies may either use EPA, NCC's IBM ES-9000 computer, or a PC version available from EPA, Office of Water, Assessment and Watershed Protection Division.

United States Geological Survey

The USGS maintains an extensive database for water known as the National Water Information System (NWIS), which is maintained at the USGS's central computer facility at Reston, Virginia (Maddy *et al.*, 1991). Within NWIS is the Automated Data Processing System, the Ground-Water Site Inventory System, the Water-Quality System and the Water-Use Data System. NWIS allows data processing over a network of terminals at USGS offices throughout the United States. Within the Water-Quality System is a database known as WATSTORE for the storage of data collected during investigations of surface and underground water resources (Hutchison, 1975). These investigations include industrial, domestic and agricultural water requirements; research to develop better understandings of hydrologic principles; and publication of results and data from these investigations. In addition to it's storage and retrieval capabilities, WATSTORE can also produce tables, graphics, statistical analyses of data, and digital plots. NWIS accepts input from local terminals and returns output either to the user's screen or to disk files. Information contained in WATSTORE is available to other Federal agencies and cooperators, and water quality information is periodically transferred to EPA's STORET system (Hutchison, 1975; U.S. EPA, 1992).

6.2. A State Agency Approach to Water Quality Data Management

LDEQ, OWR maintains a Water Quality Data Management System (WQDMS), which was developed over a number of years in order to meet requirements outlined in section 6.1.1. WQDMS was developed by the Technical Services Section of the LDEQ in cooperation with the OWR, which represents the end user. The system consists of a series of menu driven options which lead the user to a variety of pre-programmed reports to meet most user needs. Flexibility in the system permits changes to be made as required by new data demands or improvements in technology and methods. WQDMS will be presented here as an example of a successful system currently in use by a state agency. The discussion will be developed from basic technical parameters of the system, through data entry, verification/correction, and recovery.

Handling of water quality network data, which comprises a large percentage of the entry/recovery requirements, will be described in order to familiarize the reader with use of the system. However, references will be made to special project data which comprise the second most extensive database used for strictly environmental purposes. Omitted from the discussion will be related systems developed and used for water discharge permit tracking, water discharger inventories, discharge monitoring reports, and toxic release inventories. By working through the steps needed to enter, verify, correct and recover water quality network data, it is hoped the reader will gain an understanding of how to develop their own environmental data management system. While this system was developed specifically for water quality data, it should be adaptable to other environmental data requirements.

6.2.1. MAIN FRAME REQUIREMENTS[1]

Computer processing facilities for 10 divisions of LDEQ are maintained by the Technical Services Division of the LDEQ. A Digital Equipment Corporation VAX (DEC VAX) cluster of 6000-610 and 6000-420 processors is used for all mainframe data management (Joni DeVilbiss, 1993, personal communication). Twenty-eight gigabytes of disk space are available on the DEC VAX cluster. A VMS operating system is used. Five languages FOCUS, COBOL, FORTRAN, Basic, and SAS are used to maintain the Water Quality Data Management System's entry, retrieval and report writing capabilities. Each of these 5 languages, along with Lotus 1-2-3 and 20/20, are available for customized data analysis and report preparation.

In addition to the in-house uses of the DEC VAX system, the Technical Services Division also maintains a link to EPA's NCC IBM ES-9000 computer at Research Triangle Park, North Carolina. A DEC SNA gateway is used to maintain this link which

---

[1] Mention or use of firm and trade names for computer hardware or software does not constitute endorsement by the Louisiana Department of Environmental Quality.

is used for transfer of water quality data to EPA's STORET system described in section 6.1.2. Data transfer to STORET is a requirement of grant commitments between LDEQ, OWR and EPA.

6.2.2. MENU SYSTEM OF DATA ENTRY, VERIFICATION AND RETRIEVAL

Data entry is provided by a series of menu driven options in order to simplify the process and help ensure accurate entry of a wide variety of data. However, because the Water Quality Data Management System menus are available to all personnel with access to the DEC VAX system, quality control and data protection must be provided by limiting most water quality data system users to read only privileges. Only those personnel with the responsibility of entering and maintaining databases are provided with privileges to write and edit data in the system.

Most data entry, verification, correction, and retrieval is achieved using a series of menu driven options beginning with the main menu shown in Fig. 6.1. Throughout this discussion, menu headings of interest will be highlighted in the text and in each figure to facilitate the following of working progression through each menu. The first menu is opened by typing **"MENU"** at the prompt after initial logon procedures have been completed. At this point, the desired menu option is typed on the enter option line. Option **"WQ"** is used to access water quality data entry screens. The remaining options access a variety of databases for enforcement actions, permitting, vehicle and property control, and other specialized storage and retrieval uses.

Water Pollution Control Division Menu of General Access Systems

CG	OWR Contracts and Grants
CSLOG	Surveillance Complaints and Spills Log
DMR	OWR Enforcement DMR Tracking System
DISCHARGE	OWR Discharger Inventory Database
ENF	OWR Enforcement Actions Tracking System
EWOCDS	OWR Early Warning Organic Compound Detection System
FISH	OWR Fish Population Information
INS	OWR Inspection Tracking System
INV	Department Property Control System
INVVEH	OWR Vehicle Log Recording System
MRTRAVEL	Mississippi River Time of Travel Program
PTS	OWR Permit Tracking System
REQS	Requisitions
VOC	OWR Mississippi River VOC Data
WQ	**OWR Statewide Water Quality Monitoring Network**
TRIS	Toxic Release Inventory System
EXIT	Exit Menu

Please enter option: **WQ**

Figure 6.1. Main menu of General Access Systems within the Water Quality Data Management System used by Louisiana Department of Environmental Quality, Office of Water Resources, Baton Rouge, Louisiana

Water Quality Network Data Entry Procedures
After entering **"WQ"**, the next screen which appears is shown in Fig. 6.2. This screen is particular to the OWR, Water Quality Management Division (WQMD) and provides for all data entry, maintenance and retrieval related to the WQN. In order to begin entry of WQN, data option **"WQN"** is used. Within WQN, 4 screens are available for data entry, maintenance, and retrieval. The 3 successive screens after screen 1 are accessed by typing **"N"** in the option desired space near the bottom of each screen. Figures 6.3-6.6 show each of the menu screens available within WQN. Screens 1-3 provide for pre-programmed reporting of all available data as described by each menu option. Screen 4 allows for 9 maintenance options to enter, delete, change and verify WQN data. To continue WQN data entry, enter **"WQNMNT"**. At this point, the user is prompted to enter station number, sample date and sample depth for which data is to be entered. A data entry screen then appears, listing all water quality parameters used for the WQN (Fig. 6.7). In order to simplify data entry, this screen mirrors field/labsheets used to record WQN data. The "TAB" or "ENTER" keys are used to move between parameters. In order to limit entry errors, valid ranges are established for each parameter. If an invalid entry is attempted, the user is prompted to correct the error.

Water Quality Information System

MAIN MENU

COL	SWQMN Coliform Data
PRJ	Projects
PRM	Parameter Codes
SPECIAL	Special Report Requests
STOR	STORET Transfer Options
STS	Sample Sites
WB	Water Bodies (Subsegments)
WQN	**SWQMN Water Quality Data**
WQS	W.Q. Surveys
305b	305B Report and Data Summary Publication Menu
DOC	Documentation

Please enter option desired: **WQN**

(O) Output menu (N) Next Menu (P) Previous Menu (T) Top Menu

(EX) Exit to FOCUS (FIN) Finish using FOCUS (TABLETALK)

Menu-Driven Reporting

Figure 6.2. Base menu for Water Quality Information, within the Water Quality Data Management System used by Louisiana Department of Environmental Quality, Office of Water Resources, Baton Rouge, Louisiana

Water Quality Information System

SWQMN DATA MENU

WQNFIELD	Reports field parameters for a user specified station and time frame.
WQNMETAL	Reports metals parameters for a user specified station and time frame
WQMETALL	Reports metals parameters for all stations and specified time frame.
WQNNUTRI	Reports NUTRIENTS for a user specified station and time frame.
WQNPHYS	Reports PHYSICAL/INORGANIC parameters for a user specified station and time frame

Please enter option desired: ....

(O) Output menu (N) Next Menu (P) Previous Menu (T) Top Menu
(EX) Exit to FOCUS (FIN) Finish using FOCUS (TABLETALK)
Menu-Driven Reporting

Figure 6.3. Screen 1 retrieval menu for State Water Quality Monitoring Network, within the Water Quality Data Management System used by Louisiana Department of Environmental Quality, Office of Water Resources, Baton Rouge, Louisiana

Water Quality Information System

SWQMN DATA MENU

WQNSTN	Reports a current listing of all station numbers in Water Quality Network and their descriptions
* WQNREP	Command procedure to request a formal report of all or selected parameters for a requested time frame
* WQNBREP	BATCH Command procedure to request a summary report of ALL parameters for a requested time frame for many stations
* WQNBREPS	BATCH Command procedure to request a summary report of ALL parameters for a requested time frame for many stations.

* These programs must be printed using the VMS PRINT commands

Please enter option desired: ....

(O) Output menu (N) Next Menu (P) Previous Menu (T) Top Menu
(EX) Exit to FOCUS (FIN) Finish using FOCUS (TABLETALK)
Menu-Driven Reporting

Figure 6.4. Screen 2 retrieval menu for State Water Quality Monitoring Network, within the Water Quality Data Management System used by Louisiana Department of Environmental Quality, Office of Water Resources, Baton Rouge, Louisiana

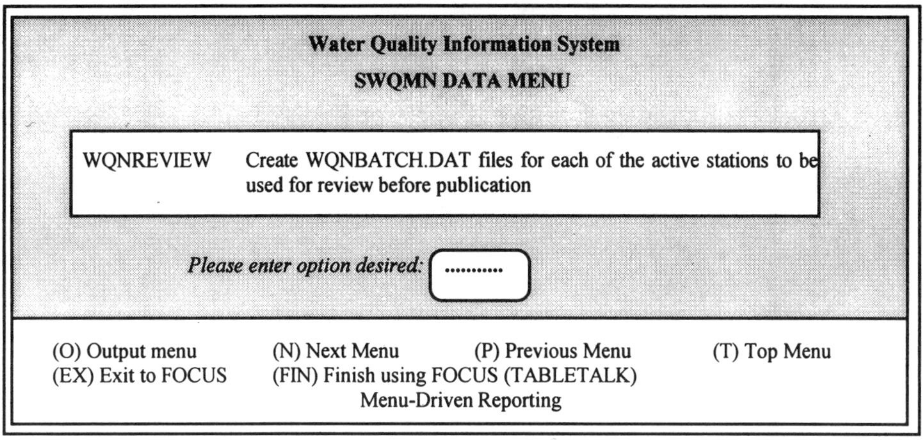

Figure 6.5. Screen 3 retrieval menu for State Water Quality Monitoring Network, within the Water Quality Data Management System used by Louisiana Department of Environmental Quality, Office of Water Resources, Baton Rouge, Louisiana

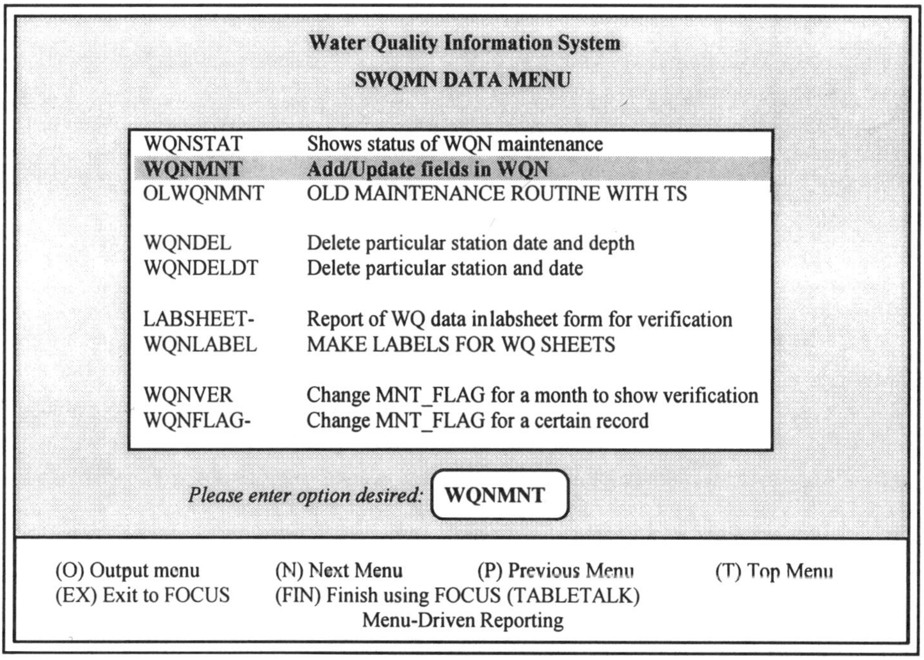

Figure 6.6. Screen 4 entry/correction menu for State Water Quality Monitoring Network, within the Water Quality Data Management System used by Louisiana Department of Environmental Quality, Office of Water Resources, Baton Rouge, Louisiana

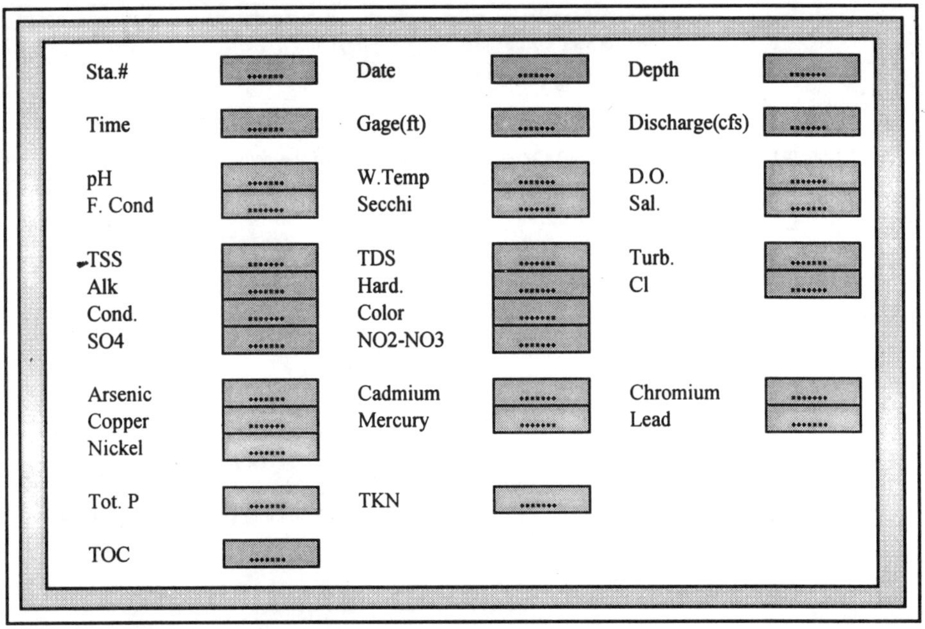

Figure 6.7. Water Quality Monitoring Network data entry screen within the Water Quality Data Management System used by Louisiana Department of Environmental Quality, Office of Water Resources, Baton Rouge, Louisiana

Following initial entry, all data are reviewed on screen prior to writing new data to the WQN file. Any errors can be corrected immediately by moving to the pertinent entry block and making the correction. This completes the first verification procedure. A series of maintenance flags are used to track the verification process. After initial entry and verification, flags are set to "0" indicating "entered". When one month's data have been entered into the computer, it is printed in a form similar to the field/labsheets for the second verification procedure. The printout is compared to each station's field/labsheet to identify and correct entry errors. Corrections are made by again using "WQNMNT" with existing station, date, and depth parameters. This brings up the applicable station and date information, allowing corrections to be made as in the initial data entry. Any unusual entries are referred to an Environmental Quality Specialist or water quality chemist for cross checking with related parameters. Illegible labsheet entries are checked with the original labsheets maintained at the laboratory. Following this third verification and correction procedure, maintenance flags are set to "1" indicating "verified". This flag permits data to be sent to EPA's STORET system. Transfers to STORET are facilitated by creation of a compatible file format, which is sent to STORET via LDEQ's DEC VAX SNA gateway hardware and software (Joni DeVilbiss, 1993, personal communication). Transfers to STORET are usually performed on a quarterly basis. After transfer to STORET, the maintenance flags are automatically set to "2" indicating "sent to STORET". At this point, the data are fully verified and ready to use for water quality assessments and publication in the Water Quality Data Summary prepared biennially.

Water Quality Surveys Data Entry Procedures

The OWR, WQMD also oversees entry of Water Quality Survey (WQS) data. WQS projects are assigned a unique 4 digit project number based on the year and sequence in which the project began. For example, the first project begun in 1993 would be assigned project number 9301, the second project would be 9302, and so on. Project sites are assigned unique 8 digit numbers based on a project and site number. For example, site 0001 of project 9301 would be designated 93010001. Different projects can have the same sites (i.e., 0001), but project site designations will be different (i.e., 58010001, 93010001).

Laboratory results for WQS projects come from a variety of sources including LDEQ's laboratory facilities, university labs, contract labs, and consulting firms which work for industries involved in LDEQ permitting or enforcement issues. As a result, laboratory data are received in a variety of forms, precluding the possibility of establishing a single entry screen, as is done for WQN data. However, entry screens have been developed for all of the priority pollutants and standard field parameters. These compounds are divided among nine subroutines for general survey results, and 4 specialized routines for long term projects. They can be reached by first selecting option **"WQS"** found in the WQDMS base menu shown in Fig. 6.2.

Following selection of option **"WQS"**, a report generation menu for the Calcasieu Toxics Inventory, an ongoing project, appears (Fig. 6.8). This screen can be bypassed by entering **"N"** in the space provided. The next screen (Fig. 6.9) allows selection of a variety of options for data entry, loading to files, and corrections. Entering **"WQSMNT"** begins a BASIC program which allows users to write data into a file established for each project. The nine subroutines are divided, based on broad chemical classifications (Fig. 6.10). Because most laboratory data are provided in groups based on these classifications, data entry is simplified by first selecting the correct chemical classification screen. As an example, the routine for adding metals data will be followed. Routines for other compound classifications are similar.

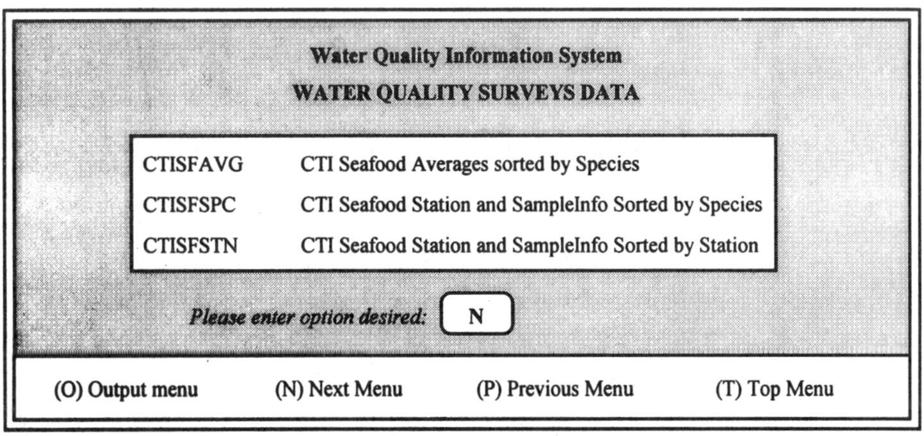

Figure 6.8. Screen 1 retrieval menu for Water Quality Survey data, within the Water Quality Data Management System used by Louisiana Department of Environmental Quality, Office of Water Resources, Baton Rouge, Louisiana

Water Quality Information System

WATER QUALITY SURVEYS DATA

WQSMNT	**Run BASIC program to write the data to a file**
WQSLOAD	Run command file to load the data into WQS.FOC
WQSINVNT	Produce an inventory of dates and sample #s for user provided IF test
WQSCVAL	Allows you to change value and remark fields
WQSCFISH	Allows you to change fishcode and species name fields
WQSDELS	Allows you to delete an entire sample number and all fishcodes and pcodes associated with it.
WQSDELSB	where sample number is blank
WQSDELSM	where sample number is missing

Please enter option desired: **WQSMNT**

Figure 6.9. Screen 2 entry/correction menu for Water Quality Survey data, within the Water Quality Data Management System used by Louisiana Department of Environmental Quality, Office of Water Resources, Baton Rouge, Louisiana

This is a list of parameter files that exist at this time:

For information on creating new parameter files, see MENUDISK:[MENU.WQDATA]WQSPARM.FILES

FIELD	Standard FIELD parameters
METAL	**Priority METALS**
VOC	Priority VOLATILE ORGANICS
BNE	BASE/NEUTRAL EXTRACTABLE compounds
AE	ACID EXTRACTABLE compounds
PEST	PESTICIDES
CHLOR	Chlorinated Organic compounds
PCB	PCB's (polychlorinated biphenyls)
TSCAN	Total Scan of PRIORITY POLLUTANTS
CTI	Calcasieu Toxics Inventory (Project # 8706) - Water phase VOC
CTISF	Calcasieu Toxics Inventory (Project # 8706) - Tissue Organics
PWS	Produced Water Study (Project # 8902)
9005	Flat River Dredging Study

Enter the appropriate file code or return to exit?: **METAL**

Figure 6.10. Screen 3 compound classification entry menu for Water Quality Survey data, within the Water Quality Data Management System used by Louisiana Department of Environmental Quality, Office of Water Resources, Baton Rouge, Louisiana

The data entry classification screen (Fig. 6.10) allows selection of the compound classification for which one has data. Entering **"METAL"** will begin a series of questions designed to specify project, site, date, and other sample information (Fig. 6.11). As each question appears in sequence, the user is asked to provide the appropriate information. Response delimiters are included to aid in data entry quality control. Following completion of the first set of questions, an information verification screen appears, repeating the project sample information (Fig. 6.12). Should corrections be necessary they can be made at this time. If all sample information is correct, option **"0"** is entered which brings up the second sample information screen (Fig. 6.13).

Please enter sample information requested

Project number	(4 digit number)	?.....
Site Number	(4 digit number)	?.....
Sample Date	(mmddyy)	?.....
Sample Time	(4 digit number)	?.....
Sample Media	(W)ater (S)ediment (T)issue	?.....
Sample Type	(G)rab (C)omposite (I)n-situ (S)ediment/Water Interface	?.....
Sample Number	(If applicable)	?.....
Sample Depth	(meters)	?.....
Analyzed by DEQ	(Y,N)	?.....

Figure 6.11. First sample delineation question screen for Water Quality Survey data entry within the Water Quality Data Management System used by Louisiana Department of Environmental Quality, Office of Water Resources, Baton Rouge, Louisiana

Sample Information Verification

1	Project Number :	....
2	Site Number	....
3	Sample Date :	....
4	Sample Time :	....hours
5	Sample Media :	....
6	Sample Type :	....
7	Sample Number :	....
8	Sample Depth :	....
9	DEQ Sample :	....

Choose option -> (#) of field to edit or (0) to continue ......?

Figure 6.12. First sample delineation verification screen for Water Quality Survey data entry within the Water Quality Data Management System used by Louisiana Department of Environmental Quality, Office of Water Resources, Baton Rouge, Louisiana

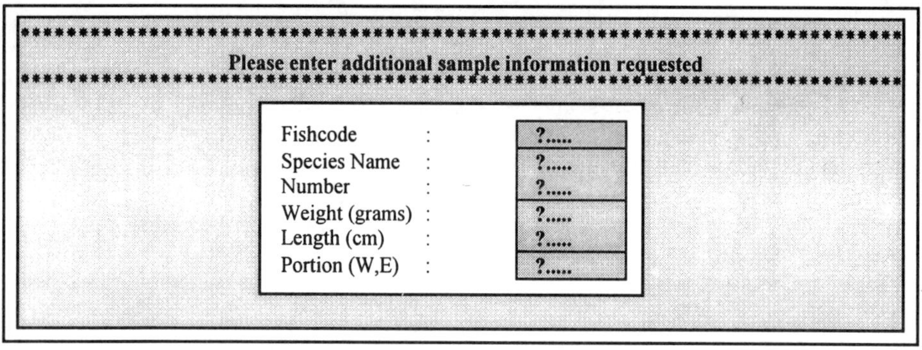

Figure 6.13. Second sample delineation question screen for Water Quality Survey data entry within the Water Quality Data Management System used by Louisiana Department of Environmental Quality, Office of Water Resources, Baton Rouge, Louisiana

The second sample information screen directs the user to provide information concerning the composition of the sample itself. Following completion of this series of questions, users are again provided a verification screen similar to that shown in Fig. 6.12. Once again, the user may correct entry errors or accept the information as shown and move on to the next screen. At this point, data entry personnel are asked if they have data on heavy metals to enter. Answering "**N**" for "no" moves the user out of the metals entry routine and brings up the option screen shown in Fig. 6.14. This allows the user to continue entering data on the same sample without reentry of initial sample information, or leave the data entry routine. Answering "**Y**" for "yes" brings up the metals data entry screen (Fig. 6.15).

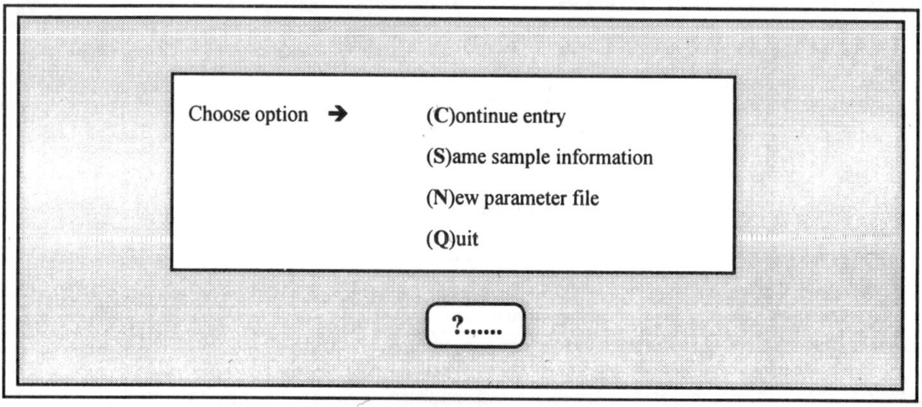

Figure 6.14. Sample continuation option menu for Water Quality Survey data entry within the Water Quality Data Management System used by Louisiana Department of Environmental Quality, Office of Water Resources, Baton Rouge, Louisiana

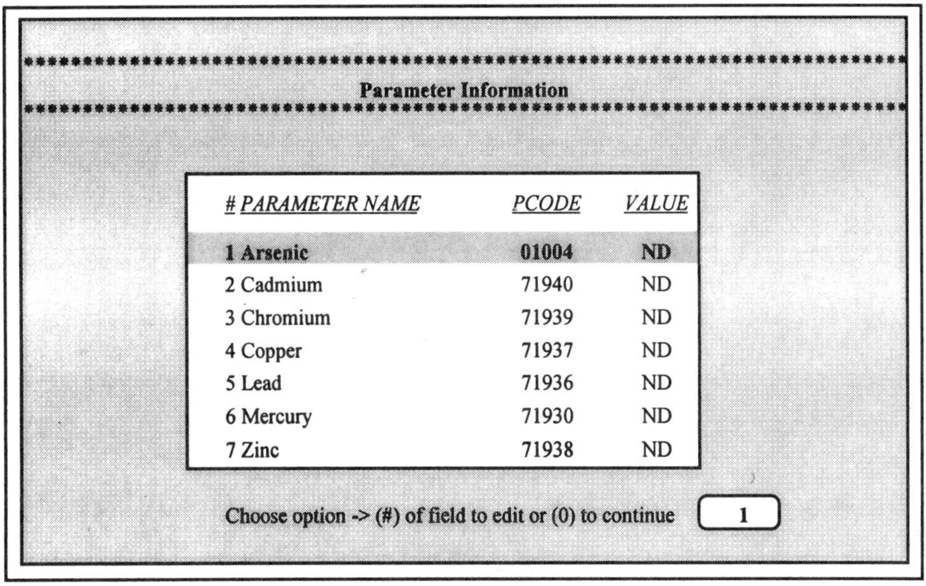

Figure 6.15. Metals entry screen for Water Quality Survey data entry within the Water Quality Data Management System used by Louisiana Department of Environmental Quality, Office of Water Resources, Baton Rouge, Louisiana

All compound entry screens are designed in the same fashion as the example metals screen. A default value of ND (non-detect) is provided, as most sample analysis results in non-detection. ND for all compounds is currently recorded in the database as 0.1 parts per million (PPM) based on the most common laboratory detection limits. However, this value can be adjusted if necessary by changes in laboratory analysis techniques. If all of the metals results are ND, then the analyst will enter **"0"** for "continue". If compound detections are present, then the analyst will enter the appropriate compound number; in this example, **"1"** is entered to allow entry of arsenic data. At this point, the analyst is asked to change the remark code or leave it blank (Fig. 6.16). No remark code is equivalent to using the actual value provided on the data sheet. A remark code of "K" is used for values less than the reported value, and a remark code of "L" is used for values greater than the reported value. This system of remark codes, along with the associated five-digit parameter codes (PCODE), is a requirement of EPA's STORET system and allows LDEQ data to be transferred to EPA as needed. Next, the actual concentration value is recorded (Fig. 6.16). Following recording of all available concentrations for metals, the analyst is asked to verify the data as was done with the first two sample information screens (Fig. 6.12). Answering **"0"** for "continue" brings up an option for writing the data to file or forgetting this sample number (Fig. 6.17). After answering this question, users are offered the sample continuation options found in Fig. 6.14.

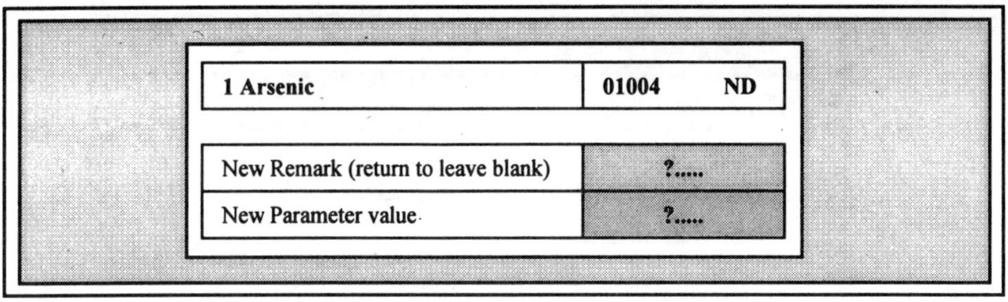

Figure 6.16. Arsenic entry window for Water Quality Survey data entry within the Water Quality Data Management System used by Louisiana Department of Environmental Quality, Office of Water Resources, Baton Rouge, Louisiana

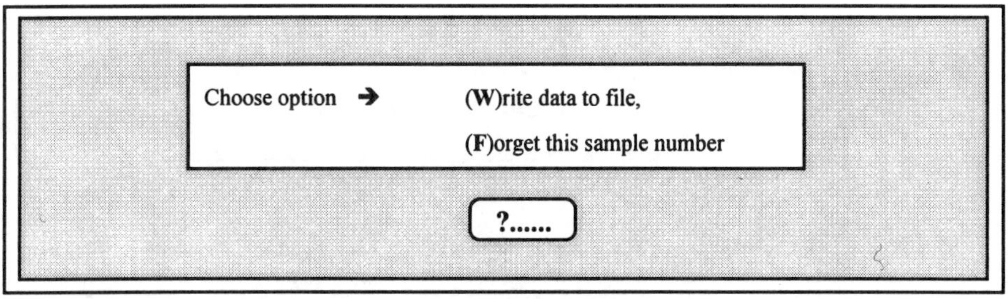

Figure 6.17. Data saving option menu for Water Quality Survey data entry within the Water Quality Data Management System used by Louisiana Department of Environmental Quality, Office of Water Resources, Baton Rouge, Louisiana

As can be seen, data entry verification is provided for during the actual entry procedure; however, an additional verification is provided following data entry. After an entry session has been completed and the analyst indicates option **"QUIT"** on the entry screen shown in Fig. 6.14, all data recorded during that session are saved in a file called "NEWDATA" which is placed in the users directory. This file can then be reviewed for accuracy at a later date. Corrections to the entered data can be made directly by modifying the preliminary "NEWDATA" file created after data entry. After this verification and correction, the file is loaded into a permanent database using option **"WQSLOAD"** found on the menu shown in Fig. 6.9. Should errors be detected after data has been loaded, provisions have been made to correct them by using the **"WQSCVAL"**, **"WQSCFISH"**, **"WQSDELS"**, **"WQSDELSB"**, or **"WQSDELSM"** options found on the same menu with **"WQSLOAD"** (Fig. 6.9). These options allow for changes to remarks, values, fish codes, or names. Users can also delete entire sample entries if they are wrong or duplicated.

Water Quality Network and Survey Data Recovery Procedures

Data retrieval from both the WQN and WQS databases can be accomplished in two ways. First, a series of menus has been developed, which provides for the most commonly requested types of information. These menus can be reached by entering **"WQN"** or **"WQS"** at the base menu (Fig. 6.2). For recovery of WQN data, a series of three menus is available, which are shown in Figs. 6.3-6.5. Choosing any of these options leads to a query which prompts the user to specify what type of information they wish to retrieve. These queries include what station, time frame, or parameters are desired. Recovered data may be directed to the screen, saved in a file, or sent directly to a printer. For WQS data retrieval, Fig. 6.8 shows a menu designed to recover Calcasieu Toxics Inventory (CTI) data. CTI is a specific long-term project, for which a simplified menu driven retrieval system was deemed appropriate.

Both WQN and WQS data, as well as all other information stored on the WQDMS, may be retrieved by writing customized FOCUS data retrieval programs. Data retrieved by customized programs can be output directly to the screen for quick review or written to WordPerfect compatible files for later manipulation or printing. Output can also be written to ASCII files for importing to a spreadsheet or SAS program. While FOCUS does permit limited analysis of raw data, importing files to spreadsheets or SAS permits a wide range of analyses to be performed on water quality, fish tissue, or sediment data.

Currently a combination of FOCUS and SAS programs are utilized biennially by the LDEQ, WQMD to determine water quality status as required by EPA under section 305(b) of the Clean Water Act. In addition to this biennial review of water quality data, studies are frequently established to determine water or fish tissue quality at specific sites throughout the state. In order to perform these studies, either existing WQN data or new data developed for the study must be retrieved from the database and transferred to a spreadsheet or SAS for analysis. Finally, requests for data come from industry, citizen groups, and other state and federal agencies. These requests must be met rapidly and accurately in order to meet the demand while not taking personnel away from ongoing projects. These requests are met by providing the desired data either as hardcopy or in ASCII or WordPerfect format on diskettes.

Flexibility of LDEQ, OWR Data Management System

As with any system in active use, changes in procedure or advances in knowledge necessitate modification of the system to meet new demands. For example, when laboratory techniques improve, detection levels will be reduced. This necessitates the need to change default detection limit values. When changes in assessment protocols or stream standards change, the ability to modify assessment programs must be easily available, either to the original programmer, or ideally to the user and the decision maker. Each of these situations can be readily resolved by the WQDMS. All of the data retrieval and data assessment menus available on the system are controlled by one or more programs. Each of these programs can be modified, either by Technical Services personnel, or by end-user personnel provided with proper access and writing protection capabilities. After a program is modified and saved, it once again becomes part of the WQDMS and is available for use. In this manner, routine changes in assessment protocols can be quickly handled by the end-user without tying up Technical Service personnel with responsibility toward the entire Department.

Should sampling stations be added to or dropped from the WQN or WQS networks, menus are available to easily make these changes. Stations which are added must have information provided on station number, projects associated with the station, exact location, reason for addition of the station, and time frame during which the station will be used. All of this information may be added by answering a series of queries provided by the menu driven system.

If larger changes to the system are needed, such as an ability for users to personally modify or delete incorrect data, new programs must be created by Technical Services personnel familiar with the system. These capabilities, while placed on the general WQDMS, must be protected by limiting writing access to those responsible for the data. By requesting correction programs of this type, end-users are again able to free-up valuable Technical Service personnel from making routine corrections to the databases.

In summary, it is important for end-users and Technical Service programmers who developed the WQDMS to maintain close ties. This enables modifications to the system to be implemented quickly and easily, enabling both parties to continue daily operation of the data management system.

6.3. Summary of Environmental Data Management

This chapter has provided a brief description of two environmental data management systems currently in use throughout the United States. While limited to two agencies, selection of systems used by EPA and USGS permit the reader to see the wide range of capabilities and the enormous size of systems being used and developed in the United States. In addition, the chapter has provided a detailed discussion of one data management system currently used by the Louisiana Department of Environmental Quality, Office of Water Resources. While this chapter has been geared toward water quality data, information provided should allow the reader to begin development of a data management system tailored to the needs of any environmental resource agency.

Use of a menu based system allows users to quickly become familiar with the requirements of data entry and retrieval. Menus also permit more experienced users to input a wide variety of data more rapidly and accurately, and allow for easy verification and correction as needed. Retrieval of commonly required data is also facilitated by use of menus designed to accommodate frequent data needs. While the use of menus allows for an easy to use, user friendly system, the availability of customized reporting procedures is a necessity. This allows for more a detailed retrieval and analysis as required for environmental decision making or research.

References

Maddy, D.V., L.E. Lapp, D.L. Jackson, R.H. Coupe, T.L Schertz, and K.T. Garcia (1991) *National Water Information System User's Manual*, Volume 2, Chapter 2, Water quality system, Version 91.1, United States Geological Survey,Reston, Virginia.

Hutchison, N.E. (1975) WATSTORE user's guide, National water data storage and retrieval system, United States Geological Survey.

United States Environmental Protection Agency (1993) PC water body system user's manual (Version 3.1), Office of Water, Assessment and Watershed Protection Division, July 1, 1993.

United States Environmental Protection Agency (1992) Office of Water, Environmental and program information systems compendium, FY 1992, Office of Water (WH-556), 800-B92-001, 152pp.

CHAPTER 7

DATA ANALYSIS

N.B. Harmancioglu, S.D. Ozkul and O. Fistikoglu

Abstract. This chapter considers the analysis of observed environmental data specifically in the case of water quality. Although data analysis procedures for air quality and solid waste properties may differ in the types of data they require and use (i.e., spatial and temporal frequencies, variables monitored, etc.), the basic approach remains the same as that used for analyzing water quality. Therefore, the approach described in the following sections may as well be used to investigate other environmental processes. The methods reviewed are the most common techniques employed in statistical data analyses. The emphasis here is on how these techniques perform in the case of messy water quality data with short sample sizes, missing values, and gaps within the series.

7.1. Selection of the Appropriate Data Analysis Methodology

There are several methodologies, basically statistical in nature, that are used to analyze the properties of observed water quality data. The principles underlying these methodologies are available in general statistical literature and in publications devoted particularly to the case of water quality. Among numerous studies, which would be too exhaustive to cite here, one may refer to Chapman (1992) for a general summary of methods used for water quality data analyses and to Hipel and McLeod (1994) for an extensive and highly detailed review of environmental data analysis techniques. It is not intended herein to restate the mathematical background of such techniques; rather, a general critical overview of the data analysis procedures is presented, including mathematical details when deemed necessary for treatment of water quality data. Some of the most frequently used methods are discussed with respect to their application to messy water quality data such as those observed in the majority of Turkish river basins. The selection of a particular data analysis methodology for investigating water quality depends basically on two factors:

a) the type of information sought;
b) the nature of available data.

In the case of water quality, there are essentially three types of information to be derived by data analyses:

N.B. Harmancioglu et al. (eds.), Environmental Data Management, 141-196.

a) information on the *mean values* of water quality variables;

b) information on the *extreme values* of water quality variables;

c) information on *trends* (spatial or temporal) in water quality.

Each of these properties needs different data analysis techniques so that it can be reliably described. Such techniques are further classified according to their suitability to the nature of available data. Some methodologies require regularly collected data; whereas some can better adapt to the sporadic nature of water quality observations.

Since water quality variables can be treated as hydrologic variables, all techniques of hydrology currently in use can be employed to investigate water quality. Yet, this statement is true in the theoretical sense. In practice, the shortcomings of available water quality data, which basically stem from the monitoring practices applied, preclude the use of classical hydrologic techniques as they are. Often, these techniques need to be revised and modified to extract the maximum amount of information from irregular water quality data.

7.2. The Nature of Water Quality Data

Hydrologic processes, including water quality, may be analyzed as univariate series in the form of either time series (as a function of time) or as line series (as a function of distance) when one of the dimensions, time or space, is kept constant. However, information is often needed on both the temporal and the spatial distribution of such processes so that one has to consult to multivariate analysis techniques for a full understanding of how they evolve over time and space. In this sense, runoff data are probably the least problematic as they are regularly observed within a systematically operated network. Water quality data, in contrast, pose significant difficulties in multivariate analyses due to the monitoring practice applied.

Three basic features of sampling affect the resulting information gain about water quality:

a) variables sampled,

b) the frequency of sampling, and

c) sampling sites.

With respect to the first feature, the difficulty is that the quality of water, even at a single site, has to be described by a large number of variables in contrast to streamflow which is represented by a single variable at a point in space. Accordingly, the analysis of water quality for a single site becomes a multivariate one where the relationships between several variables have to be investigated. There is no problem when all variables are monitored regularly at the same time points. However, if different frequencies are applied for each variable, such relationships may be pretty difficult to be reliably described.

The second feature of water quality monitoring, i.e., the temporal frequency of sampling, is the most problematic aspect about water quality with respect to data analysis. Water quality variables are often sporadically observed at irregular time intervals. Furthermore, their data series have several gaps and missing values as there may be long intervals where observations are not made. Another problem is that periods of water quality observations are often pretty short. With these characteristics, the nature of water quality data is often described to be "messy" (Hipel and McLeod, 1994). Consequently, the application of classical techniques of time series data is often made difficult by this messy character of water quality observations.

The third feature of water quality monitoring relates to the adequate spatial representation of water quality. Even if there exist sufficient numbers and locations of sampling sites, information transfer between water quality variables in the space domain is often poor. This is because temporal sampling frequencies for a single variable at different sites do not match, or because different variables are monitored at different sites. Thus, a full spatial description of water quality along a water course cannot be realized efficiently.

It follows from the above that the multivariable, multisite and messy character of water quality data complicate their analysis so that researchers are in continuous search of appropriate techniques to identify the space/time distributions of water quality.

7.3. Analysis of Water Quality Means

In water quality management, it is often required to identify the mean value of a water quality variable at a particular site. Such information is sought for management purposes such as general surveillance or treatment needs to control river water quality. For example, the design of a treatment plant to regulate instream quality is based on the knowledge of the mean values of particular variables monitored at a site. The design criteria are based on the true means to be estimated from observed data. Obviously, one or two random observations are not sufficient to decide upon the true mean value. A series of data should be available in adequate amounts so that the mean water quality concentrations can be reliably estimated. Then the question is how many samples should be taken to determine the true mean with a certain level of confidence.

Sanders and Adrian (1978) have proposed a method for estimating the mean value of a water quality variable from a series of monitored data. Essentially, they have developed this methodology to determine the required sampling frequencies in time if the information sought is the true mean value of a water quality variable at a specified level of statistical confidence. The method depends on the assumption that the primary objectives of future water quality monitoring networks are the determination of ambient water quality conditions and an assessment of yearly trends. The purpose of the method is to derive a sampling frequency criterion from standard statistical procedures that are used to determine the relationship between sampling frequency and the expected half width of the confidence interval of the random component of an annual mean variable concentration (Sanders and Adrian, 1978; Sanders *et al.*,1983; Sanders,1988). It must be noted here that, upon lack of sufficient water quality data, the method was demonstrated by Sanders and Adrian (1978) for the case of river flows so that the annual statistic used was the mean log river flow.

For a series of random events, the confidence interval of the mean decreases as the number of samples increases. Thus, the accuracy of the estimate of the mean is a function of the number of sample observations. Therefore, a sampling frequency, as number of samples per year, can be determined for a specified confidence interval of the mean. Unfortunately, most hydrological time series are not random but significantly correlated and nonstationary, which makes standard statistical analyses difficult. Thus, the method can be applied only after removing the serial correlation and nonstationarity from the series.

The Student t-statistic is selected to estimate the relationship between sampling frequency and the confidence interval of the mean of the random component. If the observations x_i (i=1, ..., n) are stationary, independent and identically distributed, the variable t of Eq. (7.1) can be defined by a Student t-distribution:

$$t = \frac{\bar{x} - \mu}{S / \sqrt{n}} \tag{7.1}$$

where,

$\bar{x}$ = the calculated mean of the independent residuals,
μ = the theoretical population mean,
S^2 = the sample variance of x_i, and
n = the number of independent observations (Sanders and Adrian, 1978).

For a specified level of significance, the variable t will lie in a confidence interval defined by known constants. This means that the probability that the random variable t is contained within the interval is equal to the level of significance (1-α), and the probability that the variable t is not contained within the interval is equal to α. This situation can be written by using the common statistical notation:

$$P_r = \left\{ t_{\alpha/2} < \frac{\bar{x} - \mu}{S / \sqrt{n}} < t_{1-\alpha/2} \right\} = 1 - \alpha \tag{7.2}$$

where, $t_{1-\alpha/2}$ and $t_{\alpha/2}$ are constants defined from the Student t-distribution for a specified level of significance and the number of samples.

By using the equality $t_{1-\alpha/2} = -t_{\alpha/2}$, the confidence interval of the theoretical residual mean can be written as:

$$\bar{x} - \frac{t_{\alpha/2} S}{\sqrt{n}} < \mu < \bar{x} + \frac{t_{\alpha/2} S}{\sqrt{n}} \tag{7.3}$$

and the width of the confidence interval of this mean of the random sequence $[x_i]$ is:

$$2R = \frac{2\, t_{\alpha/2}\, S}{\sqrt{n}} \tag{7.4}$$

where R represents half the expected confidence interval of the mean (Sanders and Adrian, 1978). Figure 7.1 shows the sampling distribution of the mean $\bar{x}$ together with the confidence interval bounded by μ_{upper} and μ_{lower} limits for a specified level of significance (1-α). 2R, then, is the confidence interval between the limits defined. Thus, R is a function of the standard deviation of the observed residuals, the square root of the number of the data, and the constant from the Student t-distribution. Consequently, to determine the temporal sampling criterion, a plot of half of the expected confidence interval of the residual mean versus the sampling frequency is sufficient since the confidence interval is symmetric about the mean.

Sanders and Adrian (1978) showed the application of the method for the case of streamflows due to the lack of sufficient water quality data for statistical analysis. In their procedure, they first removed all series components that cause nonstationarity (trends, periodicity and serial correlations). Next, the sample variance of residuals S_a^2 are computed and plotted against the sampling interval as shown in Fig. 7.2 (Sanders and Adrian,1978).

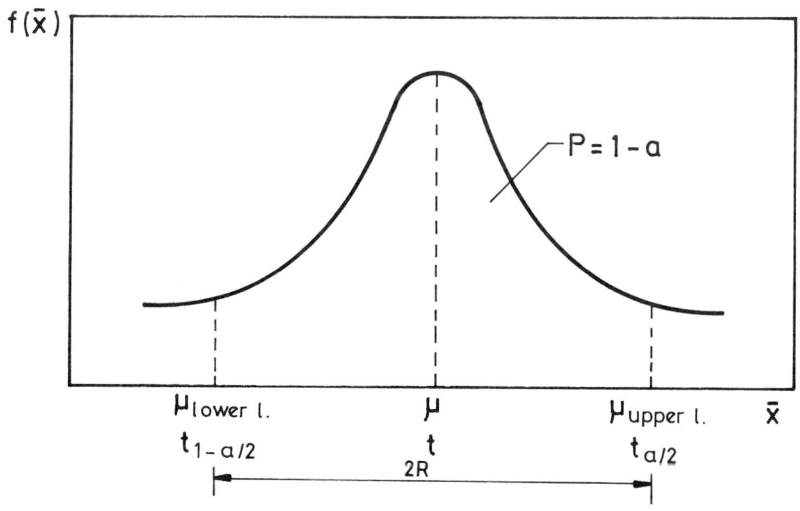

Figure 7.1. Sampling distribution of the mean x and the confidence interval 2R for a specified level of significance (1-α) (Sanders and Adrian,1978)

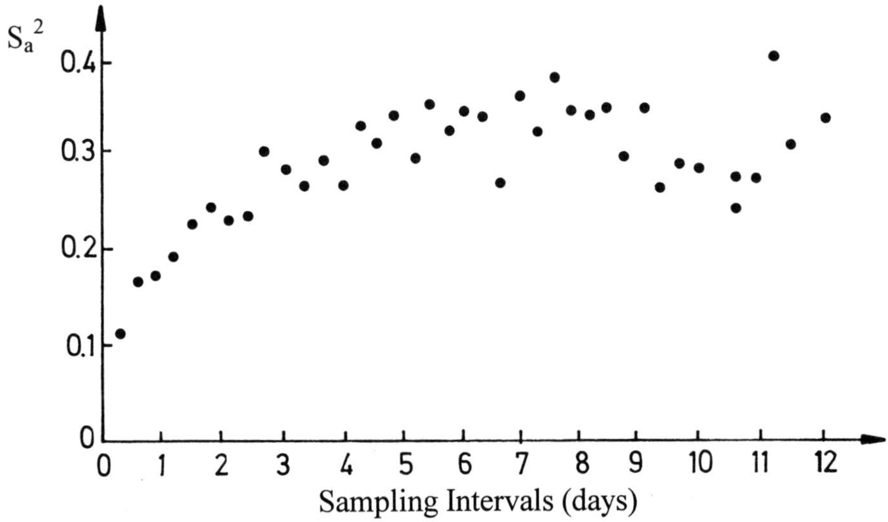

Figure 7.2: Sampling variance of residuals versus the sampling interval (Sanders and Adrian,1978)

The S_a^2 values stabilize after a certain sampling interval and approach a limiting value. After a certain sampling interval for which S_a^2 stabilizes, the variance becomes almost constant and is independent of the sampling interval. Sanders and Adrian (1978) stated that this is a necessary condition so that the analysis of the relationship between R and n becomes theoretically valid. Next, for the streamflow series used, they derived the plots of R versus n (number of samples per year) for specified levels of significance (1-α) as shown in Fig. 7.3.

Sanders and Adrian (1978) used daily streamflows in their analysis so that the required sampling frequency is found by dividing the number of days in a year by the number of samples per year:

$$sampling\,frequency = \frac{365}{n} \tag{7.5}$$

To determine the sampling frequency by this method, one has to specify the level of significance first. Then, using the relationship shown in Fig. 7.3, the number of samples per year (n) can be determined for a particular value of R.

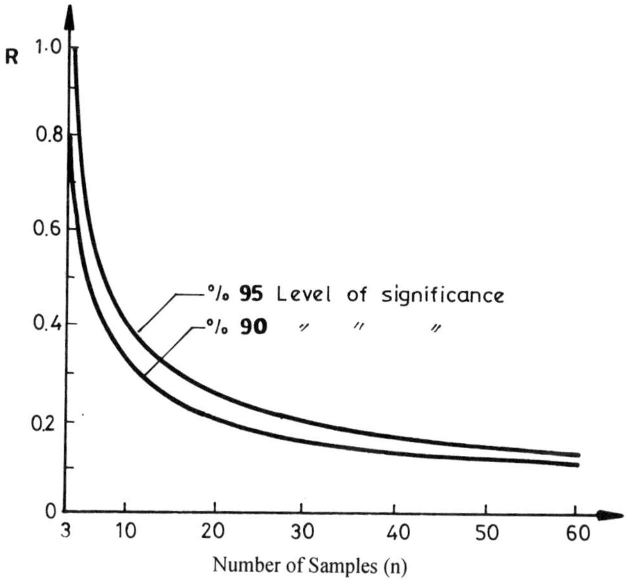

Figure 7.3. Relationship between the expected half width of the confidence interval R and the number of samples per year (Sanders and Adrian,1978)

Tokgoz (1992) tested the applicability of the above methodology on the most complete set of water quality data available in Sakarya River basin in Turkey. The monthly data record covered a period of 6 years with a number of missing observations and irregularly spaced gaps within the series. Application of Sanders' method has been realized with a number of difficulties due to the sampling intervals and the total observation period of available data. First, although Sanders and Adrian (1978) propose that the sampling frequency can be determined by the required number of samples per year, it has not been possible to investigate within-the-year frequencies by considering one year as the major cycle. The reason for this difficulty is that available data are observed on a monthly basis so that the method could only be applied to investigate the required frequencies within the total period of observation, i.e., 72 months. In this case, Eq. (7.5) becomes:

$$\text{sampling frequency} = \frac{\text{no. of months in total observation period}}{n} \tag{7.6}$$

To investigate different sampling frequencies, the total number of data are divided into 2, 3, 4, ..., 12 to obtain samples with different observation frequencies such as monthly, bimonthly, every three months, every four months, etc. Such a procedure inevitably produces smaller sample sizes for each step; thus, the reliability of the statistical analysis is reduced.

Next, the relationships between Sa^2 and sampling intervals are obtained as Sanders and Adrian (1978) require that the method is valid only when Sa^2 values stabilize at large sampling frequencies. The relationship shown in Fig. 7.4 do not agree with that of Sanders and Adrian (1978) in Fig. 7.2 since the Sa^2 values do not seem to stabilize or to approach a limiting value as the sampling interval increases. For almost all variables except pH (and PV to a certain extent), all the Sa^2 values are highly scattered. This means that the theoretical basis for the method is not validated here for small samples of monthly observed data which contain missing values and gaps within the series. Then, it can be stated that the evaluation of the statistical results remain vague and unreliable to determine the required sampling frequency.

The above mentioned difficulty with the instability of Sa^2 values are further reflected in Fig. 7.5 which shows the R-n relationships. Tokgoz (1992) has observed that only pH gives smooth curves whereas the others show significant fluctuations at small numbers of n. Almost all variables give results that can be evaluated only beyond 20 or 30 number of samples. Sample sizes smaller than 20 have fluctuations in the R values so that one cannot infer reliably about the relationships between R and n. In general, results for the water quality data used (except pH) may be that at least 20 to 30 samples are required for each variable so that one can roughly estimate the number of samples needed to obtain a specified value of R. This may mean that each variable has to be sampled at least every 3 or 4 months. On the other hand, this result can be expressed only when the total observation period is considered. Due to the already selected sampling frequencies of the available data in the Sakarya River basin, one can investigate neither within-the-year frequencies nor frequencies higher than a month.

The application of Sanders' method has shown that, in general for the water quality variables investigated, there are significant statistical difficulties in both applying and evaluating the method. In this case, the current monitoring practices of water quality stations do not seem to permit a valid and reliable statistical analysis. Thus, the available data appear to be insufficient for purposes of this specific method. Basically, Sanders' method was previously applied to daily observed streamflow data. The results had then shown that the method could be very well applied to determine required number of samples for a reliable estimation of the mean value at particular levels of statistical significance. In the application by Tokgoz (1992), however, the method is applied to monthly observed small samples of actual water quality data. In the application, significant difficulties are encountered with respect to statistical computations so that the results need to be evaluated with caution. In fact, the basic theoretical requirement for the proposed method could not be validated so that the eventual results are not considered to be very reliable. Within this respect, one concludes that the available data produce very little information as to the required sampling intervals and the true mean values of the variables investigated.

As demonstrated in the study by Tokgoz (1992), the methodologies for estimation of mean values are perfectly valid in the statistical sense. However, their application to short duration irregularly observed water quality data does not always produce reliable results since the underlying assumptions of such methods are often not met by water quality series. This may not be the case for a number of developed countries where data banks are already filled up with regularly observed data. However, in a great majority of countries, including the developing ones, reliability of such statistical approaches may be pretty low.

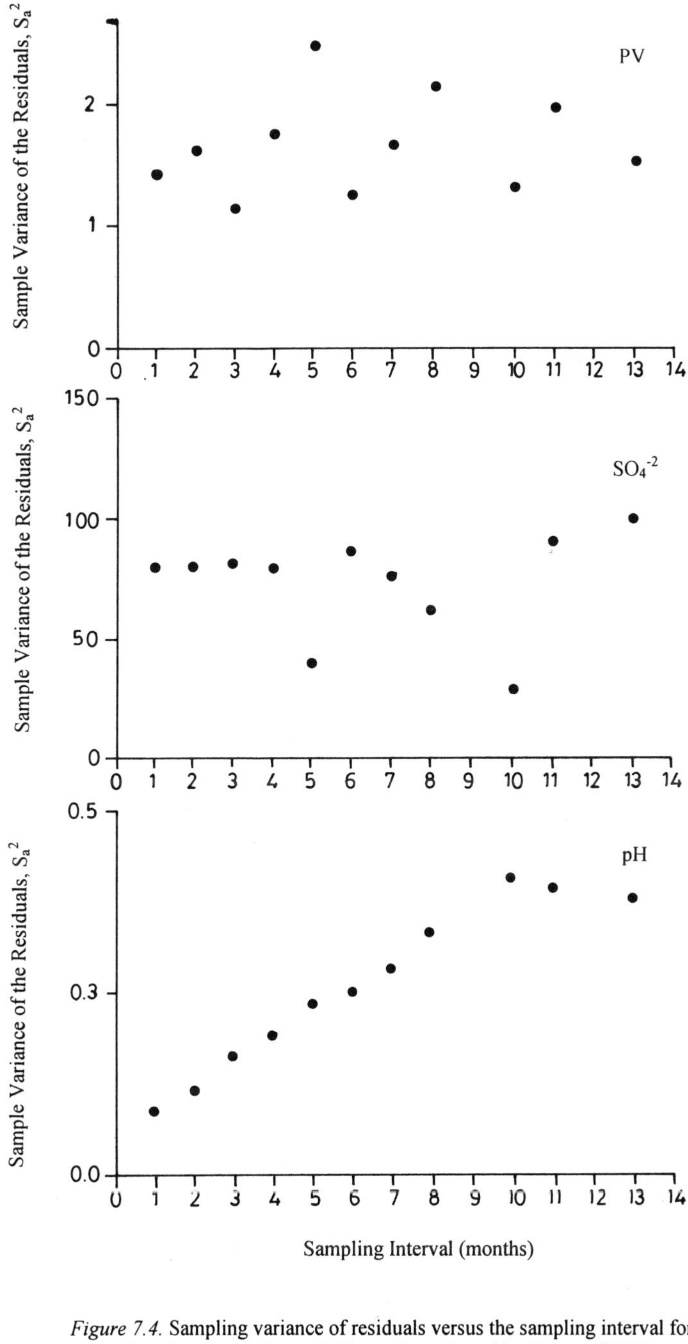

Figure 7.4. Sampling variance of residuals versus the sampling interval for selected water quality variables as investigated by Tokgoz (1992)

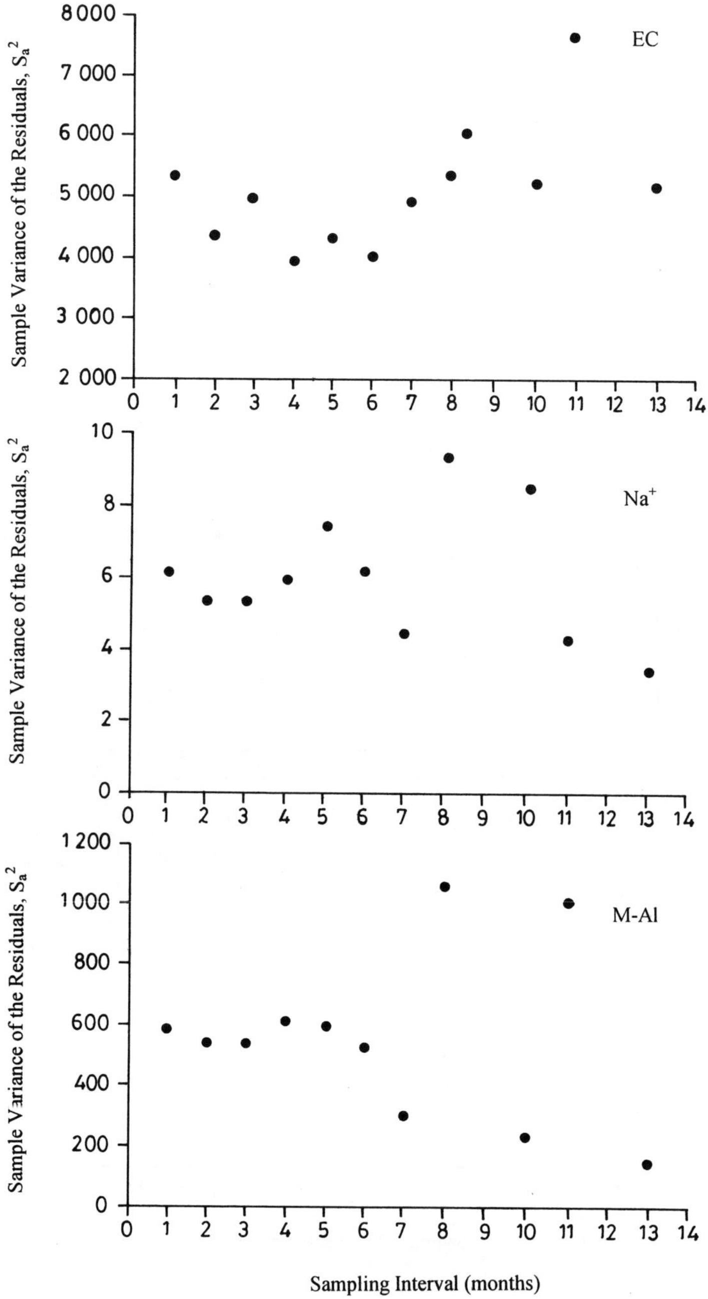

Figure 7.4. (cont.)

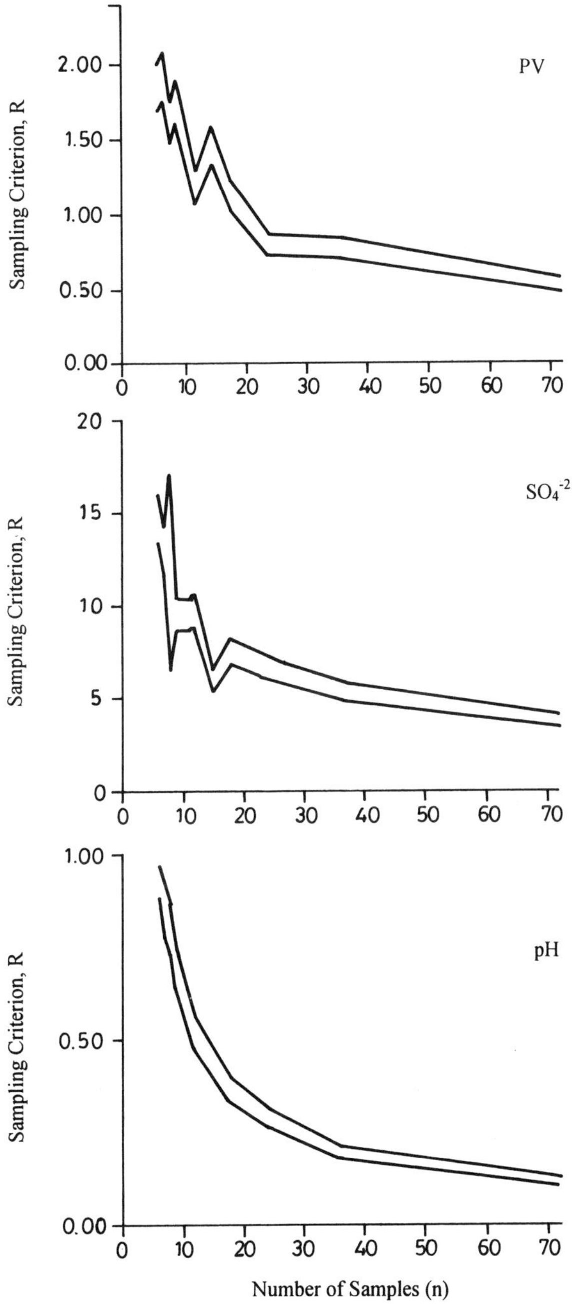

Figure 7.5. Relationship between the expected half width of the confidence interval R and the number of water quality samples per year Tokgoz (1992)

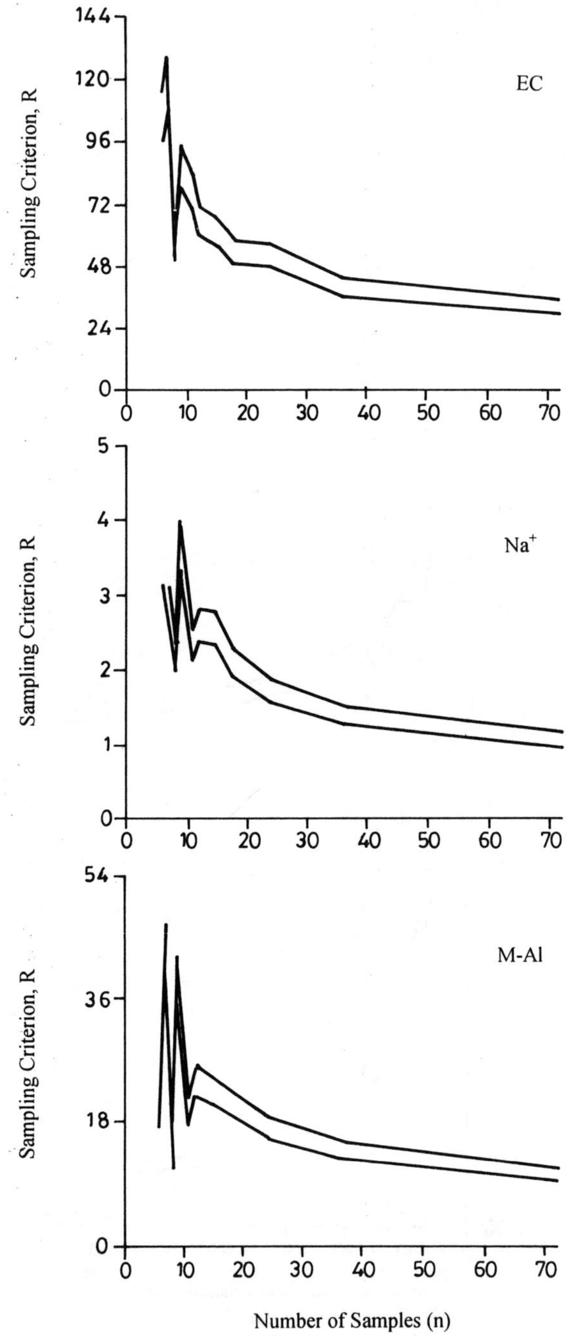

Figure 7.5. (cont.)

7.4. Determination of Extremes in Water Quality

It is not only the means but the extreme values of water quality conditions that are of interest to managers. Knowledge on extremes is required for regulatory purposes such as detecting standard violations. A major difficulty associated with assessment of standard compliance is that it is based on sampling. In this context, it is highly affected by sampling errors and the resulting uncertainties. For example, the actual quality of water at a time and space point may be exceeding a critical value, but it may not be noticed if it is not sampled at that time. Or, if a sample taken at a certain time shows that the quality is good, it is assumed until the next sampling that it will remain good. In each case, our decisions carry a risk of failing to observe the actual quality of water (Alpaslan *et al.*, 1993).

The major problem in assessment of compliance stems from the selected monitoring frequencies. Essentially, continuous monitoring is required if one wants to detect extreme values which may lead to standard violations. Recently, advances in measurement capabilities permit us to observe much better the variability and the uncertainty in the behavior of water quality processes. By means of continuous monitoring, we can now identify not only the average concentrations of pollutants, but also the occurrence of extreme events in the form of shock loadings, which are similar to flood events in the case of water quantity (Beck and Finney, 1987). On the other hand, continuous monitoring is often costly in time, labor, and money in addition to being highly sensitive to system failures, e.g., equipment failures (although internationally the number of stations with continuous monitoring is increasing (Mulder, 1994)). Then the question is how frequently should water quality be sampled or how many samples should be taken so that extremes do not go unnoticed. The answer to this question is basically treated by probabilistic approaches, which are valid for water quality variables since they are random in nature.

An extreme condition regarding water quality can be described as the probability of exceedance $P(X>x_{cr,h})$, or the probability of nonexceedance $P(X<x_{cr,l})$, with X representing the random variable, and $x_{cr,h}$ and $x_{cr,l}$, the critical high or low values of the variable shown in Fig.7.6, where x_m represents the mean value or the median of the variable. The exceedance (or nonexceedance) probabilities may be determined by either a parametric or a nonparametric approach. The former requires the fitting of probability distribution functions, as in Fig.7.6, to describe the random variable, and is therefore subject to modeling uncertainties (errors). The nonparametric approach to quantification of natural and/or impact risks covers the use of the well-known plotting position formulas. The relationship between parametric and nonparametric probabilities results in the following formula, which is commonly used in risk analyses to describe the probability of at least one failure within a record of n samples (Harmancioglu and Alpaslan, 1992; Harmancioglu *et al.*, 1993):

$$R_{(1)} = 1 - [1 - P(X) x_{cr})]^n \tag{7.7}$$

where $R_{(1)}$ represents the risk of being equaled or exceeded once or more in a record of n samples.

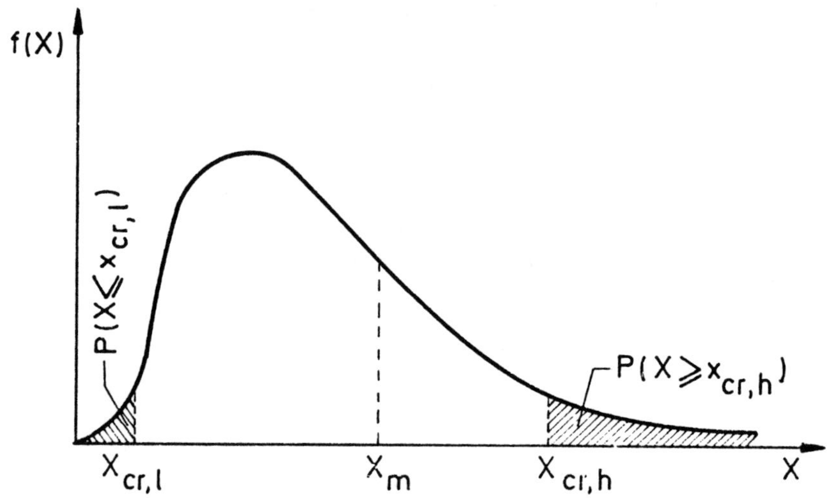

Figure 7.6. Probabilistic description of extreme values in water quality in the form of exceedance $P(X>x_{cr,h})$ and nonexceedance $P(X<x_{cr,l})$ probabilities

The condition of compliance or violation entails a risk factor in the form of a water quality variable exceeding (or nonexceeding) a critical value, $x_{cr,h}$ or $x_{cr,l}$, set as a standard. In this case, exceedance or nonexceedance probabilities shown in Fig. 7.6 represent the risk of violation of a standard.

Sometimes the frequency (f) of occurrences of undesirable outcomes (compliance failures) may be considered more significant such that it does not exceed a critical value f_{cr}. This condition may be expressed as the probability (Harmancioglu and Alpaslan, 1992):

$$P\left[f \geq f_{cr}\right] \leq \alpha_f \tag{7.8}$$

which is required to remain below an acceptable or specified level of risk α_f. If, in addition to frequency of failures, the degree of failures is also of concern (Dendrou and Delleur, 1979), a similar requirement can be expressed using quantities. For example, the concentration C_l of a pollutant load may be desired not to exceed a particular critical level $c_{l,cr}$. Then, the probability of such exceedance can be assessed to remain below a specified risk level as:

$$P\left[C_l \geq c_{l,cr}\right] \leq \alpha_{l,cr} \tag{7.9}$$

In this case, the standard is set as the critical level $c_{l,cr}$, and the risk factor here refers to that of samples failing to meet the standard. Defining $\alpha_{l,cr}$ means that an acceptable level of risk or compliance failure is a priori determined so that a particular percentile of the samples are expected to satisfy the condition in Eq. (7.9). For example, if $\alpha_{l,cr}$ is selected as 5%, it indicates an acceptable failure rate of 5% so that 95% of the samples is required to meet the standard. Such an evaluation foresees the determination of percentile values from an observed statistical sample using parametric or nonparametric approaches. Within this respect, Crabtree *et al.* (1987) claim that the parametric approach serves better to obtain the maximum amount of information from available data. However, this procedure is subject to modeling uncertainties so that tests for goodness of fit need to be applied. Nevertheless, Crabtree *et al.* (1987) have found that the "best" estimate of the 95 percentile from a data set must be based upon a fitted probability distribution. Among the 334 water quality data sets they have analyzed, approximately half of them could be fitted by a normal, lognormal or Pearson Type 3 distribution, the latter providing the most flexibility in the case of water quality. Crabtree *et al.* (1987) have suggested that nonparametric techniques can be used if failure occurs in fitting a parametric distribution.

Interpretation of standards as percentile values avoids assessment of compliance by using a standard that is unnecessarily rigid in some circumstances and not rigid enough in others. Similarly, it avoids the use of a standard that is inconsistent with the capability of existing treatment facilities or with the cost of possible improvements (Crabtree *et al.*, 1987). Thus, the standards are now described on a probabilistic basis to take into consideration the local circumstances and river quality objectives. As described earlier, 95 percentile class limits are adopted to describe water quality standards in most countries. This implies an acceptable 5% risk of compliance failure (Warn, 1988; Crabtree *et al.*, 1987).

Sanders *et al.* (1983) also discuss that the use of a statistical or probabilistic basis for compliance assessment requires fitting of a distribution function to observed series of water quality data. For example, water quality observations may be assumed to be independent and identically distributed to follow a normal distribution function $N(\mu,\sigma)$ with μ the population mean and σ the standard deviation of the variable. Figure 7.7 shows such a distribution where the standard is described as the $\mu + 2\sigma$ concentration. In this case, 95/100 samples will have concentrations within an interval bounded by $\mu - 2\sigma$ and $\mu + 2\sigma$; whereas 5/100 samples will remain outside these boundaries to be less than $\mu - 2\sigma$ or greater than $\mu + 2\sigma$. If the stream standard for a single sample were set at $\mu + 2\sigma$, the probability P of any one sample exceeding the standard is the shaded area in the figure to the right of the standard, which is approximately 0.0227 (Sanders *et al.*, 1983).

If the standard were set at $\mu + 3\sigma$, this will decrease the probability of a single sample concentration to exceed the standard. In this case, the risk of violation would be 0.14%. If the standard were specified as $\mu + 3\sigma$ and never to be exceeded, that would necessitate continuous monitoring of the variable. The same standard can be set as the limit not to be exceeded by more than a single sample within a specified time interval. Then, the frequency of sampling has to be determined since the number of available samples affects the interpretation of compliance (Sanders *et al.*, 1983).

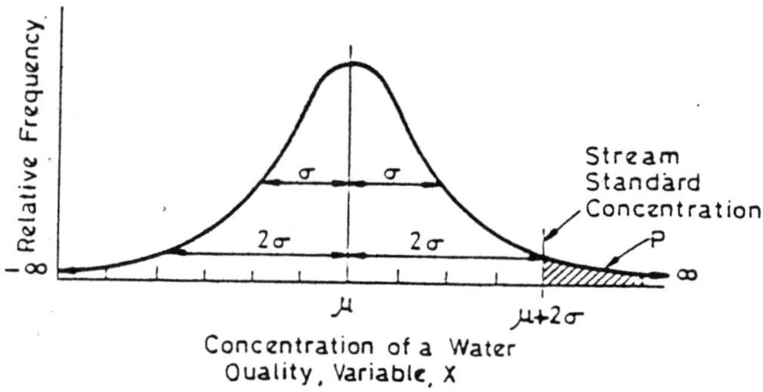

Figure 7.7. Assumed frequency distribution of a water quality variable which is normal with mean μ and standard deviation σ (Sanders *et al.*, 1983)

As denoted earlier, the probabilistic basis for standard compliance as described above indicates another difficulty associated with compliance as it is based on sampling. In this context, it is highly affected by sampling errors and the resulting uncertainties. Warn (1988) stresses that these errors can be minimized by defining confidence limits which describe the range within which the true value of the standard (expressed as percentiles) lies. Loftis and Ward (1981) also present methods for defining confidence limits around standard violation probabilities. Such an approach aids in the assessment of uncertainty of the standard violation information obtained from collected series of water quality data (Sanders *et al.*, 1983).

A major difficulty in comparing a particular percentile value of water quality with a standard is that, by chance, one may have a set of bad samples which shows noncompliance, or on the contrary, a set of good samples that indicate compliance although the sampled site actually violates the standard (Warn, 1988). This uncertainty in assessment occurs due to sampling error when cases of true failure or true compliance cannot be observed due to discrete sampling. Such sampling errors lead to the risk of making wrong decisions regarding compliance with standards. This risk can be expressed in the form of confidence limits to describe the range within which the true value of the 95 percentile is expected to lie (Warn, 1988).

Warn (1988) defines 95% confidence limits around a 95 percentile standard and refers to them as optimistic (lower limit) and pessimistic (upper limit) confidence limits. Then three cases may occur in assessment of compliance using the 95 percentile value:

a) if the pessimistic confidence limit is less than the 95 percentile standard, then one can be at least 95% confident that the standard is not exceeded;
b) if the optimistic confidence limit exceeds the 95 percentile standard, one can be at least 95% confident that the standard is not met;

c) if the 95 percentile standard lies between the optimistic and pessimistic confidence limits, it is not possible to assess compliance or failure with at least 95% confidence.

Warn (1988) further discusses that such an assessment may be worked out for different levels of confidence limits (50%, 80%, 99%, 99.99%, etc.) for interpreting compliance at different rates of assessment risks. The selection of the risk level depends on the planner's evaluation of the significance pertaining to the problem considered. To get a complete picture of effects of sampling errors, more than one risk level may be investigated by describing confidence limits at different levels of probability. Warn (1988) states that such an approach provides a sound basis for interpreting the performance of treatment plants where sampling is realized infrequently.

To illustrate the above considerations, samples of BOD, COD, and TSS (total suspended solids) concentrations, observed daily for three years in effluents from the treatment plant of a paper factory in Turkey, are evaluated for assessment of compliance with relevant preset standards (Alpaslan *et al.*, 1993). Sampling distributions of the three variables are analyzed and basic statistical parameters are computed. As shown in Fig.7.8, all three variables are assumed to follow a normal distribution. 95 percentile values are estimated using these distributions, and 95% confidence limits are computed for the 95 percentile of each variable.

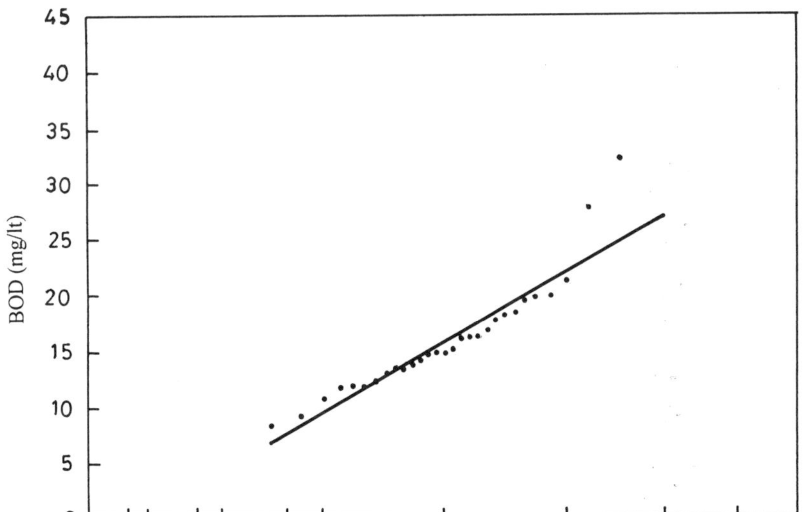

Figure 7.8. Frequency distribution and fitted normal probability distribution function of samples of BOD, COD and TSS observed in treated effluents of the paper factory investigated

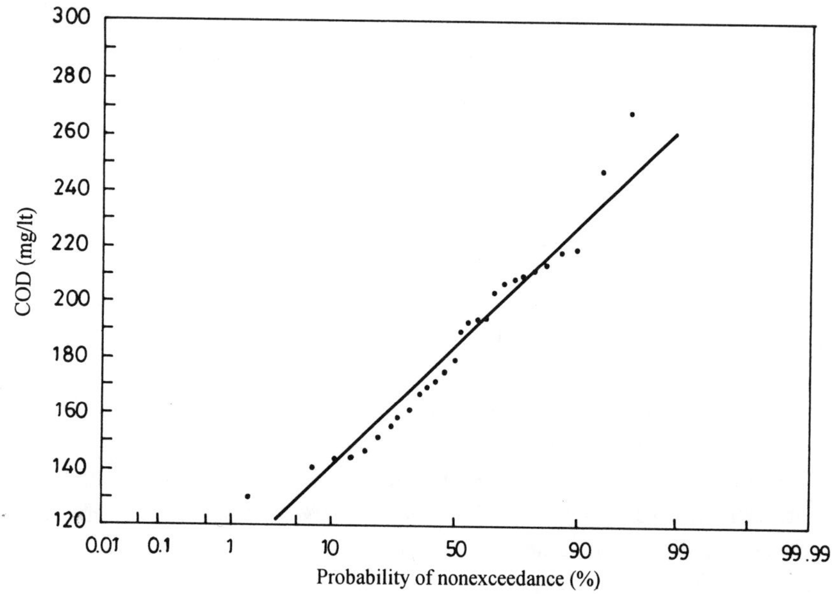

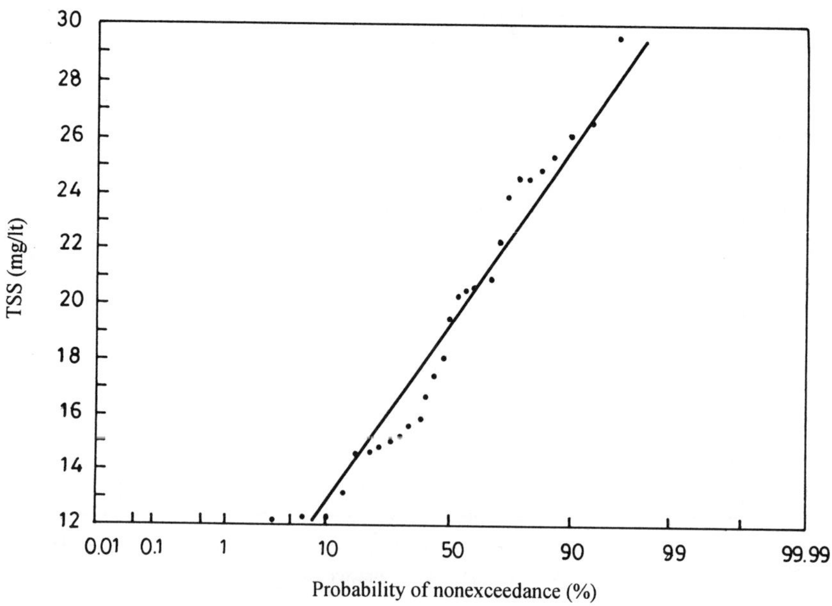

Figure 7.8. (Cont.)

Table 7.1 shows the results of such computations. It may be observed from the above table that, for all of the three variables, one may be at least 95% confident that the standards were met as the pessimistic confidence limits are much lower than the foreseen standards. The same situation occurs even when the confidence limits are specified at 99.99% probability level. In this particular example, standards are set according to the local conditions and resources which would permit a particular level of regulation and enforcement. If there were a more stringent standard, say for example 20 mg/lt. for BOD, this would imply 95% confidence of compliance failure.

As it may be followed from the above discussion, assessment of extremes in water quality requires the fitting of a probability distribution, which is often made difficult by the nature of collected water quality data. Then one has to consult to nonparametric methods. A similar situation exists in assessment of trends in water quality.

TABLE 7.1. Assessment of compliance in case of BOD, COD and TSS samples collected from the treated effluents of a paper factory

Variable	Mean	Standard Deviation	95 Percentile	Standard	Optimistic Confidence Limit	Pessimistic Confidence Limit
BOD	15.88	5.04	24.2	350	21.39	27.01
COD	184.15	33.26	238.9	800	220.00	257.80
TSS	19.09	4.89	27.1	80	24.36	29.83

All figures given are in mg/lt.

7.5. Determination of Trends in Water Quality

7.5.1. OBJECTIVES OF TREND ASSESSMENT

Within the framework of water quality control efforts, emphasis has been given to treatment of effluent discharges from point and nonpoint sources. To this end, monitoring practices have been initiated to evaluate the consequences of pollution control efforts, to determine the current status of surface water quality, and to detect possible changes, if any, in water quality processes with respect to time and space. Essentially, in respect of these objectives, monitoring efforts have been intensified in almost all developed countries where significant amount of research is devoted to investigation and evaluation of trends in surface water quality on a regional or national basis. It is often difficult to determine the true means or extremes in water quality due to temporally varying characteristics of quality variables so that it becomes imperative to investigate possible trends in water quality with respect to time.

On the other hand, water quality observations often constitute intermittent series of messy data so that the application of classical methods for trend analysis has generally produced unsatisfactory results in the case of water quality. Thus, efforts have intensified to develop, apply and test more effective techniques for trend analyses. In these studies, nonparametric methods are proposed since, by means of such techniques, problems related to probability distributions of water quality variables, short observation periods, and the sporadic character of quality data are effectively handled. Since the 70's, several nonparametric techniques based on order statistics of observed data are developed particularly for the analysis of trends in water quality. The goodness of fit of these techniques depends on the series structural properties and the type of temporal trend investigated.

7.5.2. PARAMETRIC METHODS OF TREND DETECTION

Parametric statistical methods for trend detection are techniques that use the numerical values of observed data directly and which, therefore, require the probability distribution of the process be known (Lettenmaier *et al.*, 1991). Parametric methods basically cover the well-known parametric t tests; other parametric techniques that have been developed for serially correlated data include time series analysis and intervention analysis (Belle and Hughes, 1984; Hipel and McLeod, 1994; Icaga, 1994).

The classical statistical tests require an assumption that an observed data vector X is independently distributed as:

$$f x_1 x_2 \ldots x_n (x_1, x_2, \ldots x_n) = f x_1 (x_1) \,.\, f x_2 (x_2) \ldots f x_n (x_n) \tag{7.10}$$

In addition, parametric tests (those which assume a knowledge of the form of the probability density function) usually require that the elements of the data vector must have identical probability density functions. The best known parametric tests are the t tests, which assume that the probability density function f is normal (Lettenmaier, 1976).

Two types of trends may be considered by parametric methods: step trends which consist of a step change in the mean level of a process at the midpoint of the data series; and linear trends which consist of a process with mean level that varies linearly throughout the data record. These two types of trends in water quality observations refer to sudden changes (such as improvement in stream quality due to establishment of a new treatment plant) and gradual changes (increases in nonpoint polluting contributions to a river due to urbanization or land use) (Lettenmaier, 1976).

The trend detection problem by parametric methods is basically a statistical hypothesis testing problem as shown in Table 7.2. The null hypothesis H_0 is that an event A has not occurred, and the alternative hypothesis H_1 is that A has occurred. A test statistic T is used to test H_0 and H_1. The probability of choosing H_0 when H_1 is true is the confidence level of the test or $(1 - \alpha)$. The probability of choosing H_1 when H_1 is true is the power of the test denoted by $(1 - \beta)$. The probability of a Type 2 error or β is a function of sample size, the population, the confidence level α and the alternative hypothesis H_1. In the trend detection problem, H_0 is the hypothesis that there is no trend in the underlying population. H_1 states either that there is a trend in the data (two-sided test) or that a positive or negative trend exists in the data (one-sided test) (Lettenmaier, 1976).

TABLE 7.2. Hypothesis testing and error probabilities (Lettenmaier, 1976)

State of Nature	Test Indication	
	H_0	H_1
H_0	No error ($P = 1 - \alpha$) confidence	Type 1 error ($P = \alpha$)
H_1	Type 2 error ($P = \beta$)	No error ($P = 1 - \beta$) power

Testing for trends is basically the determination of the sample size $n = n_1 = n_2$ needed for a particular power of the test, where two population means are compared (Walpole and Myers, 1990). The hypotheses may be:

$$H_0 : \mu_1 = \mu_2 \qquad H_1 : \mu_1 \neq \mu_2 \tag{7.11}$$

where population standard deviations σ_1 and σ_2 are known. For a particular alternative such as $\mu_1 - \mu_2 = \delta$, the power of the test is given by

$$1\text{-}\beta = 1 - P\left[-z_{\alpha/2} - \frac{\delta}{\sqrt{\left(\sigma_1^2 + \sigma_2^2\right)/n}} \langle z \langle z_{\alpha/2} - \frac{\delta}{\sqrt{\left(\sigma_1^2 + \sigma_2^2\right)/n}} \right] \tag{7.12}$$

where the statistic:

$$\frac{\bar{x}_1 - \bar{x}_2 - \delta}{\sqrt{\sigma_1^2 + \sigma_2^2)/ n}} \tag{7.13}$$

is a standard normal variable with $\bar{x}_1$ and $\bar{x}_2$ being the two sample means. Using Eq. (7.12):

$$-z_\beta \cong z_{\alpha/2} - \frac{\delta}{\sqrt{\left(\sigma_1^2 + \sigma_2^2\right)/n}} \tag{7.14}$$

is obtained so that the required sample size becomes:

$$n \cong \frac{\left(z_{\alpha/2} + z_\beta\right)^2 \left(\sigma_1^2 + \sigma_2^2\right)}{\delta^2} \tag{7.15}$$

When the population variances are not known, the statistic of Eq. (7.13) is assumed to follow the Student t-distribution (Walpole and Myers, 1990).

The power of the test is important in the trend detection problem because it gives, at a fixed confidence level, the probability of the trend detection. The power of the test depends upon the sample size, trend magnitude and the marginal probability distribution of the dependent data series, which is assumed to be normal. For dependent series, it varies also with the form of the dependence of data.

To avoid an assumption of the distribution type, Lettenmaier (1976) proposes the use of nonparametric tests such as Spearman's rho test for step trends and Mann-Whitney's test for linear trends, which require the use of Monte Carlo simulated time series. Otherwise, Student t test is used, assuming that the independent data series are normally distributed.

The parametric test described above can be applied to detect step trends in the following way. The data set x_i of size n, divided into two parts of equal size, has the means μ_1 and μ_2 so that the hypotheses:

$$H_0 : \mu_1 = \mu_2$$
$$H_1 : \mu_1 \neq \mu_2$$

are tested as given in Eq. (7.11).

For a confidence level $(1 - \alpha)$, the test statistic is defined as:

$$t = \left\{ \left[\left| \bar{x}_1 - \bar{x}_2 \right| n^{\frac{1}{2}} \right] / 2S \right\} - t_{1-\alpha/2,\nu} \tag{7.16}$$

where $t_{1-\alpha/2,\ \nu}$ is the quantile of the Student t-distribution at probability level $1-\alpha/2$, the degrees of freedom ν are n-2, and S is the sample standard deviation of the data set:

$$S^2 = \frac{1}{n-2} \left[\sum_{i=1}^{n/2} \left(x_i - \bar{x}_1 \right)^2 + \sum_{i=n/2+1}^{n} \left(x_i - \bar{x}_2 \right)^2 \right] \tag{7.17}$$

Hypothesis H_0 is accepted when $t \leq 0$ or rejected when $t > 0$.

In parametric methods, a test criterion N_T is defined as a population statistic assuming that the population trend and standard deviation are known. With $T_r = |\mu_1 - \mu_2|$ representing the absolute value of the true difference between the two means, N_T is defined as:

$$N_T = T_r \sqrt{n} / 2\sigma \tag{7.18}$$

The power of the test is then:

$$1 - \beta = F\left(N_T - t_{1-\alpha/2,\ \nu} \right) \tag{7.19}$$

where F is the cumulative distribution function of the standard Student t-distribution with ν = n-2 degrees of freedom (Lettenmaier, 1976).

In case of a linear trend, a similar approach is used to obtain the test criterion N_T' as:

$$N_T' = \frac{[n\,(n+1)\,(n-1)]^{1/2}\ T_r'}{n\,(12)^{1/2}\ \sigma_\varepsilon} \tag{7.20}$$

for the well-known regression model:

$$y_i = \tau\, x_i + \gamma + \varepsilon_i \tag{7.21}$$

where ε_i is normally distributed with zero mean and variance σ_ε^2. τ is the trend magnitude and γ is the base level constant. In Eq. (7.20), σ_ε represents the standard deviation of ε_i and $T_r' = n\,\tau$. The test criterion N_T' for linear trends is similar to N_T of Eq. (7.18) for step trends except for the constants. The power of the test is again computed by Eq. (7.19).

Lettenmaier (1976) developed curves as shown in Fig. 7.9 to describe the relation between the power of the test (1-β) and N_T or N_T'. He described that the power curves are essentially functions only of N_T or N_T' for sample sizes greater than 20.

He further developed this figure into the relationship between the detection power and the sampling interval Δ for different specified values of T_r. These relationships are shown in Fig. 7.10. Here, the number of samples n is replaced by T/Δ, where T is the total observation period and Δ is the sampling frequency. Thus, by substituting n in Eqs. (7.18) and (7.20) by T/Δ, a direct relation between the detection power and sampling frequency is obtained. The figure is valid for both step and linear trends (T_r or T_r').

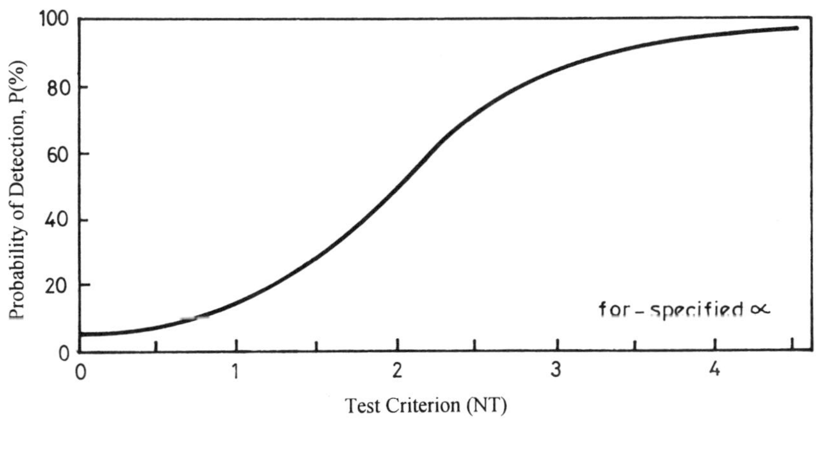

Figure 7.9. Relation between the detection power and the test criterion N_T (or N_T') (Schilperoort *et al.*,1982)

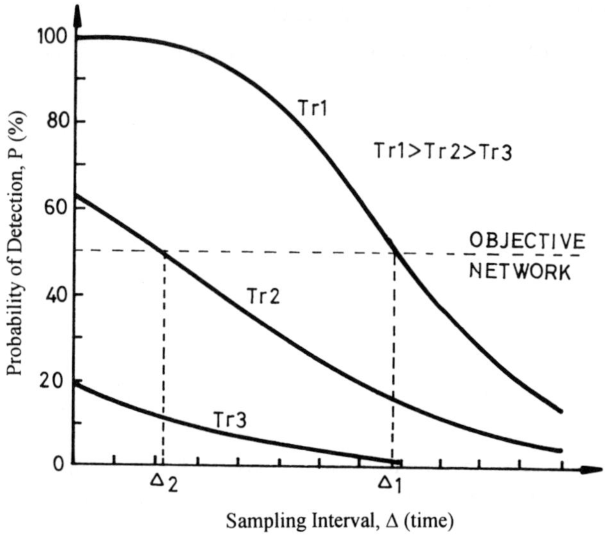

Figure 7.10. The probability of detection as a function of the sampling interval and the trend size (T_r or T_r')

For the case of dependent series, Lettenmaier (1976) replaces n by the effective sample size n_e which is a function of the dependence type. For an AR(1) process, n_e becomes:

$$n_e = n \frac{1 - \sigma_1}{1 + \sigma_1} \tag{7.22}$$

where σ_1 is the sample first-order autocorrelation coefficient. Then, the test criterion N_T' for dependent series can be computed as in Eq. (7.18) or Eq. (7.20) by using n_e instead of n. With the above method, Schilperoort *et al.*(1982) used the data from an existing network to investigate two issues:

a) which trend over a certain observation period (T) can be detected with the present sampling interval;

b) which sampling interval is necessary to detect a specified trend T_r (as a certain percentage of the mean concentration) over the period T.

Lettenmaier's technique on trend detection has the advantage that it is an objective-based approach to selection of sampling frequencies. Furthermore, it can be used for small sample sizes to determine what information the available data bring at particular levels of detection power. This technique is demonstrated on actual water quality data by Lettenmaier (1976) and Schilperoort *et al.*(1982). Their results show that the method works pretty well under the given assumptions.

7.5.3. APPLICATION OF PARAMETRIC TESTS TO MESSY WATER QUALITY DATA

Tokgoz (1992) applied Lettenmaier's method of trend detection to the same water quality data analyzed by Sanders' method in section 7.3. To investigate the applicability of the method, only linear trends were tested by defining the test criterion N_T' of Eq. (7.20) and using the Student parameter for hypothesis testing. The method was applied to two sets of variables, the first set comprising series with no serial dependence and the second with significant dependent components. For the first group, the total number of observations (n) was used in computing N_T', whereas for the second, effective sample sizes n_e were used instead of n to account for serial dependence. All tests were realized on two levels of significance at 90% and 95%. Furthermore, four trend levels T_r' were selected at 10%, 20%, 50% and 80% of the mean value for each variable.

The results of the study are presented in detail for Cl^- which represents the group of independent variables and for DO which represents the dependent variables. Figure 7.11 (a through d) shows the application in case of Cl^- for four specified levels of trends T_r' as 10% (a), 20% (b), 50% (c) and 80% (d). Each trend is investigated for two levels of significance as 90% (lower curve) and 95% (upper curve). Next, these four figures are merged as in Fig. 7.12 (a and b) to show the relationship between probability of detection and the sampling interval (Δ) for four levels of T_r' and at 90% and 95% levels of significance.

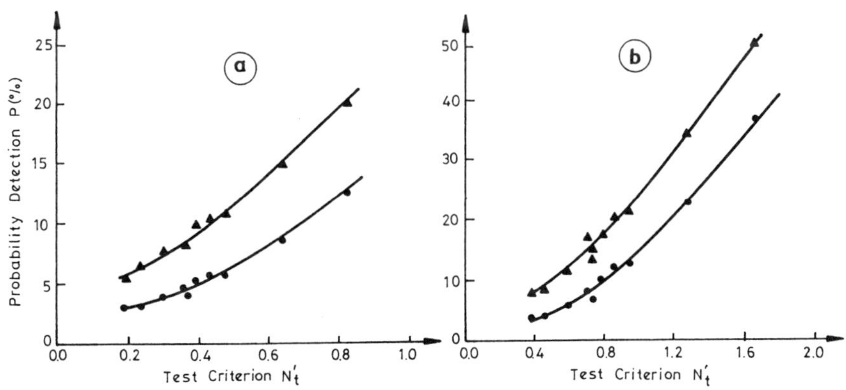

Figure 7.11. Relationship between probability of detection and the test criterion N_T' for Cl^- at 90% (lower curves) and 95% (upper curves) levels of significance for T_r': (a) 10%; (b) 20%; (c) 50%; (d) 80% of mean $\overline{x}$

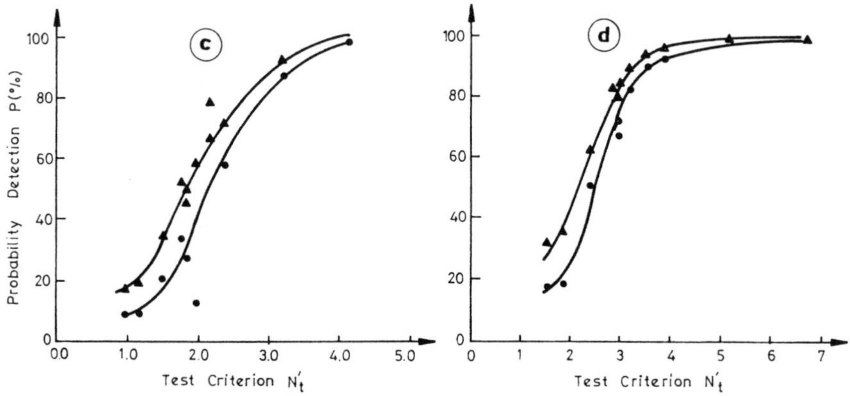

Figure 7.11. (Cont.)

The same analyses are carried out for the serially dependent DO where the test criterion N_T' is computed in terms of the effective sample size n_e. Figure 7.13 shows the relationships between the probability of detection and the test criterion at 90% and 95% levels of significance and for four levels of T_r' as done for Cl^-. Figure 7.14 further shows the relationships between the probability of detection at the sampling interval (Δ) again for different levels of specified trends.

Application of this method has been statistically more convenient and has not created significant difficulties due to shortage of data. The relationship between probability of detection and the test criterion N_T have produced fairly smooth curves both for independent variables as shown in Fig. 7.11 for Cl^-, and for dependent variables like DO as presented in Fig. 7.13. From these figures, one can infer on the probability of detection or the power of hypothesis testing with respect to the test criteria. In addition, this can be applied for various levels of specified trends. The curves of Fig. 7.11 and 7.13 agree well with the theoretical ones proposed by Lettenmaier (1976).

The basic advantage of the method is further observed in Figs. 7.12 and 7.14 where one can infer on the information produced by available data to evaluate the presence of particular trend levels. In these figures, it is possible to determine the probability of trend detection for different intervals. For example, in Fig. 7.12, one can observe that for detecting trends in the order of 10% of the mean value, the maximum probability of detection is attained at monthly intervals. However, this probability is still very low such that it is less than 20% at 90% level of significance and is about 20% at 95% level of significance. These results indicate the extent of information that can be produced by the existing data on Cl^- about particular trends.

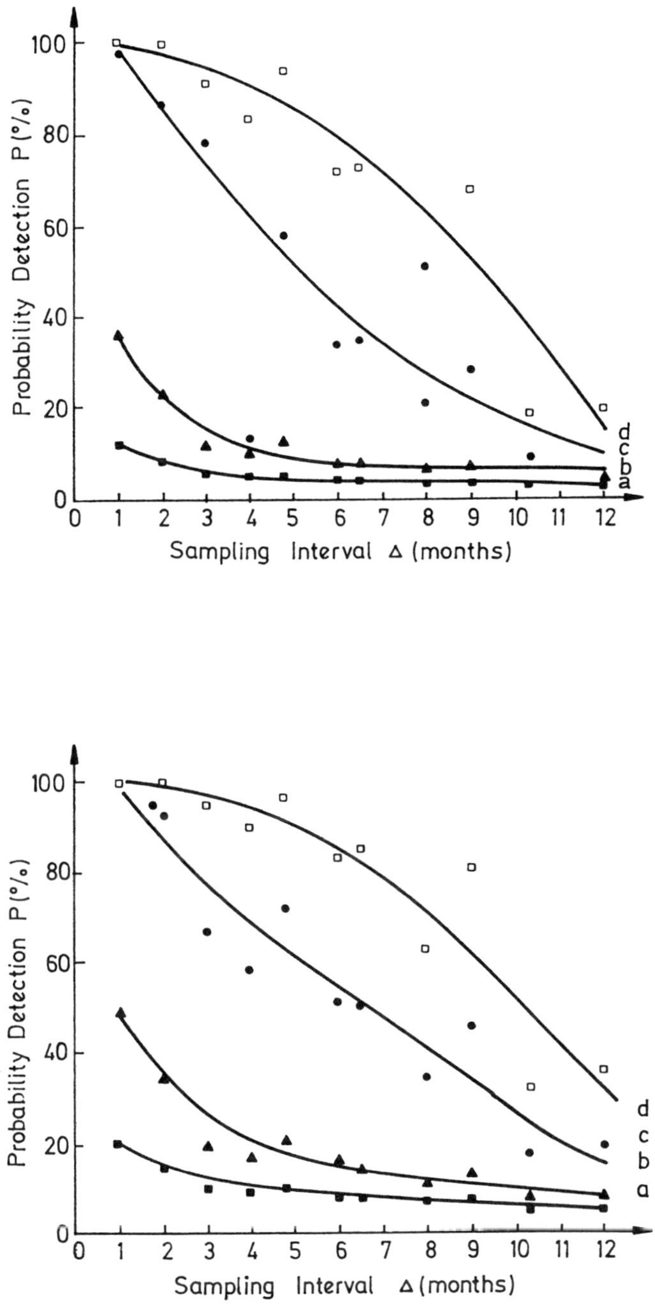

Figure 7.12. The probability of detection as a function of a sampling interval and the trend size at 90% and 95% levels of significance for T_r' as (a) 10%; (b) 20%; (c) 50%; and (d) 80% of mean Cl^- value

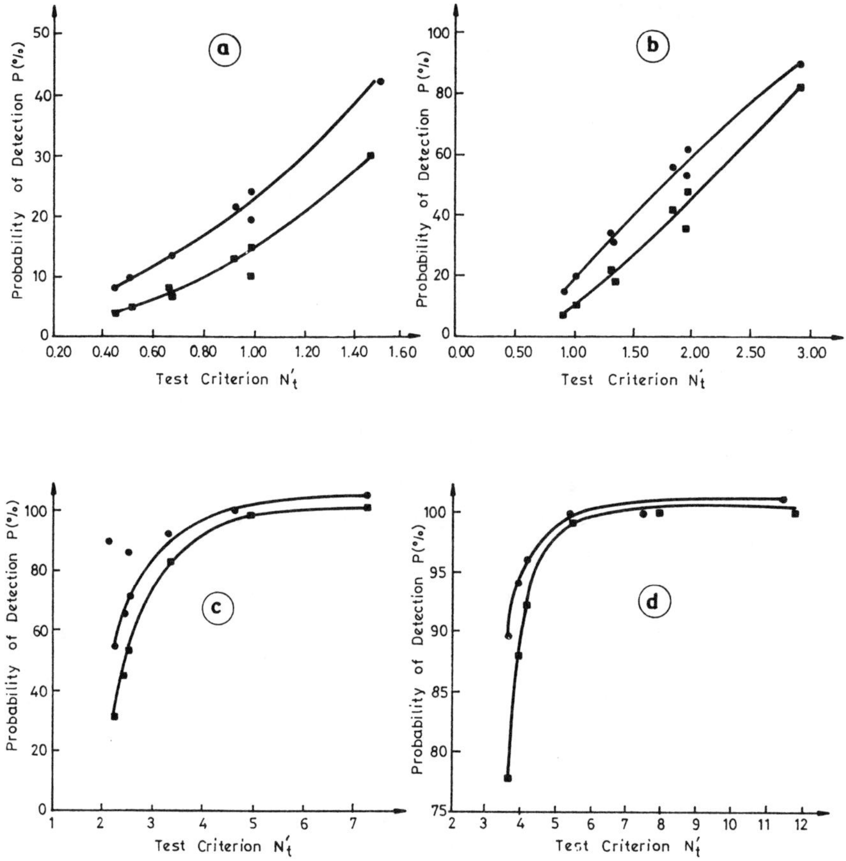

Figure 7.13. Relationship between probability of detection and the test criterion N_T' for DO at 90% (lower curves) and 95% (upper curves) levels of significance for T_r': (a) 10%; (b) 20%; (c) 50%; (d) 80% of mean $\overline{x}$

Probabilities of detection versus sampling interval show similar curves for all variables Tokgoz (1992) analyzed, except that the probabilities of detection are different for each. They seem to assume the highest values in case of DO only, as shown in Fig. 7.14. Thus, only for this variable, one may consider to increase the sampling intervals beyond monthly practices.

Lettenmaier's method appears to give a good fit to small sample sizes such as those available in the above the study. It has the advantage that is takes into consideration the utility of available data with respect to particular monitoring objective. Furthermore, one can infer, at particular levels of detection probability, about alternative sampling frequencies when the purpose is to detect specified trends. The general result for all

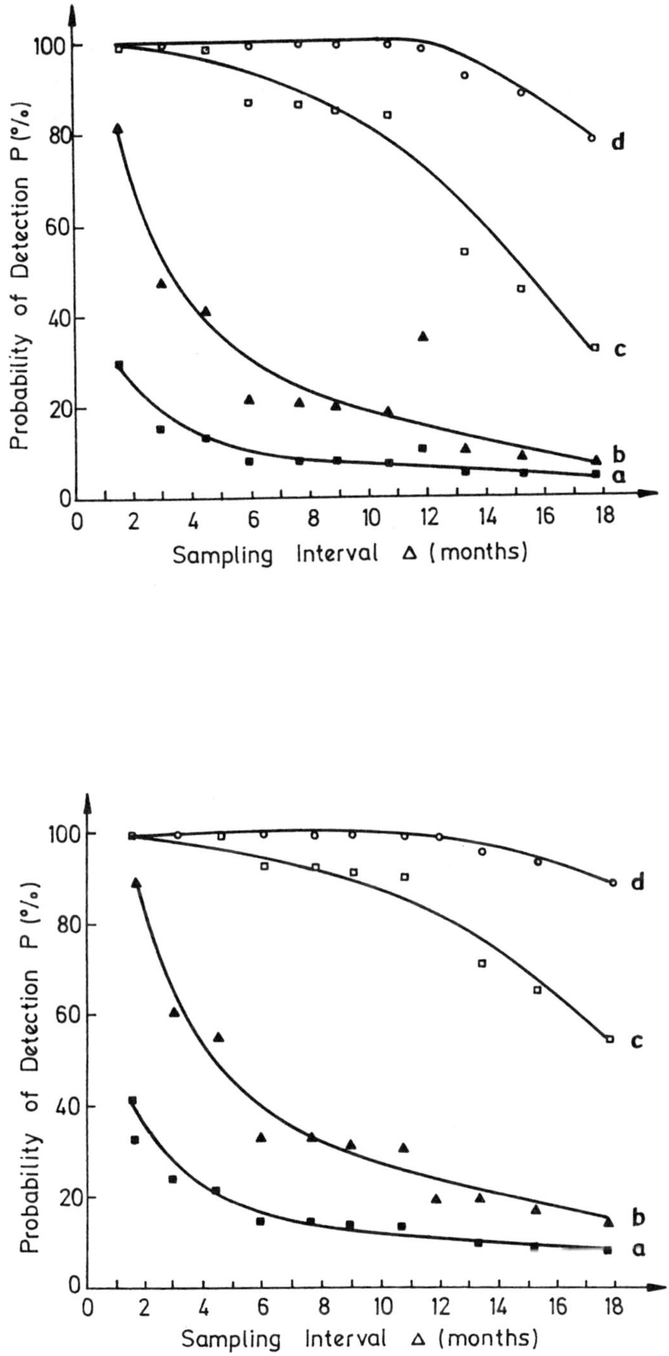

Figure 7.14. The probability of detection as a function of a sampling interval and the trend size at 90% and 95% levels of significance for T_r' as (a) 10%; (b) 20%; (c) 50%; and (d) 80% of mean DO value

variables analyzed by Tokgoz (1992) is that even monthly sampling intervals are not sufficient to detect trends in the order of 10 % or 20 % with an acceptable probability of detection. For larger trend values, the sampling intervals may be increased depending on the desired levels of detection probability. All these evaluations depend on the specified probability of detection and the basic objective of the network. Lettermaier (1976) proposes that the method is better fit to small sample sizes when nonparametric hypothesis testing approaches are used in determining the test criterion N_T^1. Although only the parametric tests are used here, the results are satisfactory in the sense that available data can be evaluated with respect to the information they produce.

On the other hand, the method is data dependent such that all statistical evaluations are based on the actual frequencies and sample sizes of available data. Thus, it is difficult, for example, to evaluate the case of frequent intervals such as daily or weekly when the existing sampling programs are based on monthly intervals.

It follows from the above that parametric trend tests based on the t-statistic have particular requirements, i.e., knowledge on the probability distribution of the process, independence of the observed series, and normality of the process.

Apart from the standard technique described, other parametric methods for trend assessment cover time series analysis and intervention analysis (Hipel *et al.*, 1975; Lettenmaier, 1988; Hipel and McLeod, 1994).

7.5.4. PROBLEMS ASSOCIATED WITH PARAMETRIC METHODS IN CASE OF WATER QUALITY

Many existing water quality databases have been found unsuitable for analysis by standard parametric methods as available data series do not fulfill the requirements of such methods. Basically, the application of traditional statistical techniques to spatially and temporally correlated, nonnormal water quality data is problematic. Other techniques developed for serially correlated data, such as time series and intervention analyses, are not suitable for water quality data because of missing data, censored data and changing laboratory techniques (Belle and Hughes, 1984).

Difficulties associated with water quality data in assessment of trends by parametric methods are described by various researchers as (Lettenmaier, 1988; Montgomery and Reckhow, 1984; Hirsch and Slack, 1984; Hirsch *et al.*, 1982; Berryman *et al.*, 1988):

a) there are usually many variables;
b) there are gaps and/or missing data in most of the records;
c) some concentrations are below detection limits and cannot be assigned a numerical value in the laboratory;
d) there is strong seasonality in some variables;
e) the data may be correlated in time and between constituents;
f) data records are often short;
g) techniques and sensitivities of analytical methods have changed over the years;
h) sampling location and frequency have changed in many cases;
i) natural background variability often hides water quality trends;

j) water quality time series often have nonnormal distributions;

k) water quality variables are often correlated with flow so that it is difficult to identify true trends in water quality.

The above characteristics often make trend detection in water quality data complicated because they do not comply with the assumptions of the classical methods. In particular, Hirsch and Slack (1984) define three basic reasons why the use of traditional techniques becomes problematic in case of water quality:

a) water quality data are often nonnormally distributed. Especially when the data sets are small, the tests for normality only reveal the most extreme violations;

b) there are missing values in most observed series. The parametric procedures for trend detection, used when serial dependence exists, require uniform sampling. Techniques are available to estimate missing values when they are few. However, if there are lots of missing values and long gaps, the effect of data filling on the identification of the stochastic process and trend testing leads to several problems;

c) water quality data are censored. When "less than detection limit" observations occur in a data set, parametric methods require substituting some numerical value for such data. However, such substitution makes the parametric test inexact and violates the assumption of normality.

It is due to the above problems that recent research on trend detection has proceeded in favor of nonparametric methods for the case of water quality variables.

7.5.5. NONPARAMETRIC METHODS

Recently, several nonparametric tests for trends in water quality have been proposed. These tests have been developed because the assumptions of classical parametric methods (i.e., normality, linearity, independence) are usually not met by water quality data. Additional properties of the data, such as missing values, censored data, and seasonality complicate the problem as discussed in section 7.5.4. The nonparametric methods are more flexible and can handle these problems more easily.

A nonparametric test is a method for testing a hypothesis where the test does not depend on the form of the underlying distibution of the null hypothesis. Therefore, nonparametric methods are sometimes referred to as distribution-free methods.

In response to the need for nonparametric procedures, authors like Lettenmaier (1976), Hirsch *et al.* (1982), Hirsch and Slack (1984), Van Belle and Hughes (1984) and a number of other researchers have made significant contributions to the development and application of nonparametric techniques in water resources. Research on this topic is still continued due to the wide range and great number of water quality problems encountered.

Lettenmaier (1976) claimed that Mann-Whitney's test for step trends and Spearman's rho test for linear trends perform very well in comparison to parametric t-tests. On the other hand, since these two tests also require independent data, Lettenmaier focused on detection of trends in water quality from data records with dependent observations. He considered that, for dependent time series, the power of the trend test varies with the form of the dependence of the observations. Accordingly, he developed a method of trend detection in case of dependent data by establishing an equivalence between power curves for dependent and independent observations.

Hirsch *et al.* (1982) presented techniques that are suitable in the presence of complications related to water quality data and proposed them for the exploratory analysis of monthly water quality data for monotonic trends. The first procedure they described is a nonparametric test for trend detection applicable to data sets with seasonality, missing, or censored values: the seasonal Kendall test. For stochastic processes with seasonality, skewness, and serial correlation, the seasonal Kendall test performs better than its parametric alternatives although it cannot be considered an exact test in the presence of serial dependence. The second procedure proposed by Hirsch *et al.* (1982) is an estimator of trend magnitude. It is an unbiased estimator of the slope of a linear trend and has a higher precision than a regression estimator where data are highly skewed. It gives lower precision in case of normally distributed series. The third procedure described by Hirsch *et al.* (1982) provides a means for testing for change over time in the relationship between water quality concentrations and flow, thus avoiding the problem of identifying trends in water quality that result from particular discharge series observed. In this method, a flow-adjusted concentration is defined as the residual based on a regression of concentration on some function of discharge. These flow-adjusted concentrations, which may also be seasonal and nonnormal, are then tested for trend by using the seasonal Kendall test (Hirsch *et al.*, 1982).

Van Belle and Hughes (1984) have analyzed the relative power of various nonparametric procedures. They considered two classes of techniques:

a) intrablock methods which compute a statistic, such as Kendall's tau, for each block or season and then sum these to produce a single overall statistic;
b) aligned rank methods which remove the block effect from each observed value, sum the data over blocks and then produce a statistic from these sums.

Van Belle and Hughes (1984) discussed that aligned rank methods are asymptotically more powerful than intrablock methods; yet intrablock methods are more adaptable and may be generalized to deal with a broad range of models.

Hirsch and Slack (1984) analyzed application of nonparametric trend tests for seasonal data with serial dependence and proposed an extension of the Mann-Kendall (seasonal Kendall) test for trend. They claimed that, since the test is based entirely on ranks, it performs well in case of nonnormal and censored data. Seasonality and missing values present no theoretical or computational problems in its application. Hirsch and Slack (1984) have shown that this modified test is valid in case of serial dependence except when the data have a strong long-term persistence or when sample sizes are small (e.g., 5 years of monthly data).

McLeod *et al.* (1983) discuss that there are two major steps in statistical analysis of trends. The first step is called "exploratory data analysis" where important properties of the data are delineated by simple graphical and numerical studies. These studies include graphs of data against time, Box-and-Whisker plots, Tukey smoothing, and the autocorrelation function. At this stage, McLeod *et al.* (1983) use a data filling procedure to produce evenly spaced data series from data observed at unequal time intervals. The next step of the analysis is called "confirmatory data analysis" where the purpose is to statistically confirm the presence or absence of trends. For this step, McLeod *et al.* (1983) purpose the use of intervention analysis method to determine if there has been a significant change in the mean of the series.

Montgomery and Reckhow (1984) also discuss that exploratory and confirmatory data analysis procedures should be applied to detect trends in water quality. Their trend detection methodology involves the following in a step-wise manner:

a) hypothesis formulation (statement of the problem to be tested);
b) data preparation (selection of water quality variables and data);
c) data analysis by exploratory techniques;
d) statistical tests (tests for detecting trends).

Lettenmaier *et al.* (1991) used the nonparametric seasonal Kendall's test and its multivariate extension to analyze 403 water quality monitoring stations in the U.S.A. for possible trends for the period 1978-1987. The results of their study showed that, for all groups and individual constitutents, trends were present only for a minority of stations at 10% significance level. Furthermore, analysis of possible relationships between trends and land use and population did not give strong evidence of possible causes.

Hirsch *et al.* (1991) reviewed in detail methods for the detection and estimation of trends in water quality. They considered that the steps involved in the selection of a trend detection method include:

a) determination of the type of trend hypothesis to analyze (step versus monotonic trend);
b) selection of the general category of statistical methods to use (parametric versus nonparametric);
c) selection of water quality data to analyze (concentration versus flux);
d) selection among various data manipulation alternatives related to the use of mathematical tranformations and the removal of natural sources of variability (discharge, seasonality) in water quality;
e) the choice of trend detection technique for water quality records with censored data.

With respect to (b), Hirsch et al (1991) discuss that parametric procedures for trend testing are regression in the case of a monotonic trend and the two sample t-test for step trends. In these methods, estimators of trend magnitude are the regression slope and the difference in the means. Nonparametric alternatives of these methods are the Mann-Kendall test and the Rank Sum test, respectively. Hirsch *et al.* (1991) further indicate that the decision as to which procedure should be used is based on considerations of power and efficiency of the test required by the available data. Power is the probability of selecting the null hypothesis (of no trend) given a particular type and magnitude of actual trend, and efficiency is a measure of estimation error. As indicated by Hirsch *et al.* (1991), a procedure's relative efficiency can be measured by the ratio of the mean square error of an alternative procedure to the mean square error of the particular procedure considered. Hirsch *et al.* (1991) discuss also that, for any significance level, the most powerful test is the parametric procedure if residuals are normally distributed. Similarly, the relative efficiency of these procedures is higher when residuals are normally distributed. In case of nonnormal water quality variables, Hirsch *et al.* (1991) propose the use of seasonal Kendall test for monotonic trends and Seasonal Rank Sum test for step trends.

El-Shaarawi and Damsleth (1988) claimed that ignoring serial dependence can have serious effects on the performance of the t, sign, and Wilcoxon tests. Accordingly, they presented modifications of these tests and further suggested an estimate of serial correlation for binary data.

Other studies on nonparametric methods include those by Hipel *et al.* (1988), who used Seasonal Mann-Kendall test to analyze trends in lake water quality; Hughes and Millard (1988) who suggested a tau-like test for trend in the presence of multiple censoring points; Lettenmaier (1988) who extended the use of nonparametric trend tests to the multivariate case; and Hirsch (1988) who investigated the magnitude of step trends in water quality by nonparametric tests.

Berryman *et al.* (1988) presented an extensive review of nonparametric tests as applied to water quality data and evaluated current studies on the subject. They further proposed a methodology on how to select the most appropriate test for a given time series.

According to Berryman *et al.* (1988), the common types of statistical methods used for trend detection and modeling can be summarized as in Table 7.3 together with the characteristics of water quality data that limit their use.

Berryman *et al.* (1988) indicated that time series analysis is difficult to use when observations are taken at irregular intervals. It is possible to model trends by time series analysis, but such a procedure does not, by itself, detect trends that are considered significant. Only graphical and statistical tests can be used to detect such trends. On the other hand, statistical tests can be used together with time series analysis. For example, tests that cannot be used on periodic series can be applied after seasonality is removed from the series by means of time series analysis models. However, Berryman *et al.* (1988) also note that recent developments in tests for water quality data allow trend detection in a great variety of water quality time series without having to decompose the series into its components before testing it for trend.

TABLE 7..3. Statistical methods for trend detection and modeling and the characteristics of water quality data that limit their use (Berryman *et al.*, 1988)

Objective	Statistical Methods	Data Characteristics Limiting the Use of the Statistical Method (*)
Modeling the series	Time series analysis	D, E
Detecting trends	Graphical methods (double-mass, Cusum)	A, B, C
	Parametric tests	A, B, C, E
	Nonparametric tests	A

(*) A : dependence of the error terms.
B : error terms with nonnormal distributions.
C : error terms with nonconstant variance.
D : observations irregularly spaced in time.
E : observations below the limit of detection.

According to Berryman *et al.* (1988), only graphical methods and statistical tests can be used to detect significant trends. The rule here is to consider a trend as significant when its magnitude is large, compared to the variance of the process, so that the probability of its occurrence by chance only is minimal. Usually, a trend is considered significant when its probability of occurrence only by chance is below 5%.

Berryman *et al.* (1988) emphasize that parametric tests for trend require observations for which the error terms are independent, have constant variance, and come from normal populations. Most nonparametric methods have only the first of these requirements. In case of nonnormal distributions, parametric tests can be less powerful than their nonparametric alternatives. Berryman *et al.* (1988) have shown that for some nonnormal distributions, nonparametric tests are more powerful than their parametric counterparts, while for normal distributions, they are almost as powerful. Accordingly, the use of nonparametric tests is advised by Berryman *et al.* (1988), as well as by other researchers discussed above, whenever the normality of the population's distribution is doubtful.

Berryman *et al.* (1988) listed 12 tests for monotonic trends (e.g., Kendall, Spearman, intrablock tests, aligned ranks tests), 7 tests for step trends (e.g., median, Mann-Whitney, Kolmogorov-Smirnov), and 3 tests for multistep trends (e.g., Kruskall-Wallis). Among these, Spearman's and Kendall's are the most powerful when time series do not contain seasonal variations. Intrablock and aligned ranks tests can be used when data are affected by cycles. In intrablock tests, the data are blocked into seasons or months where the seasonality effect is homogenous. Then, all blocks are subjected to "treatments" that are given values of the independent variable, which is time (Berryman *et al.* 1988). Intrablock and aligned rank tests measure the relationship between time and the variable analyzed.

Berryman *et al.* (1988) have concluded that Mann-Whitney, Spearman and Kendall tests are the best methods for trend detection in water quality time series. Among these, Mann-Whitney is the most widely used two-sample test when the assumptions of its parametric equivalent, the t-test, are not met. Berryman *et al.* (1988) have finally developed a list of nonparametric tests that are appropriate for water quality time series. These methods are summarized in the following section.

7.5.6. NONPARAMETRIC TESTS PROPOSED FOR WATER QUALITY TIME SERIES

According to Berryman *et al.* (1988), nonparametric tests are good tools to detect trends in water quality data; however, this can be complicated by serial dependence which often occurs in water quality observations. Such serial dependence in due to three factors: seasonal variations, persistence, and monotonic trends. In recent years, nonparametric methods have been proposed to account for seasonal variations and persistence. Intrablock and aligned rank techniques were applied to series with seasonality (Hirsch *et al.*, 1982; Van Belle and Hughes, 1984); furthermore, modifications to the Spearman, Mann-Whitney and the intrablock tests were proposed to account for persistence (Lettenmaier, 1976; Hirsch and Slack, 1984). Berryman *et al.* (1988) have developed a list nonparametric methods proposed by various researchers. This list is presented in Table 7.4.

TABLE 7.4. Proposed nonparametric tests for detection of trends in water quality time series (Berryman *et al.*, 1988)

Trend Type	No. Of Stations	Series Characteristics: Persistence	Series Characteristics: Periodicity	Test	Adaptation in Water Quality Reference	N Minimum (*)
Step	1	No	No	Mann-Whitney	-	5 (**)
	1	Yes, Markovian order 1	No	Mann-Whitney	Lettenmaier (1976)	20
	1	No	Yes	Intrablock test	Berryman (1984a)	12
Monotonic	1	No	No	Spearman	-	11
	1	No	No	Kendall	-	9
	1	Yes, Markovian order 1	No	Spearman	Lettenmaier (1976)	20
	1	No	Yes	Intrablock	Hirsch *et al.* (1982)	24 (2 years, 12 months)
	1	No	Yes	Aligned rank test	Farrell (1980)	20 (5 years, 4 seasons)
	1	Yes	Yes	Intrablock for persistent data	Hirsch and Slack (1984)	120 (10 years, 12 months)
	>1	No	Yes	Analysis of variance and chi-square test	Van Belle and Hughes (1984)	240 (10 years, 12 months, 2 stations)
Homogeneity of monthly trends	1	No	Yes	Chi-square	Van belle and Hughes (1984)	120 (10 years, 12 months)
	>1	No	Yes	Analysis of variance and chi-square test	Van Belle and Hughes (1984)	240 (10 years, 12 months, 2 stations)

(*) : Minimum number of observations to apply method.
(**) : Ten observations are needed to conduct a test with $\alpha = 0.05$, and five are required if $\alpha = 0.1$.

Table 7.4 includes the techniques proposed so far by researchers for water quality trend analyses. The list includes the test proposed by Van Belle and Hughes (1984) to investigate the homogeneity of monthly (or seasonal) trends. Such tests have been used to show how nonparametric analysis of variance coupled with the chi-square test allows the evaluation of homogeneity of trends between sites, between seasons, and between site-seasons. Depending on the results of this analysis, data can then be tested for within season trend, within site trend, within site-season trend, or for overall trend (all stations and all seasons).

On the basis of Table 7.4, Berryman *et al.* (1988) have indicated that many tests are available for water quality trend analysis, but that they are applicable only to specific types of trend and time series. Accordingly, Berryman *et al.* (1988) proposed a methodology for selecting the appropriate test for given series.

According to Berryman *et al.* (1988), the choice of a test depends on the type of trend to detect and on the series characteristics. The objective of the analysis or a simple graphical review of data usually indicate the kind of trend to be investigated. If there is still doubt, then both step and monotonic trends can be investigated.

Dependence and the sources of dependence (trend, periodicity, or persistence in random series) are the most important characteristics used for the selection of the appropriate test. The dependence and the sources of dependence are investigated by first removing the effect of trend, then removing seasonality and by checking for residual dependence after each of these operations. Reduced dependence after deseasonalization indicates periodicity, and significant dependence remaining after this operation indicates persistence. This approach proposed by Berryman *et al.* (1988) is formulated into a flow chart as shown in Fig. 7.15. The methodology in the figure is given for monotonic trends; however, a similar approach can be used to select the appropriate test for step trend detection.

The approach summarized in Fig. 7.15, based on the identification of the sources of serial dependence in the time series, is developed as a general procedure by Berryman *et al.* (1988), who indicated that their approach is not a strict one. Thus, they essentially left the solution to the user's judgment. Berryman *et al.* (1988) also emphasized that, in a trend analysis, several tests should be used and compared since understanding of periodicity and persistence is significant and since the measures of these statistical characteristics are somewhat subjective.

Icaga (1994) and Icaga and Harmancioglu (1995) have applied the approach proposed by Berryman *et al.* (1988) to series of water quality data from several basins in Turkey. Figure 7.16 shows in a stepwise manner the basic procedure used in their studies. The procedure starts with the selection of water quality data to be tested (step 1) and their visual inspection via plots of raw data (step 2). Plots of data with respect to time provide information on both the presence of trends and the series properties such as seasonality. The next step involves flow adjustment (step 3), followed by exploratory data analysis (steps 4, 5, 6). Data analysis includes three steps to approximate trends, identify seasonality and to analyze the presence of serial dependence. The results of these steps indicate the type of nonparametric test technique to be selected. The methods shown in step 7 of Fig. 7.16 are applied on the basis of series properties of selected water quality data.

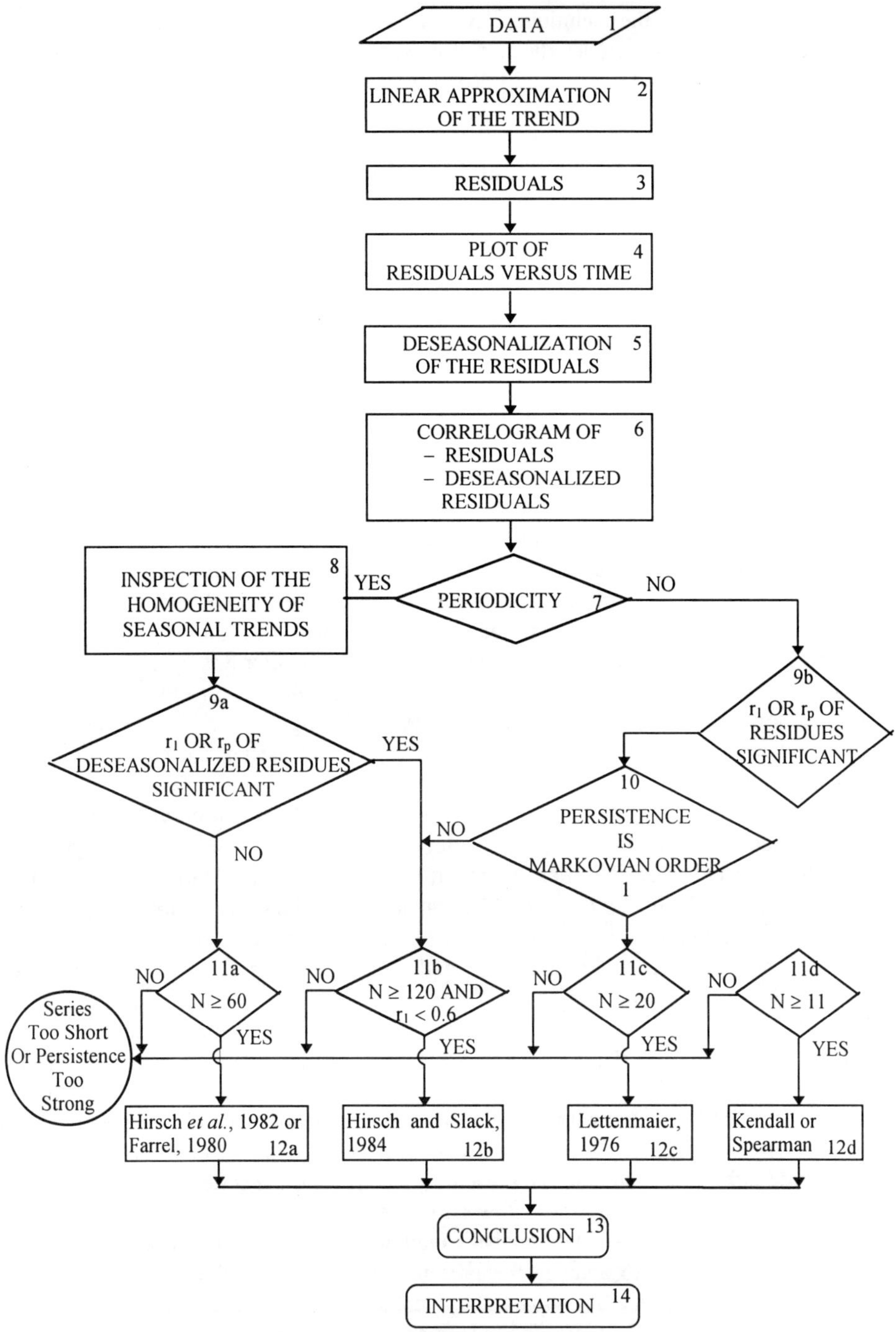

Figure 7.15. Flowchart for the selection of the appropriate statistical test for monotonic trends, proposed by Berryman *et al.* (1988)

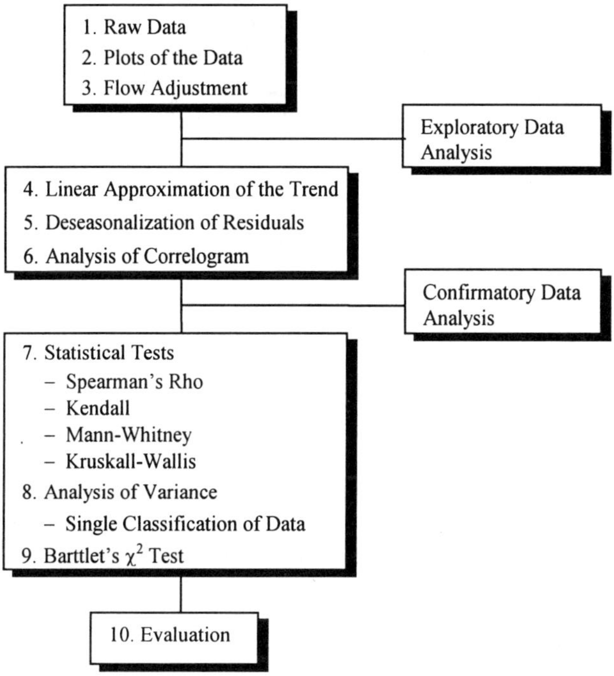

Figure 7.16. Basic steps used by Icaga (1994) and Icaga and Harmancioglu (1995) for detection of trends in water quality data with short sample sizes and significant numbers of missing values

The results of the studies by Icaga (1994) and Icaga and Harmancioglu (1995) have shown that there are significant difficulties in the first two steps of trend analysis, i.e. flow adjustment and exploratory analyses (i.e., linear approximation of the trend, modeling of periodicity and serial dependence). These difficulties have stemmed again from the messy character of available data, i.e., short sample sizes and significant numbers of missing values and gaps in data records. For flow adjustment, the problem is often that simultaneous measurements of discharge and water quality variables are not taken; or, if they are, the basic problem lies in the short duration of records such that a reliable relationship cannot be developed between discharge and water quality. Description of series properties such as periodicity and serial dependence have been found to be problematic due to the sporadic nature of available data on water quality variables. Nevertheless, the approach proposed by Berryman *et al.* (1988) may be satisfactorily used in case of regularly collected data if made available by consistent monitoring practices.

7.6. Data Correlations

7.6.1. ANALYSIS BY REGRESSION AND ENTROPY-BASED MEASURES

Information on water quality may also be procured by transferring information among water quantity and water quality, or among water quality variables themselves. Transfer of information may be carried out in time or in space. Yevjevich and Harmancioglu (1985) and Harmancioglu and Yevjevich (1986) investigated such information transfers by bivariate correlations between water quality variables observed along the Upper Potomac River Estuary. The results of their studies have shown that only a few variables have significant correlations with discharge. Furthermore, they have also delineated that information transfer between water quality variables themselves is pretty poor. Harmancioglu and Yevjevich (1986) used the informational entropy concept in their study to define the amount of transferable information to be compared with the amount of information actually transferred by regression. Such comparisons have shown that it is basically such series properties as periodicity and dependence that account for the relationship between water quality variables. Otherwise, the correlations between the random components of water quality time series are poor.

For the above studies on the Upper Potomac River Estuary, four different water quality data sets were used: WAQ (data of the Washington Aquaduct Division of the U.S. Army Corps of Engineers), USGS-PES (data collected by USGS as part of the Potomac Estuary Study), DES (data collected by the Department of Environmental Sources, Washington, D.C.) and STORET (water quality data of U.S. EPA). These data sets differ significantly in accuracy, frequency of observation, length and period of records, variables observed and the units reported (Harmancioglu and Yevjevich, 1986). The water flow of the Potomac River is systematically observed. However, in almost all data sets used, observations of water quality variables were highly irregular with respect to time. The data requirements for the types on analyses foreseen for the study included water quality inputs as short interval time series. Thus, all data were used as daily series by special care paid to missing daily observations.

The WAQ set was analyzed for the extent of information transfer among water quality variables by means of bivariate correlation analyses (Yevjevich and Harmancioglu, 1985). The purpose of this study was to investigate whether the number of variables to be observed could be reduced if significant relationships existed among the variables.

The WAQ data set comprised observations of 12 water quality variables for the period between December 1964 and February 1984, with the majority of the variables observed daily except on weekends. The WAQ data set may be considered to be an excellent set since it has considerably few missing values.

Table 7.5 presents the correlation coefficients, r, of both linear (first lines in each row) and nonlinear (the second lines) regression analyses between the observed series of WAQ variables. The figures here indicate that the correlation coefficients are, in general, small and that the application of nonlinear regression does not improve this result significantly. Among the 66 pairs of variables considered, only 5 pairs have correlation coefficients above 0.50. 2 pairs, pH-CO_2 and alkalinity (MO)-total hardness (TOT) show highly correlations in the order of 0.80.

Table 7.6 presents the coefficients of determination for the linear (first lines of each row) and nonlinear (second lines of each row) bivariate regressions between the observed values of water quality variables. The figures for the linear regression indicate that the extent of information transferred in this case is pretty low, or even negligible for most of the variable pairs. Only two pairs as in Table 7.5, MO-TOT and pH-CO_2, seem to permit information transfer up to 64%.

TABLE 7.5. Correlation coefficients, r, of linear and nonlinear regression between original WAQ data

VAR	TURB	CLD	MO	NCH	TOT	PH	$C0_2$	DO	COD	BOD_5	$N0_2$	NO_3
TURB	1.0000	0.5462	-0.3766	-0.1328	-0.3700	-0.2722	0.2823	-0.0141	0.3571	0.1325	-0.0409	0.0446
	1.0000	0.4696	-0.4723	-0.1883	-0.5218	-0.3078	0.3204	-0.0910	0.3726	0.0989	0.0619	0.0256
CLD		1.0000	-0.1222	0.2387	0.0349	-0.2596	0.2598	-0.1736	0.3704	0.2886	0.0107	-0.2485
		1.0000	-0.1005	0.2338	0.0590	-0.2184	0.1805	-0.1755	0.3535	0.2707	0.1409	-0.3428
MO			1.0000	0.2024	0.7995	0.5427	-0.2902	-0.1856	-0.0616	0.0269	0.0040	-0.2514
			1.0000	0.1540	0.8028	0.5719	-0.3226	-0.1858	-0.0312	0.0310	-0.0828	-0.2654
NCH				1.0000	0.6932	0.0775	-0.0003	-0.0420	-0.1434	0.0939	-0.2776	-0.2993
				1.0000	0.6289	0.1011	-0.0539	-0.0225	-0.1476	0.1104	-0.2823	-0.3115
TOT					1.0000	0.4029	-0.2151	-0.1626	-0.1364	0.0766	0.0960	-0.2101
					1.0000	0.4401	-0.2648	-0.1494	-0.1230	0.0711	-0.0152	-0.2586
PH						1.0000	-0.8031	-0.1546	-0.0491	0.0255	0.0448	-0.2765
						1.0000	-0.8435	-0.1121	0.0211	0.0497	-0.1292	-0.2765
$C0_2$							1.0000	0.0756	0.0157	0.0210	0.0326	0.1085
							1.0000	0.0717	-0.0040	-0.0873	0.0918	0.1760
DO								1.0000	-0.4777	0.1037	-0.0927	0.3911
								1.0000	-0.5184	0.0957	-0.0502	0.4278
COD									1.0000	0.1867	0.0481	-0.3269
									1.0000	0.0161	0.3022	-0.1303
BOD_5										1.0000	0.0914	-0.0490
										1.0000	0.1391	-0.0626
$N0_2$											1.0000	0.0919
											1.0000	0.1593
$N0_3$												1.0000
												1.0000

TABLE 7.6. Coefficients of determination, r^2, for linear and non linear regressions between original values of the WAQ variables

VAR	TURB	CLD	MO	NCH	TOT	PH	$C0_2$	DO	COD	BOD_5	$N0_2$	NO_3
TURB	1.0000	0.2984	0.1418	0.0176	0.1369	0.0741	0.0797	0.0002	0.1275	0.0176	0.0017	1.0020
	1.0000	0.2205	0.2231	0.0355	0.2722	0.0947	0.1027	0.0083	0.1388	0.0098	0.0038	0.0006
CLD		1.0000	0.0149	0.0570	0.0012	0.0674	0.0675	0.0301	0.1372	0.0833	0.0001	0.0618
		1.0000	0.0101	0.0546	0.0035	0.0477	0.0326	0.0308	0.1249	0.0733	0.0198	0.1175
MO			1.0000	0.0410	0.6391	0.2945	0.0842	0.0344	0.0038	0.0007	0.0000	0.0632
			1.0000	0.0237	0.6445	0.3271	0.1040	0.0345	0.0010	0.0002	0.0069	0.0704
NCH				1.0000	0.4805	0.0060	0.0000	0.0018	0.0206	0.0088	0.0771	0.0896
				1.0000	0.3955	0.0102	0.0029	0.0005	0.0218	0.0122	0.0797	0.0970
TOT					1.0000	0.1623	0.0463	0.0264	0.0186	0.0059	0.0092	0.0441
					1.0000	0.1937	0.0701	0.0223	0.0151	0.0051	0.0002	0.0668
pH						1.0000	0.6450	0.0239	0.0024	0.0006	0.0020	0.0764
						1.0000	0.7115	0.0126	0.0004	0.0025	0.0167	0.0765
CO_2							1.0000	0.0057	0.0002	0.0004	0.0011	0.0118
							1.0000	0.0051	0.0000	0.0076	0.0084	0.0310
DO								1.0000	0.2282	0.0108	0.0086	0.1529
								1.0000	0.2687	0.0092	0.0025	0.1830
COD									1.0000	0.0348	0.0023	0.1069
									1.0000	0.0003	0.0913	0.0170
BOD_5										1.0000	0.0084	0.0024
										1.0000	0.0194	0.0039
NO_2											1.0000	0.0084
											1.0000	0.0254
NO_3												1.0000
												1.0000

Considering the total of 66 pairs of variables analyzed, the average r^2 for linear regression is found to be 0.0747, as an overall representative figure. This average becomes 0.5882 when only the three highest r^2 values (those for NO-TOT, NCH-TOT, and pH-CO_2) are considered. The same figures are 0.0806 and 0.5838, respectively, for nonlinear regression. These results are confirmed by the entropy-based measures of information, not presented here.

It was concluded then that the number of variables in the water quality data set could not be reduced by estimating the values of one variable via its relationship with any other variable. In this case, all variables need to be monitored if, according to other considerations, information is required on each process.

The time series of water quantity and quality of the other data sets were also investigated for crosscorrelations to determine the amounts of transferred information between the pairs of variables. Linear and nonlinear (power function nonlinearity) regression was used for transfer of information. The water quality time series were also analyzed for their structural properties, i.e., trends, seasonality, and serial correlations. Correlation analyses were carried out after each step of structural analysis to find out what structural components accounted for correlations between the variables. The coefficient of determination, r^2, was used as the measure of transferred information by regression. Transferable information was measured by the entropy coefficient of information R_0 defined as:

$$R_0 = \sqrt{1 - e^{-2T_0}} \tag{7.23}$$

where T_0 is the transinformation (mutual information) representing the upper limit of transferable information between two or more variables (Harmancioglu and Yevjevich, 1987). Transinformation is an entropy-based measure of common information between two discrete random variables X and Y and is defined as:

$$\begin{aligned} T(X,Y) = {} & \sum_{i=1}^{n} p(x_i) \log \frac{1}{p(x_i)} + \sum_{j=1}^{n} p(y_j) \log \frac{1}{p(y_j)} \\ & - \sum_{i=1}^{n} \sum_{j=1}^{n} p(x_i, y_j) \log \frac{1}{p(x_i, y_j)} \end{aligned} \tag{7.24}$$

with $p(x_i)$ and $p(y_i)$ representing the probabilities of events x_i and y_i, and $p(x_i, y_i)$, the joint probabilities of X and Y.

If probability values in Eq. (7.24) are estimated from samples by the corresponding frequencies (Harmancioglu *et al.*, 1986), the transinformation T_0 obtained represents the upper limit of transferable information between the variables.

T_0 and the corresponding R_0^2 values were computed for water quality variables of the three data sets after each step of structural analysis, i.e., after removing trends, seasonality, and serial dependence in a sequential order. Comparisons between r^2 and R_0^2 revealed how much information the assumed regression function transferred with respect to the upper limit of transferable information. Another measure used to evaluate the rate of information transfer is the t_i ratio:

$$t_i = \frac{T_0 - T_1}{T_0} \tag{7.25}$$

with T_1 representing the transinformation defined for the particular type of regression:

$$T_1 = -\frac{1}{2} \ln\left(1 - r^2\right) \tag{7.26}$$

t_i basically described the relative T_0 - T_1 portion of untransferred information.

7.6.2. ANALYSIS OF USGS-PES DATA SET

The general result of analyses of the USGS-PES data set was that only four pairs out of 36 showed significant correlation between the original series. Removal of deterministic components, particularly serial dependence, reduced correlation so much that independent series appear to permit only a very low transfer of information. When the average of 36 pairs are considered, r^2 drops from 0.16 for the original data to 0.12 for the deseasonalized and to 0.05 for the independent series. The overall figure is 0.11 for the USGS-PES set. When the average of the highest correlation for three pairs are considered, the respective values of r^2 are 0.61, 0.48 and 0.25 with an overall average value of 0.45. These results are shown in Fig. 7.17.

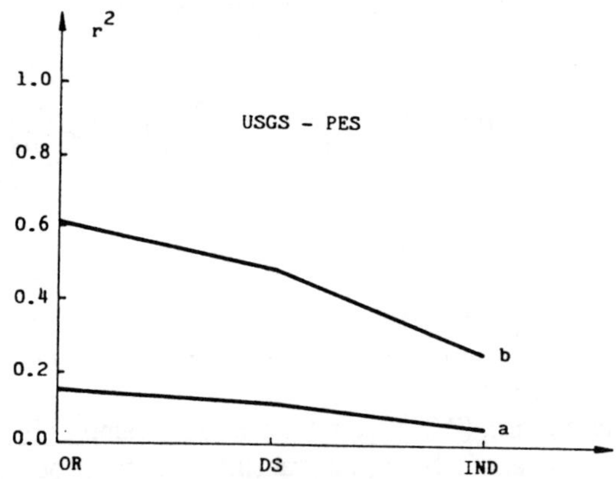

Figure 7.17. Average coefficients of determination r^2 for the pairs of water quality variables of the Potomac River, the USGS-PES data set: (a) averages for the 32 pairs, and (b) averages for three pairs with the highest correlation (for three cases: OR-original, DS-deseasonalizated and IND-independent data)

If the classical criterion that r^2 should be equal to or greater than 0.50 is assumed, only four pairs of variables, DISCHARGE-CONDUCT, DISCHARGE-SUSSED, SUSSED-TPHOS and TPHOS-TKJELN, of the USGS-PES set pass this test when the original data are used. As shown in Fig. 7.18, nonlinear regression proves to be better for the first pair, whereas linear regression transfers more information in the case of the other three pairs. The remaining 32 pairs of variables of the USGS-PES set do not seem to permit any significant transfer of information. Removal of seasonality in series parameters does not change this overall result, that is, r^2 remains below 0.50 although it shows insignificant increases or decreases for each pair. Particularly for the four pairs of variables mentioned, the result is that r^2 for DISCHARGE-CONDUCT reduces from 0.68 to 0.14 after deseasonalization. The other three pairs practically preserve the same extent of transferred information at this step as they do for their original series. However, removal of stochastic dependence from the deseasonalized series significantly reduces the correlations for all pairs. Even for the latter two pairs mentioned, r^2 is only in the order of 0.25.

The entropy coefficients of information, R_0^2, for the original USGS-PES data indicate transferability of information in the order of 0.50 to 0.77, the highest figures being obtained for SUSSED-TPHOS, DISCHARGE-CONDUCT and DISCHARGE-SUSSED. For the majority of the remaining variable pairs, R_0^2 is below 0.50. Figure 7.18 shows also the R_0^2 values for the four distinctive pairs of the USGS set. At the second step of structural analysis where the series are deseasonalized, the extent of transferable information reduces for the majority of the variables. As for the four pairs in Fig. 7.18, it practically remains the same for SUSSED-TPHOS and TPHOS-TKJELN, but decreases for DISCHARGE-SUSSED and more so for DISCHARGE-CONDUCT. At the next step, transferable information shows an increase for the independent components of variable pairs. Since r^2 falls far below R_0^2 at this step, attempts may be made to improve the regressions, especially for those pairs which indicate high transferability of information. As for the four distinctive pairs, R_0^2 increases for DISCHARGE-CONDUCT and TPHOS-TKJELN, but decreases for DISCHARGE-SUSSED and SUSSED-TPHOS. Yet, in each case, Fig. 7.18 shows considerable difference between r^2 and R_0^2. It may then be worthy of effort, especially for TPHOS-TKJELN and SUSSED-TPHOS, to improve the assumed means of transferring information.

The extent of information transferability by the assumed regressions is measured by the ratio r^2/R_0^2. This ratio varies from 0 to 90% for the original USGS variables. Deseasonalization reduces this ratio for the majority of the variables. Finally removal of dependence leads to very small amounts of transferred information measured by r^2/R_0^2 or by $1\text{-}t_i$ ratio. Figure 7.19 shows the results for the four distinctive pairs which originally reflect high percentages of transferred information, as for SUSSED-TPHOS with r^2/R_0^2 of 99%. These figures are significantly reduced for the independent series. Only in the case of DISCHARGE-SUSSED, the deseasonalized series permit 100% of information transfer.

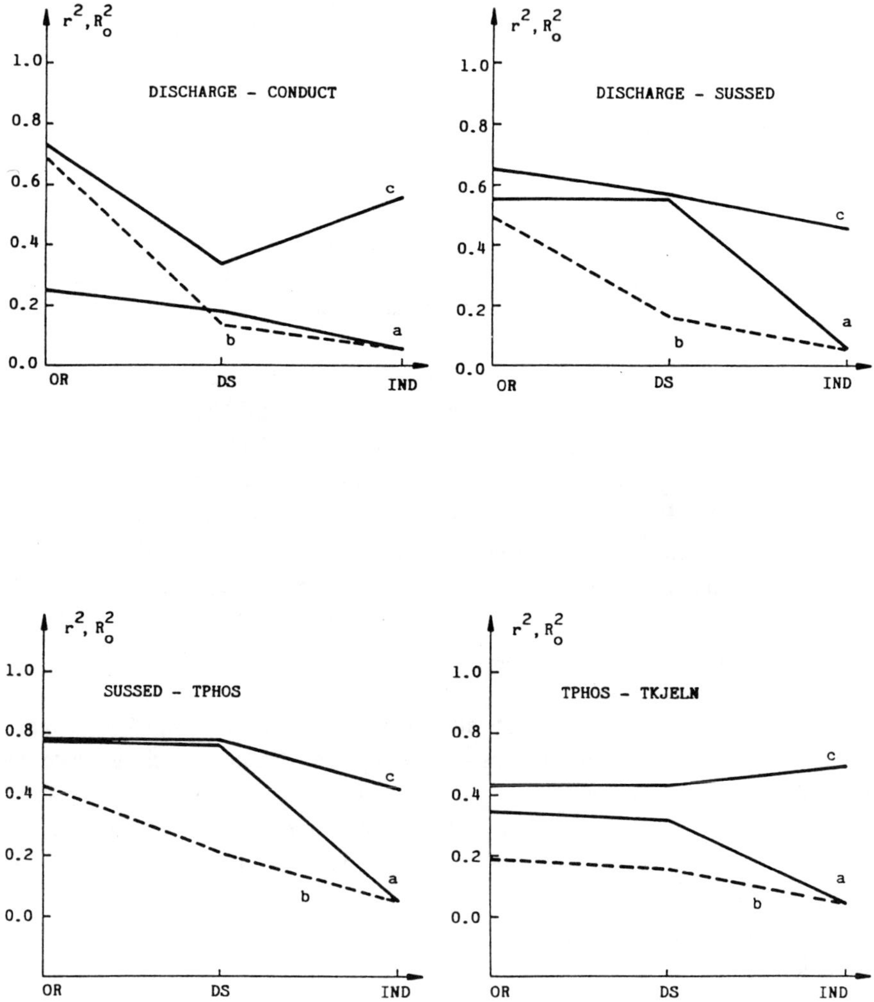

Figure 7.18. Square of correlation coefficients r^2, line (a), and square of entropy informational coefficient R^2, line (b), for the correlation between four pairs of variables in the USGS-PES data set (OR-original, DS-deseasonalizated and IND-independent data)

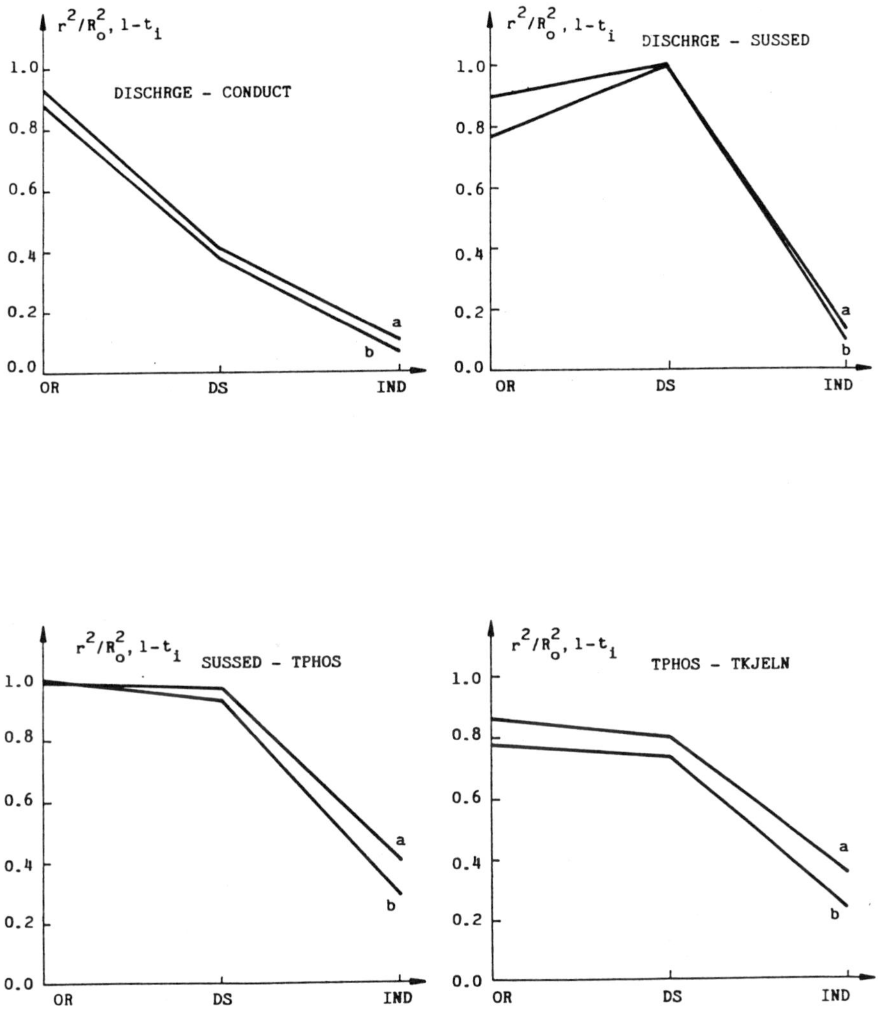

Figure 7.19. The ratio r^2/R_0^2, line (a), and the measure $1\text{-}t_i$, line (b), of transferred to transferable information, for the pairs of USGS-PES data set of water quality variables of the Potomac River (OR-original, DT- detrended and DS-deseasonalizated data)

7.6.3. ANALYSIS OF DES DATA SET

Similar analyses carried out for the DES data set shows that none of the DES variables, except TEMP-DO, appear to permit any transfer of information between each other. Yet, as shown in Fig. 7.20, the fairly high value of r^2 (0.79) obtained for linear regression between the original TEMP and DO series decreases very fast to 0.20 when the two variables are detrended. This indicates that parallel trends in the original series account for the correlation between the two. Removal of seasonal patterns causes further reduction in the detected correlation. At this point, the extent of information transferred between independent TEMP-DO cannot be investigated, since this step of structural analysis has been impossible to carry out due to the sporadicity of observations. Other than this pair, it may be stated in general that the DES variables do not permit any transfer of information. This fact is reflected in the r^2 values when expressed as an average of the 45 pairs in the data set. Figure 7.21 shows these average values which lead to an overall representative value of 0.04 for the whole set.

Transferable information R_0^2 at any step is similarly found to be too low for the DES variables. Only for the TEMP-DO pair of Fig. 7.21, R_0^2 is as high as 83% at the original state where linear regression transfers $r^2/R_0^2 = 95\%$ of transferable information (Fig. 7.22). At further steps of structural analysis, Fig. 7.20 shows decreasing R_0^2 and Fig. 7.22 shows decreasing r^2/R_0^2 and $1-t_i$ values. For the other variable pairs, R_0^2 indicates practically no transferability of information. Furthermore, the amounts of transferred information indicated by r^2/R_0^2 are pretty low at every step of structural analysis. In this case, it doesn't appear to be worthy of effort to try an improvement for the applied regressions.

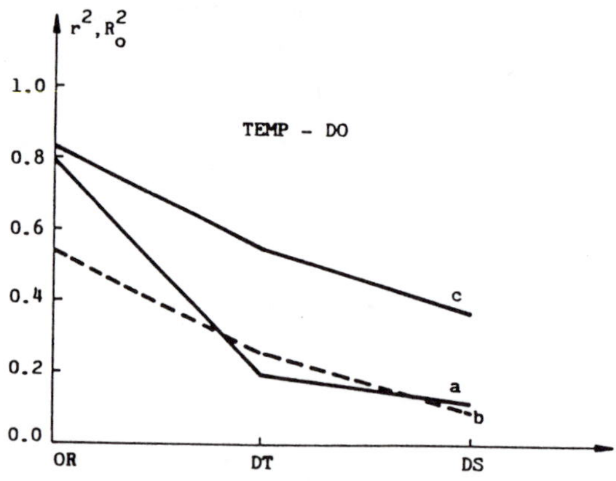

Figure 7.20. Square of correlation coefficients r^2, line (a), and square of entropy informational coefficient R_0^2, line (b), for the correlation between TEMP-DO of the DES data set (OR-original, DT- detrended and DS-deseasonalizated data)

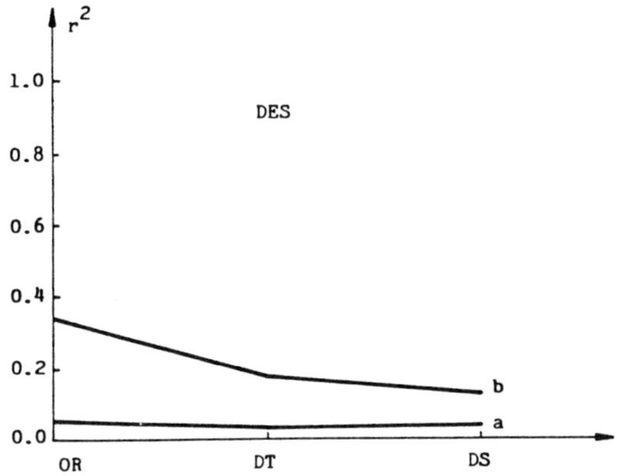

Figure 7.21. Average square of correlation coefficients r^2 of the pairs of water quality variables of the Potomac River, the DES data set: (a), averages for the 45 pairs; and (b) averages of the highest correlation pairs (OR-original, DS-deseasonalizated and IND- Independent data)

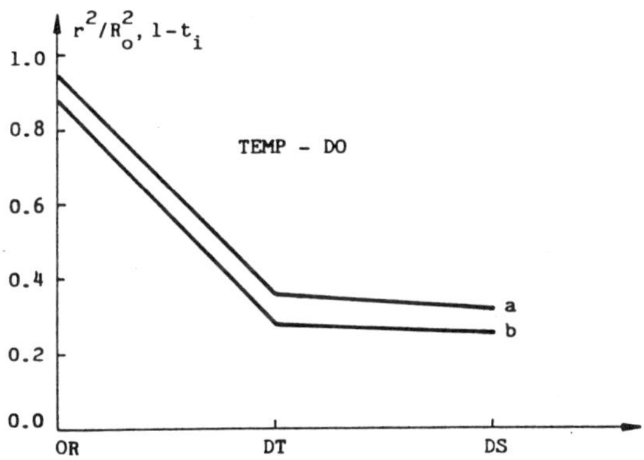

Figure 7.22. The ratio r^2/R_0^2, line (a), and the measure $1\text{-}t_i$, line (b), for the pairs of water quality variables of the Potomac River of the DES data set, and the case TEMP-DO of Fig. 7.19 (OR-original, DT- detrended and DS-deseasonalizated data)

7.6.4. ANALYSIS OF STORET DATA SET

The investigation of STORET data set basically gives results similar to the other two sets. Yet, as shown in Fig. 7.23, the overall representative figure of r^2 for STORET is about 0.20; it increases to 0.47 for the three pairs with the highest correlations. Compared to the USGS and DES sets, it may be stated that the STORET variables permit a greater percentage of information transfer between each other. Yet, the general result remains the same, namely that the parallel deterministic characteristics of the variables basically account for the detected correlations.

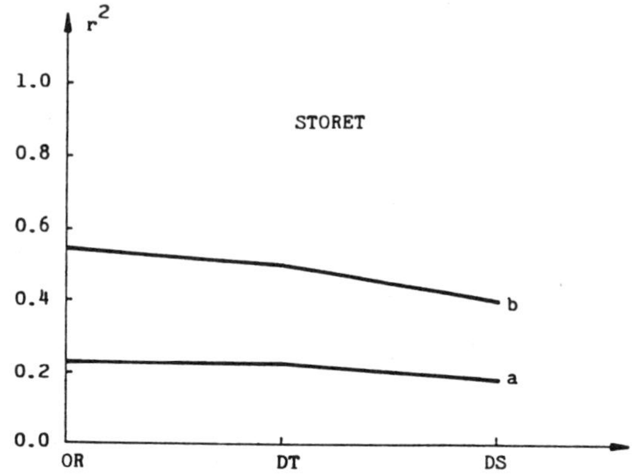

Figure 7.23. Average squares of correlation coefficients, r^2, for the pairs of water quality variables of the Potomac River, the STORET data set: (a), averages for all the pairs; and (b) averages of the highest correlation pairs (OR-original, DT- detrended and DS-deseasonalizated data)

As shown in Fig. 7.24, only three pairs in the original set, DISCHARGE-CONDUCT, DISCHARGE-TOTN, and TKN-TPHOS give r^2 greater than 0.50. Removal of trends basically has no effect on the regressions, except for DISCHARGE-TOTN and TKN-TPHOS. Removal of seasonal patterns, however, reduces the r^2 values to below 0.50 for all pairs of variables. In case of DISCHARGE-CONDUCT, r^2 decreases from 0.75 to 0.20, indicating that parallel seasonal patterns basically account for much of the transferred information. This result is the same as that obtained for the same pair in the USGS data set.

The analysis of serial dependence could be carried out for only four variables of the STORET set. However, the results even for these variables have to be evaluated with caution due to the high sporadicity of data. Therefore, analysis of information transfer between independent STORET series cannot be realized in full. r^2 values for the four independent series (DISCHARGE, TKN, DPHOS, TPHOS) indicate very low transfer of information; yet this result is questionable since the analysis of serial dependence is pretty vague due to sporadicity of data. Therefore, investigations of transinformation transfer is carried only up to the deseasonalization step as previously done for the DES set.

Analysis of transferable information for the STORET set gives results similar to those of the USGS set although a larger number of pairs indicate R_0^2 is reduced for all pairs when the series deterministic components are removed. Figure 7.25 gives similar information as to the extent of transferred information. The r^2/R_0^2 ratio and $1\text{-}t_i$ value indicate that the amount of transferred information by the applied regressions is pretty high, as for TKN-TPHOS with the ratio being 100%. However, these percentages are reduced when the series are detrended and deseasonalized, although DISCHARGE-TOTN indicate a slight increase at the last step. It may also be noted here that the decrease in r^2/R_0^2 or $1\text{-}t_i$ is rather slow for TKN-TPHOS so that even after deseasonalization, the extent of information transferability is as high as 83%.

An interesting pair in this set is TOTN-DPHOS which shows high information transferability yet with an r^2/R_0^2 ratio of 0.38. As shown in Figs. 7.24 and 7.25, both r^2 and R_0^2 for this pair increase after removal of seasonality. It may be worthwhile for this pair to attempt an improvement of the applied regression.

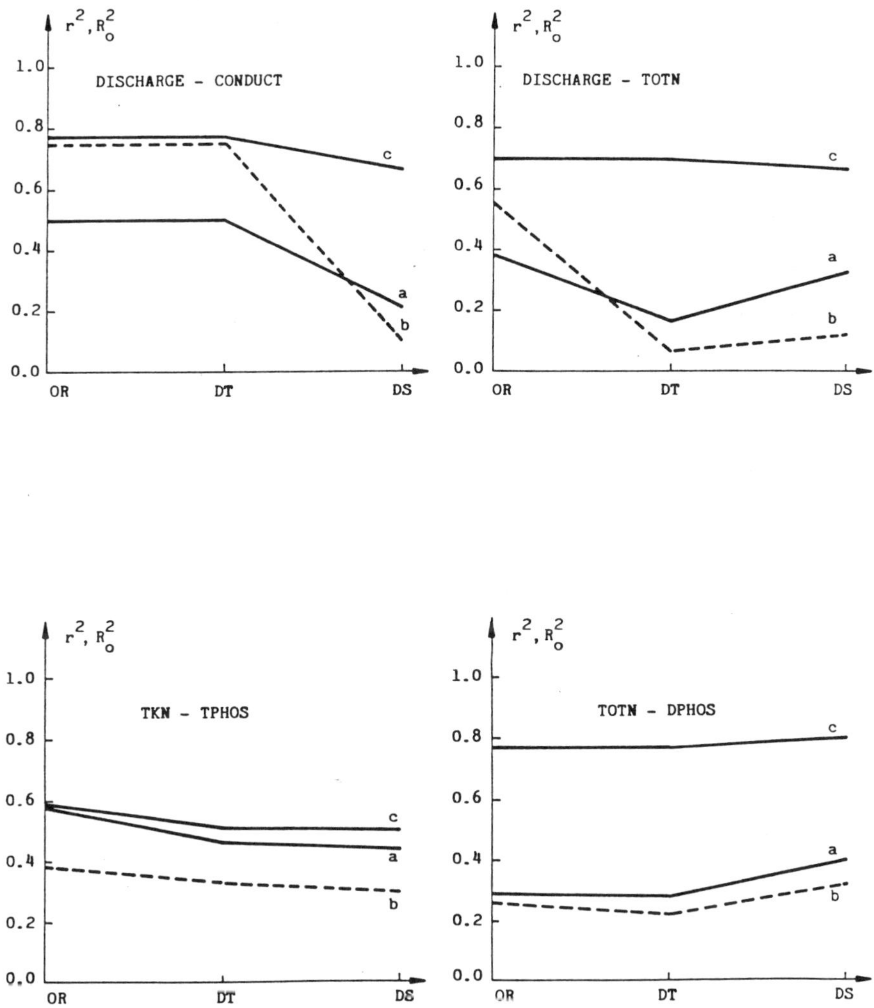

Figure 7.24. Square of the correlation coefficient of linear association, r^2, line (a), r^2 for nonlinear regression, line (b), and square of entropy informational coefficient R_0^2, line (c) for the correlations between four pairs of quality variables in the STORET data set of the Potomac River (OR-original, DT- detrended and DS-deseasonalizated data)

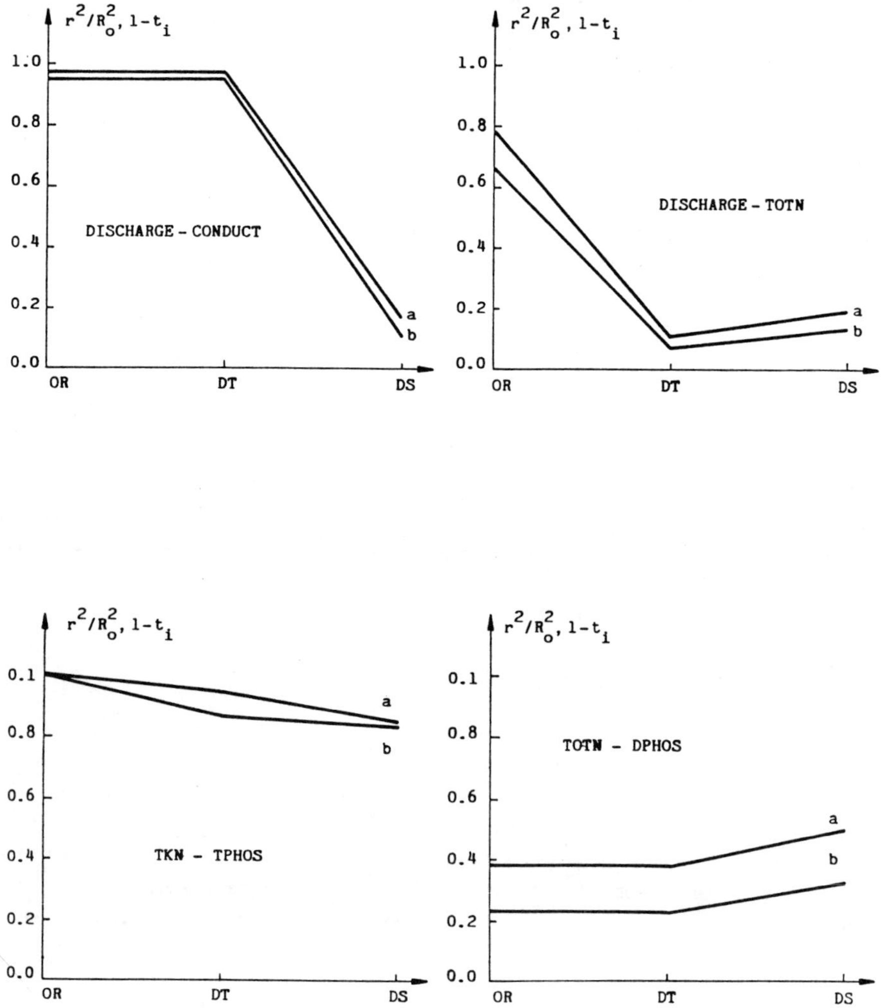

Figure 7.25. The ratio r^2/R_0^2, line (a), and the measure $1-t_i$, line (b), for the pairs of water quality variables of the Potomac River and STORET data set, and four cases of pairs of Fig. 7.23 (OR-original, DT- detrended and DS-deseasonalizated data)

7.6.5. GENERAL RESULTS OF THE CORRELATION ANALYSES

The results presented in this section basically indicate low information transferability between water quality variables of the three data sets analyzed except for a limited number of pairs. Furthermore, it is the parallelism in series deterministic components that accounts for much of the detected correlations.

Similar results have been obtained previously for the rather uniform WAQ data set. For purposes of comparison, the average r^2 values for the 66 pairs of WAQ variables are shown in Fig. 7.26 for each step of structural analysis (Yevjevich and Harmancioglu, 1985). The overall representative value here is 0.06, which increases to 0.51 for the three pairs with the highest correlations. When Fig. 7.26 is compared with Figs. 7.17, 7.21 and 7.23, it may be stated that the results for all data sets are practically the same, namely that information transferability between water quality variables is low and that much of it is due to parallelism in series deterministic characteristics.

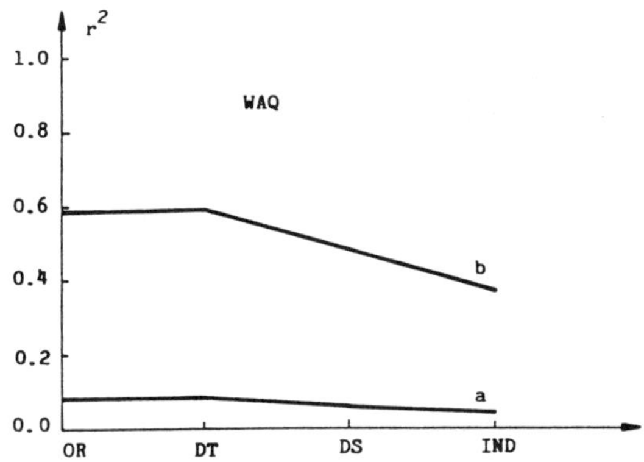

Figure 7.26. Average square of correlation coefficients, r^2, for the pairs of water quality variables of the Potomac River, of the WAQ data set: (a), averages for all the pairs; and (b) averages of the highest correlation pairs (OR-original, DT- detrended, DS-deseasonalizated and IND- Independent data)

The following general conclusions may be drawn as a result of investigations on information transferability between water quantity and water quality variables as well as among water quality variables, as demonstrated in case of the Potomac River:

1) when water quality data series have few observations with large gaps throughout the period of record, the application of time series modeling techniques becomes very difficult. Furthermore, observations made sporadically at irregular intervals, plus changes in monitoring practices, cause the structural analysis of water quality variables to be highly uncertain;
2) in the three data sets analyzed, only a limited number of pairs of variables show significant correlation. As indicated by transferred and transferable amounts of information, the majority of series do not permit a transfer of information so that these variables need to be monitored systematically;
3) the deterministic characteristics of series account for much of obtained correlations. Only a few pairs of variables retain their association after all deterministic properties of series are removed. When this is the case, such

variables do not need to be monitored systematically. Information on some of them may be obtained by transferring information from systematically observed water quality variables;

4) to model series, and eventually use them to transfer information to other variables, some water quality variables should be observed at regular times even when data must be obtained by using laboratory analyses.

7.7. Conclusions

In this chapter, a review of currently used data analysis methods is presented to focus on three basic properties of water quality data: *means*, *extremes*, and *trends*. Essentially, the methods discussed are the most common techniques employed in statistical analyses. The emphasis here has been on how these techniques perform in the case of messy water quality data; thus, their application to such data is demonstrated in case of irregularly observed water quality time series with short sample sizes, missing observations, and gaps within the series.

The examples presented clearly indicate that classical statistical methods should be applied with caution in the presence of messy water quality samples. For that matter, nonparametric approaches appear to perform much better in comparison with their parametric alternatives.

An important conclusion to be drawn is that difficulties in the application of most statistical methods stem from deficiencies in the monitoring practices applied, i.e., selection of sampling frequencies, sites, sampling duration, etc. Thus, it may be advisable at this point to reassess the current status of monitoring programs. In particular, it is essential to define a priori the specific information expected from monitoring and further to express this expectation in statistical terms. This appears to be the only way of producing data that comply with the assumptions and requirements of available data analysis methods.

The last section of the chapter focuses on the relationships between water quality variables. It is observed on the examples presented that the rates of information transfer among the majority of the variables are poor. Thus, one often cannot infer on the behavior of one variable using the information available on another variable. Accordingly, it appears that most of the water quality variables need to be monitored systematically.

References

Alpaslan, N., Harmancioglu, N.B., and Ozkul, S. (1993) Risk factors in assessment of compliance with standards, Los Angeles, California, *IAWRPC 1993 Conference on Risk, Risk Analysis Procedures and Epidemiological Confirmation*, August 1993, 13 p.

Beck, M.B. and Finney, B.A. (1987) Operational water quality management: Problem context and evaluation of a model for river quality, *Water Resour. Res.* **23 (11)**, 2030-2042.

Berryman, D., Bobee, B., Cluis, D., and Haemmerli, J. (1988) Nonparametric tests for trend detection in water quality time series, *AWRA, Water Resources Bulletin* **24(3)**, 545-556.

Chapman, D. (ed.) (1992) *Water Quality Assessments - A Guide to the Use of Biota, Sediments and Water in Environmeral Engineering*, Chapman & Hall Ltd., London, 585 p.

Crabtree, R.W., Cluckie, I.D. and Foster, C.F. (1987) Percentile estimation for water quality data, *Water Research* **21(5)**, 583-590.

Dendrou, S.A. and Delleur, J.W. (1979) Reliability concepts in planning storm drainage systems, in: E.A. McBean, K.W. Hipel and T.E. Unny (eds.), *Reliability in Water Resources Management*, Fort Collins, Water Resour. Publ., pp. 295-321.

El-Shaarawi, A.H. and Damsleth, E. (1988) Parametric and nonparametric tests for dependent data, *Water Resources Bulletin* **24(3)**, 513-519.

Harmancioglu, N.B. and Alpaslan, N. (1992) Risk factors in water quality assessment, in: *Managing Water Resources During Global Change, Proceedings of the AWRA 28th Annual Symposium*, Reno, Nevada, (November 1-5, 1992), Symposium Session S-XC, pp. 299-308.

Harmancioglu, N. and Yevjevich, V. (1987) Transfer of hydrologic information among river points, Amsterdam, Elsevier, *Journal of Hydrology* **91**, 103-118.

Harmancioglu N.B. and Yevjevich V. (1986) *Transfer of Information Among Water Quality Variables of the Potomac River, Phase III: Transferable and Transferred Information*, Report to the UDC Water Resour. Res. Cent., Washington, D.C., Int. Water Resour. Inst., June 1986.

Harmancioglu, N.B., Alpaslan, N., and Ozkul, S. (1993) Quantification of risk components in water quality assessment and management, Los Angeles, California, *IAWRPC 1993 Conference on Risk, Risk Analysis Procedures and Epidemiological Confirmation*, August 1993, 25 p.

Hipel, K.W. and McLeod, A.I. (1994) *Time Series Modeling of Water Resources and Environmental Systems*, Elsevier, Developments in Water Science, no.45, Amsterdam, 1013 p.

Hipel, K.W., McLeod, A.I., and Weiler, R.R. (1988) Data analysis of water quality time series in Lake Erie, *Water Resources Bulletin* **24(3)**, 533-544.

Hipel, K.W., Lennox, W.C., Unny, T.E., and McLeod, A.I. (1975) Intervention analysis in water resources, *Water Resources Research* **11(6)**, 855-861.

Hirsch, R.M. (1988) Statistical methods and sampling design for estimating step trends in surface water quality, *Water Resources Bulletin, AWRA* **24(3)**, 493-503.

Hirsch, R.M. and Slack, J.R. (1984) A nonparametric trend test for seasonal data with serial dependence, *Water Resources Research* **2(6)**, 727-732.

Hirsch, R.M., Alexander, R.B., and Smith, R.A. (1991) Selection of methods for the detection and estimation of trends in water quality, *Water Resources Research* **27(5)**, 803-813.

Hirsch, R.M., Slack, J.R., and Smith, R.A. (1982) Techniques of trend analysis for monthly water quality data, *Water Resources Research* **18(1)**, 107-121.

Hughes, J.P. and Millard, P.S. (1988) A tau-like test for trend in the presence of multiple censoring points, *AWRA, Water Resources Bulletin* **24(3)**, 521-531.

Icaga, Y. (1994) *Analysis of Trends in Water Quality Using Nonparametric Methods*, (Ph. D. Thesis in Civil Engineering), Dokuz Eylul University Graduate School of Natural and Applied Sciences, Izmir, 156 p., (advisor: N.Harmancioglu).

Icaga, Y. and Harmancioglu, N. (1995) Determination of water quality trends in the Yesilirmak River basin (in Turkish), *TMMOB, XIII. Technical Congress of the Chamber of Civil Engineers of Turkey*, Ankara, 20-22 December, pp.481-497.

Lettenmaier, D.P. (1988) Multivariate nonparametric tests for trend in water quality, *AWRA, Water Resources Bulletin* **24(3)**, 505-512.

Lettenmaier, D.P. (1976) Detection of trends in water quality data from records with dependent observations, *Water Resources Research* **12(5)**, 1037-1046.

Lettenmaier, D.P., Hooper, E.R., Wagoner, C., and Faris, K.B. (1991) Trend in stream quality in the Continental United States 1978-1987, *Water Resources Research* **27(3)**, 327-339.

Loftis, J.C. and Ward, R.C. (1981) Evaluating stream standard violations using a water quality data base, *Water Resources Bulletin*, **17(6)**.

Mcleod, A.I., Hipel, W.K., and Comancho, F. (1983) Trend assessment of water quality time series, *AWRA, Water Resources Bulletin* **19(4)**, 537-547.

Montgomery, R.H. and Reckhow, H.K. (1984) Techniques for detecting trends in lake water quality, *AWRA, Water Resources Bulletin* **20(1)**, 43-52.

Mulder, W.H. (1994) Water quality monitoring, forecasting and control, in: *Advances in Water Quality Monitoring*, Report of a WMO Regional Workshop, Vienna, 1994, WMO, Technical Reports in Hydrology and Water Resources, No.42, WMO/TD-No. 612, pp. 130-137.

Sanders, T.G. (1988) Water quality monitoring networks, in D. Stephenson (ed.), *Water and Wastewater System Analyses*, Elsevier, Development in Water Science No.34, ch. 13, pp. 204-216.

Sanders, T.G. and Adrian, D.D. (1978) Sampling frequency for river quality monitoring, *Water Resources Research* **14(4)**, 569-576.

Sanders, T.G., Ward, R.C., Loftis, J.C., Steele, T.D., Adrian, D.D., and Yevjevich, V. (1983) *Design of Networks for Monitoring Water Quality*, Water Resources Publications, Littleton, Colorado, 328p.

Schilperoort, T., Groot, S., Watering, B.G.M., and Dijkman, F. (1982) *Optimization of the Sampling Frequency of Water Quality Monitoring Networks*, "Waterloopkundig" Laboratium Delft, Hydraulics Lab., Delft, the Netherlands.

Tokgöz, S. (1992) *Temporal Design of Water Quality Monitoring Networks*, (Master of Science thesis in Civil Engineering), Dokuz Eylul University Institute for Graduate Studies, Izmir, (advisor: N.Harmancioglu).

Walpole, R.E. and Myers, R. H. (1990) *Probability and Statistics for Engineers and Scientists*, MacMillan Publishing Company, New York, 765 p.

Warn A.E. (1988) Auditing the quality of effluent discharges, in: *Workshop on Statistical Methods for the Assessment of Point Source Pollution*, 12-14 September, Canada Centre for Inland Waters, Burlington, Ontario, Canada.

Van Belle, G. and Hughes, J.P. (1984) Nonparametric tests for trend in water quality, *Water Resources Research* **20(1)**, 127-136.

Yevjevich, V. and Harmancioglu, N.B. (1985) *Modeling Water Quality Variables of Potomac River at the Entrance to its Estuary, Phase II (Correlation of Water Quality Variables within the Framework of Structural Analysis),* Report to D.C. Water Resources Research Center of the University of the District of Columbia, Washington, D.C., Sept. 1985, 59p.

CHAPTER 8

STOCHASTIC ENVIRONMENTAL MODELING

E. McBean, K. Ponnambalam, and W. Curi

Abstract. Alternative methodologies for use in examining the stochastic aspects of environmental modeling are examined. Some of the computational features and assumptions implicit in First-order analysis, Fokker-Planck equations, stochastic calculus and the probability density function/moment method are described.

8.1. Introduction

A mathematical model of environmental phenomena consists of a set of mathematical expressions defining the physical, biological and chemical processes that are assumed to be relevant for a particular application. The expressions usually consist of equations based on the conservation of mass and/or energy. A particular model will include then, the inputs of constituents from outside the system, the transport of constituents, and the reactions that either increase or decrease the constituent concentrations or masses. However, although the concept is general, the specific features of any model may assume many different forms, depending upon the problem, with the result that there are many different types of water quality models. For any specific situation, the appropriate model and the required data depend on the purpose of the study. In some cases, models should be relatively simple, whereas in others, much more detailed assessments are appropriate.

Although the information to follow in this chapter is equally applicable to the various environmental phenomena (e.g., air, groundwater and surface water), for ease of discussion, much of the attention is given to surface water quality modeling. Consider first, then, that many of the water quality models in use today are extensions of two simple equations proposed by Streeter and Phelps in 1925 (Streeter and Phelps, 1925) for predicting the biochemical oxygen demand of various biodegradable constituents and the resulting dissolved oxygen concentration in rivers. More complex nonlinear multiconstituent models have also been proposed and applied to predict the physical, chemical and biological interactions of many constituents and organisms found in natural water bodies.

The models referred to above, and extensions therefrom, can be used to evaluate steady-state conditions and/or transient phenomena such as nonpoint storm runoff and accidental spills. Assumptions pertaining to the mixing of pollutants in water bodies dictate the spatial dimensionality of the mathematical model. However, the noteworthy consideration herein is that these models are deterministic models. Given input values to the model, and the equations within the model, a single output value is determined.

N.B. Harmancioglu et al. (eds.), Environmental Data Management, 197-212.

Most applications of deterministic models yield estimates using only the mean values of various quality constituents. Alternatively, stochastic environmental models are much more data-demanding than deterministic models. The probabilistic models are developed with the objective of accounting for the randomness or uncertainty of various physical, biological or chemical processes. As a result, there is not a single output in response to a single input but a series of outputs, each with different probabilities of occurrence.

In a manner similar to the deterministic models, an array of stochastic model formulations exist. It is the intent herein to examine some of the alternatives with the objective of isolating some of the differentiating aspects between the models, and commenting on the data input requirements.

The increasing concern for efficiency and safety of most engineering designs, including those of waste assimilation allocation, has led researchers to develop techniques that provide the range of variability in the prediction of physical system responses due to inherent uncertainties. In some applications, this concern has been translated into a worst-case application employing deterministic models, such as, for example, the use of an extreme low flow such as the lowest seven-day average flow occurring once in twenty years. This approach involves the identification of the extreme output value arising from the uncertainty. Safety factor procedures are a form of this worst-case approach in that various forms of failure and sources of randomness are ignored. On the other hand, there are alternative approaches in which the focus on the uncertainty is less narrow. Some of these methods provide information on how the model output changes with parameter variability, while others can provide the statistical distributions for output values based on the input distribution.

8.2. Models and Interactions with Data

A random variable whose value changes through time according to probabilistic laws is called a stochastic process. An observed time series of a water quality constituent is one realization of a stochastic process. For example, if the concern is with the water quality at a particular location, a number of possibilities exist as the following:

i) the properties of a stochastic process may be determined from a time series or realization. This approach might then assume the form of a statistically-based model in which one uses a time-series based model to predict future water quality at that location; or,

ii) it may be possible to obtain inputs of pollutants to the water body, which in turn translate into downstream water quality at the location of interest. This second type of model is a mechanistically-based model in that the relevant mathematical equations are utilized to represent how the system responds in translating the water quality from one location to another.

Model application/development may then assume the form of modification to calibration, if necessary for computational reasons. Model calibration is performed using one or more observed data sets (to the extent to which the data are available) of both inputs and outputs. The model parameters and indeed the model itself are adjusted or modified so as to produce an output that is as close to the observed data as possible.

Subsequently, the validation or verification of models should be carried out. Verification requires an independent set of input and output data to test the calibrated model. A model is said to be verified if the model's predictions, for a range of conditions other than those used to calibrate the model, compare favorably with observed field data.

The validation of stochastic models is especially difficult due to the quantity of data necessary to compare probability distributions of variables rather than just their expected or mean values.

Several types of problems exist with stochastic environmental models, like:

i) sometimes the shortcoming of application of a model is a result of the computational difficulty of the model. This will be a very relevant comment in relation to some of the stochastic models discussed in the following section. The integration of stochastic differential equations will be seen to be a major concern. The assumptions required may be unacceptable for certain types of applications;
ii) sometimes the shortcoming of an application will be the availability of data. If inaccurate data are used in the input, inaccurate results will be obtained from the model. As an alternative example, for many models, the parameters cannot be measured directly; instead, the calibration of a model involves parameter adjustment so that the predictions of the model match those of the parameters which can be monitored. Obviously, the availability of monitoring data is essential to allow this procedure to be utilized.

With these types of considerations in mind, the intent in the following is to examine some of the available stochastic modeling approaches.

8.3. Examples of Stochastic Environmental Models

8.3.1. METHODS OF ANALYSIS

Over the years, a progression in the modeling of water quality phenomena has developed. At the outset, deterministic models were utilized. However, a fundamental characteristic of environmental engineering phenomena is their intrinsic stochastic nature. In addition to the frequent problem of having an inadequate number of observations, the problem of modeling the uncertainty in the observations or environmental variables, resulting from measurement errors and other random disturbances of natural origin, is a challenging problem.

Certainly, deterministic models are mathematically much simpler to deal with in contrast to the stochastic models. However, deterministic models can predict only the mean or the expected value of the process or the worst-case situation, and this is an acceptable solution for modeling processes having very small uncertainty.

As an alternative response, a number of researchers have considered uncertainty both implicitly, for example by using sensitivity analyses, and explicitly, using stochastic methods. Examples of the methods that consider uncertainty include:

i) methods based on a First-order analysis (FOA) (e.g., Burges and Lettenmaier (1975) and Tung and Hathhorn (1988));

ii) applications of Fokker-Planck equation of evolution for probability density functions of random variables (FPE) (e.g., Finney et al, 1982));

iii) methods based on the moment equations of the stochastic calculus of Ito or Stratonovich (ME) (e.g., Finney *et al.* (1982), Leduc *et al.* (1986 and 1988), Zielinski (1988), and Curi and Unny (1991)); and

iv) the probability density function/moment method (PDF/M) (e.g., Tumeo and Orlob (1989)).

Apparent from the preceding is that a considerable array of methodologies have been developed for use in examining the stochastic aspects of environmental modeling.

To allow a greater focus on the trade-offs between some of the alternative modeling approaches, a simple BOD model is considered in the following; the conclusions are equally valid, however, for more complex modeling involving a large number of differential equations.

From knowledge of the biological processes and the uncertainty in sampling and laboratory analyses, it is clear that the BOD and DO profiles have many random fluctuations (Ponnambalam and Curi, 1992; Zielinski, 1992). The reasons for this randomness are attributable to the stochastic variation of many random features including sedimentation, adsorption losses, temperature, velocity of the stream or turbulence, measurement errors, etc. Thus, if BOD values are plotted on space-time axes, they form a "random field". It is also noteworthy that, if sampling were done at closer intervals, the BOD sampling function is likely to have more random fluctuations in a given length of time. Therefore, every model that considers the uncertainty in the BOD values should be able to reproduce the "random walk" nature of the BOD sample functions, to be considered a correct representation of measured values.

Consider now the problem of simulating such random profiles so as to match the statistical characteristics of the sampling function. To examine this, assume that the rate of change of BOD can be represented by the following simple first-order ordinary differential equation (ODE) as:

$$\frac{dx(t)}{dt} = ax(t) \tag{8.1}$$

where 'a' describes the decay rate constant of this process. In deterministic BOD models, 'a' is almost always constrained as $a < 0$ (except in scour and resuspension situations).

Given the random fluctuations in the BOD profile, it follows that, when utilizing a constant and always positive decay rate, the simple model of Eq.(8.1) will never be able to reproduce the random nature of BOD sample functions. Therefore, as an increment in mathematical description, consider now the following stochastic model.

$$\frac{dx(t)}{dt} = (\mu_a + \varepsilon_t)\, x(t) \tag{8.2}$$

where μ_a is the mean value of 'a', and ε_t is assumed to be a Gaussian White noise process, with zero mean and the variance parameter σ_a^2, which is also called the intensity of the noise. Although the assumption of Gaussian White noise is an idealization and somewhat simplified, it is justified here because many random factors affect the decay rate constant; and according to the central-limit theorem, the probability distribution of the additive effect of independent random variables tends to be Gaussian distributed. In addition, when there are many random influences, it is unlikely to be practical to have sufficient data so as to prove otherwise. However, even when there are strong justifications for considering nongaussian random noises (as discussed in Zielinski (1992)), the fundamental question of whether the noise is a random variable or a stochastic process still remains, and this question needs to be discussed. It is to be noted that for pedagogical reasons, considerations herein are limited to random noise only in the parameter. The extension of this model to include noise in parameters, inputs and initial conditions is relatively straightforward.

Separating the deterministic and stochastic parts and rearranging terms in Eq.(8.2), the relationship becomes:

$$dx(t) = \mu_a x(t) dt + x(t)\varepsilon_t\, dt \tag{8.3}$$

Since Gaussian White noise is formally a derivative of Brownian motion (β_t), then $\varepsilon_t dt = d\beta_t$ with E ($d\beta_t$) = 0 and E ($d\beta_t^2$) = $\sigma_a^2 dt$, where E is the expectation operator (Jazwinski, 1970). Therefore, $\sigma_b^2 = \sigma_a^2\, dt$. Eq.(8.3) becomes:

$$dx(t) = \mu_a x(t) dt + x(t)\, d\beta_t \tag{8.4}$$

If, for simplicity, the initial condition is considered deterministic, that is, $x(t{=}0) = x_0$, the stochastic differential equation (SDE) in Eq.(8.4) can be integrated using a numerical method similar to the explicit Euler method for solving deterministic ODEs, which is the simplest numerical technique for stochastic integration (SI). The following is the resulting difference equation:

$$x(t+\Delta t) = x(t)[1 + \mu_a \Delta t] + x(t)\Delta\beta_t \tag{8.5}$$

where $\Delta\beta_t$ is $N(0, \sigma_a^2\Delta t)$ and $\Delta t = T/n$ with n as the number of discrete time steps in the interval 0 to T (Zwilling, 1989). The first term on the right-hand side of Eq.(8.5) is referred to as drift, and the second term is referred to as diffusion.

Sample functions simulated using the numerical stochastic integration described by Eq.(8.5) are presented in Fig. 8.1 for two different values of σ_a. Figure 8.1a illustrates sample functions for a value of $\sigma_a = 0.25$ with $\Delta t = 0.001$. Figure 8.1b presents the same function but further sampled only at 20 points. Similarly, Figures 8.1c and 8.1d present corresponding sample functions for $\sigma_a = 0.5$, respectively. The lines connecting the sampled BOD values in Figs. 8.1a-8.1d are linear interpolations. An important point to note is that the resulting process according to stochastic differential equation (8.4) is a Markov process, which means that, given a value of BOD at a location in the river, the BOD value of an adjacent location can only be predicted in a probabilistic sense.

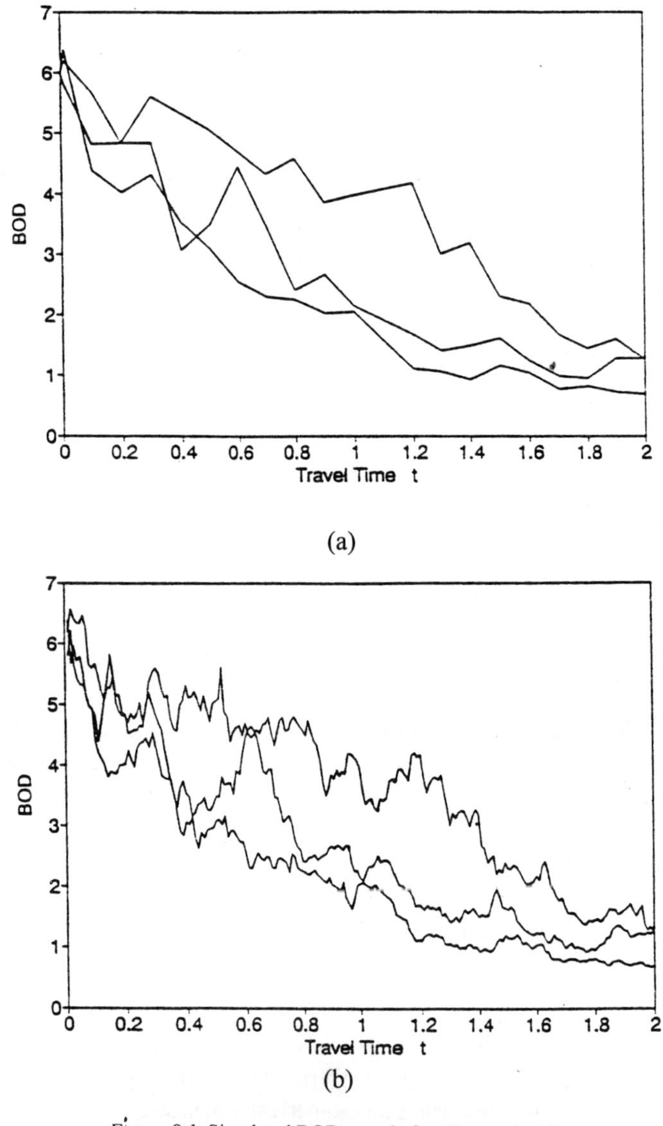

Figure 8.1. Simulated BOD sample functions using the Stochastic Integration method (SI)

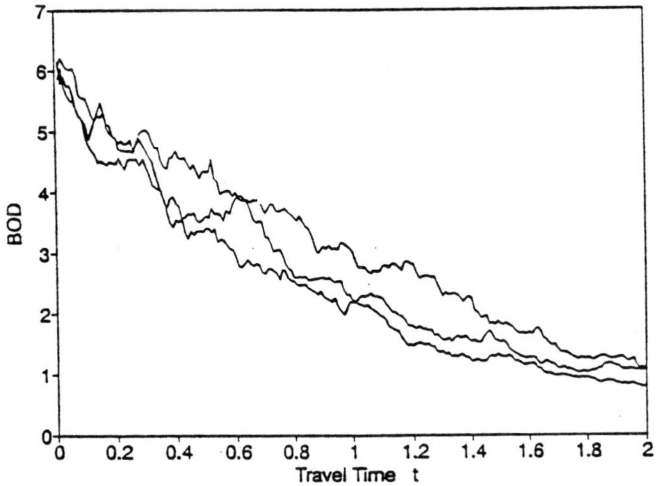

(c)

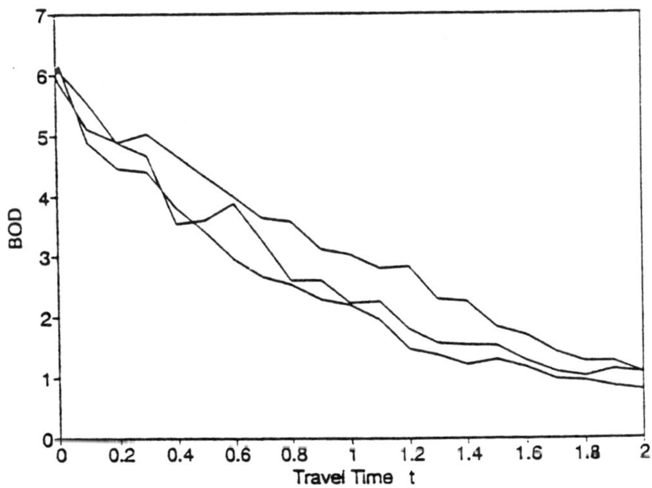

(d)

Figure 8.1. (cont.)

8.3.2. FOKKER-PLANCK AND MOMENT EQUATION METHODS

The solution process to SDE in (8.4) is a Markov process, and the evolution of such processes is provided by the Fokker-Planck equation (FPE), also called the Kolmogorov's forward equation:

$$\frac{\partial p}{\partial t} = -\frac{\partial [pf]}{\partial x} + \frac{1}{2}\frac{\partial^2 [pg^2]}{\partial x^2} \tag{8.6}$$

where p = p(x, t) is the probability density function of x, f is $\mu_a x(t)$ in the case-at-hand, g is x(t), and the SDE in (8.4) is understood in the Ito sense (Jazwinski, 1970). Note that $\int_{-\infty}^{\infty} p(x, t)dx = 1$ and the initial probability density function $p(x_0, 0)$ is assumed to be known; the boundary conditions for the partial differential equation (PDE) in (8.6) are usually taken as p $(\pm\infty, 0) = 0$ for all t. Although the FPE method is commonly used when the random noise is a white Gaussian noise process, the FPE method can be adapted to solve nongaussian random noise as in Srinivasan and Vasudevan (1971).

In theory, by integrating the PDE in (8.6) with the given initial and boundary conditions, one is able to derive the probability density function of the random functions directly. However, FPE is suitable only for limited problems before it becomes impractical to use. For example, Finney *et al.* (1982) compared the CPU time of FPE with the moment equations method (ME) described below and concluded that, even for a system of two differential equations, FPE required a factor of ten times more computational effort than the Ito's moment equations method (ME). This result is to be expected because the FPE is a very general method which does not assume any particular distribution, and therefore it describes the evolution of the stochastic processes much more completely than any other method. If it is satisfactory to have knowledge of only the lower order moments of the random outputs, generally the first and second order moments, then the ME method based on Ito's differential lemma can be used to study the evolution of SDE in (8.4).

Ito's differential lemma provides a way to generate the moment equations of a SDE as described in Appendix I. The following moment equations, which are deterministic ODEs, result when Ito's lemma is applied for the case of linear SDE in (8.4), namely:

$$d\overline{x} = \mu_a \overline{x}\, dt, \quad \overline{x} = E(x) \tag{8.7}$$

$$d\overline{x^2} = 2\mu_a \overline{x^2}\, dt + \sigma_a^2 \overline{x^2}\, dt, \quad \overline{x^2} = E(x^2) \tag{8.8}$$

By integrating the above set of deterministic ODEs, wherein the variables are the mean $\overline{x}$ and the second moment $\overline{x^2}$, the equation $V(x) = E(x^2) - [E(x)]^2$ can be used at every time step to calculate the variance V(X) of the stochastic process described by (8.4). Appendix I presents a brief description of Ito's differential lemma and the derivation of Eq.(8.8)

A difficulty with using Eqs. (8.7) and (8.8) is that, for a system of n SDEs, there are n deterministic ODEs describing the evolution of the mean and an order of n^2 (precisely there are n(n+1)/2 equations due to symmetry) deterministic ODEs describing the evolution of the second moment of the system of SDEs. However, with the present-day computers and sparse matrix techniques, even equations of the order of 10,000 (about 140 SDEs) are solvable in a reasonable CPU time. The reader is referred to Curi and Unny (1991) for a description of an automatic formulation of the moment equations, which is essential for solving large systems.

In summary, although the Fokker-Planck method is the most general method, it is not practical for large problems. The moment equations method, based on Ito's differential lemma, converts the system of SDEs into a system of deterministic ODEs and therefore is practical for implementation on problems for which it is sufficient to know only the mean and variance of the solutions.

8.3.3. FIRST-ORDER ANALYSIS

The method of first-order analysis (FOA) also attempts to approximate the mean and variance of one random variable, as a function of means and variances of the other random variables, where the variables are related by a function (Benjamin and Cornell, 1970). However, the approximation of FOA, in general, is useful only if the function relating the variables is linear or near linear. For highly nonlinear relations, the errors can become large. Following Burges and Lettenmaier (1975), Chadderton *et al.* (1982), Tung and Hathhorn (1988) and Song and Brown (1990), the FOA is applied to the following deterministic solution of ODE of Eq. (8.1):

$$x(t) = x_0 \, e^{at} \tag{8.9}$$

Using the FOA method described in Appendix II, the mean and variance relations are developed as:

$$\bar{x}(t) = x_0 \, e^{\mu_a t} \, [1 + \frac{1}{2} \, \mu_a^2 \sigma_a^2] \tag{8.10}$$

$$V[x(t)] = x_0^2 \, \mu_a^2 e^{2\mu_a t} \sigma_a^2 \tag{8.11}$$

However, if the FOA is applied on a piecewise linear solution of X, that is, on only the numerical integration formula of the explicit Euler method for deterministic ODE (Eq.(8.1)), the following relations for mean and variance of X, and the covariance of X and a can be derived (ME):

$$x(t + \Delta t) = x(t)[1 + a\Delta t], \;\; t = 0, \ldots .T, \;\; \Delta t = T / n \tag{8.12}$$

$$\bar{x}(t+1) = \bar{x}(t)[1 + \mu_a \Delta t] + \Delta t \, Cov[x(t), a] \tag{8.13}$$

$$V[x(t+1)] = [1+\mu_a \Delta t]^2 \, V[x(t)] \\ = [\bar{x}(t)\,\Delta t]^2 \sigma_a^2 + 2\,[1+\mu_a \Delta t]^2 [\bar{x}(t)\,\Delta t]\, Cov[x(t), a] \tag{8.14}$$

$$Cov[x(t+1), a] = [1+\mu_a \Delta t]\, Cov[x(t-1), a] + \Delta t\, \bar{x}(t-1)\, \sigma_a^2 \tag{8.15}$$

It is worthwhile noting that the mean, variance and covariance relations are calculated recursively, and the relations are similar to the numerical integration of moment equations (8.7 and 8.8) in the ME method. McLaughlin (1983) presented formulations similar to the above for a more complex problem. Because FOA was applied on the linear equation (8.12), the approximations calculated with Eqs. (8.13 to 8.15) should be better than those in (8.10) and (8.11). It is considered that Cov $(X_0, a) = 0$ and $X_0 = 0$ in Eqs. (8.13) to (8.15) because the initial BOD was assumed to be deterministic. The covariance terms are necessary herein because both X(t) and a are random variables, and, in Eq. (8.12), X(t) is not independent of a.

8.3.4. THE PROBABILITY DENSITY FUNCTION METHOD

The probability density function method (PDF/M) of Tumeo and Orlob (1989) can be described as follows. For the stochastic model in Eq. (8.2), the PDF/M method assumes that the variable X can be separated into deterministic and random parts as per:

$$\frac{dx'}{dt} = a'\, x_\mu + \mu_a x' + a'x' \tag{8.16}$$

where x_μ is the deterministic part of x, (the subscript t in x is dropped for convenience), a' is the random part of the random variable a, and x' is the random part of x. Separating the deterministic and random components in Eq. (8.16), the following system of differential equations results:

$$\frac{dx_\mu}{dt} = \mu_a x_\mu \tag{8.17a}$$

$$\frac{dx'}{dt} = a'\, x_\mu + \mu_a x' + a'x' \tag{8.17b}$$

In the PDF/M, using the method described in Soong (1973), the solution to Eqs. (8.17a and b) is derived as if x and x' are simply two variables. Further, the mean and the variance of the solution given below is then derived by taking expectations of the solutions. The reader is referred to Tumeo and Orlob (1989) and Zielinski (1991) for further details on their derivations.

$$\bar{x}(t) = x_0\, e^{\mu_a t}\, e^{0.5\, \sigma_a^2 t^2} \tag{8.18}$$

$$V[x(t)] = x_0^2 e^{2\mu_a t} [e^{2\sigma_a^2 t^2} - e^{\sigma_a^2 t^2}] \tag{8.19}$$

In Tumeo and Orlob (1989), the FPE is also used to derive the probability density function of the random variables. However, as noted previously, the use of FPE is not practical for large problems.

8.3.5. MONTE CARLO SIMULATION AND LATIN HYPERCUBE

In Monte Carlo sampling, an approximate probability distribution function for the system response can be attained by performing a deterministic analysis in which the input parameters are randomly chosen according to their distribution functions. The number of model runs that are required to yield the statistical properties of the system response is problem-dependent.

The method of Monte Carlo simulation was used in Padgett *et al.* (1977), Song and Brown (1990), and Zielinski (1991). In Zielinski (1991), when the parameter a was considered random with distribution $N(\mu_a, \sigma_a^2)$, the deterministic solution of Eq. (8.1) was used as follows:

$$x(t) = x_0 e^{(\mu_a + r\sigma_a) t} \tag{8.20}$$

where r is a random variable from N(0,1), and the generated sample functions are shown in Fig. 2. However, in the sample functions of Fig. 2, the 'random walk' nature of the measured BOD values is missing. Moreover, according to this method, the uncertainty is really in estimating the parameter a; and once the parameter is correctly determined, for example, by correctly determining one BOD value at a time $t = \tau$, then the values of BOD for other times $t > \tau$ can be predicted exactly. However, this understanding does not seem to correspond to reality wherein the observed BOD values are random. Incidentally, the expected value and the variance of a large number of sample functions generated by the above method equal the expected value and the variance, respectively, calculated using the method of Tumeo and Orlob (1989).

If R_i, (i = 1,2,...n) represents the average model responses for n model simulations, the sample standard deviation, S, can be calculated as:

$$S = \sqrt{\frac{\sum_{i^2}^{n} (R_i - \overline{R})^2}{n-1}} \tag{8.21}$$

where $\overline{R}$ represents the mean response of the sample. In a random sample size n, an estimate of the standard error of the sample mean $\sigma_{\overline{R}}$ (e.g., Snedecor and Cochran, 1980) is obtained by:

$$\sigma_{\bar{R}} = \frac{\sigma}{\sqrt{n}}\sqrt{1-b} \tag{8.22}$$

where b is the sampling fraction, n/N, or the fraction of the population that is included in the sample. If N is very large in comparison with n, then b approaches 0, and the estimate of the standard error is reduced to:

$$\sigma_{\bar{R}} = \frac{\sigma}{\sqrt{n}} \tag{8.23}$$

The above expression demonstrates that, for the case of the mean estimation, the required number of parameter realizations in a Monte Carlo analysis depends on the shape of the probability distribution function in each case through its standard deviation. In view of the large number of samples required to properly define the tails of the distributions, usually only the first and second moments are used to characterize the results. If the tails of the response distribution are relevant, then a larger number of samples is needed (Wallis *et al.*, 1974).

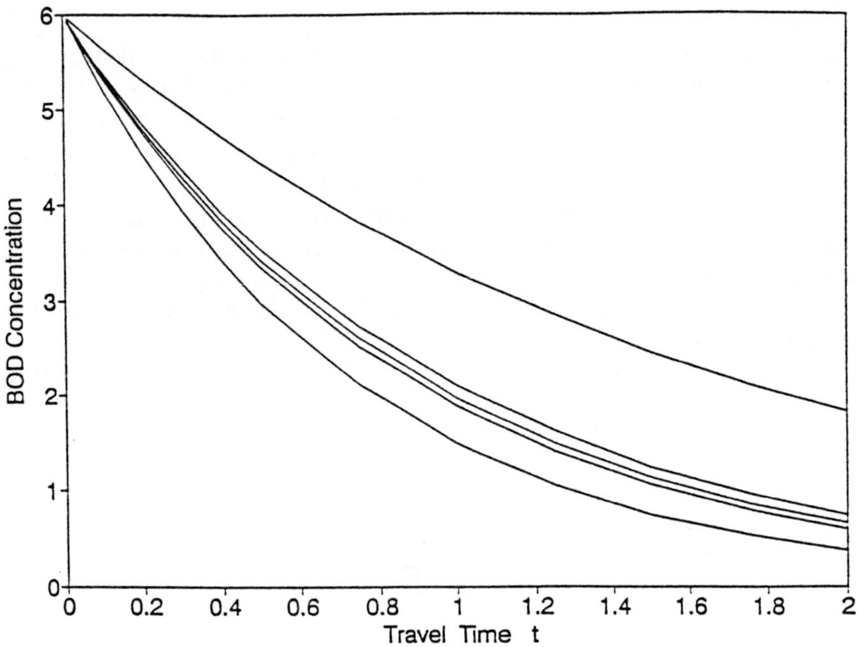

Figure 8.2. Monte Carlo (MC) simulated BOD sample functions

A more rational use of a Monte Carlo analysis is the Latin hypercube sampling technique introduced by Imam and Conover (1980). This method can provide better estimates than simple random sampling since it produces realizations from the entire probability distribution for the same number of simulations. The Latin hypercube technique involves the stratification of the range of each parameter into equally probable intervals. Then, a uniform distribution is assumed for each section, and a parameter value is randomly chosen from each section. The variance of a Latin Hypercube based sample mean is usually less than or equal to the variance of a Monte Carlo based sample mean for the same number of model simulations (McWilliams, 1987).

The use of the Monte Carlo and Latin Hypercube sampling methods to predict information about the probability density functions of the physical responses of a natural system should always consider the fact that some results may contradict feasible situations. In practice, some estimates of the system behavior are known from field measurements.

Two different approaches can be developed to deal with this problem. Basically, the approaches establish constraints that allow the filtering either of the input data or the system output. The objective of the first class of filtering technique is the reduction of uncertainty of the parameters and, consequently, a minimization of the uncertainty in estimation of the system behavior. This approach consists of the problem of parameter calibration.

An algorithm that would filter the system output is a process by which, after each realization of a Monte Carlo-type analysis, a comparison is made between the estimated result and observed results. The realization is rejected if it contradicts some established criteria. Techniques such as kriging and cokriging (e.g., Journel and Huijbrets, 1978) could be used as filters of a system output, given the fact that these methods would provide error estimates based on the statistical structure of the field measurements. An example of such an approach is Yeh *et al.* (1983) in which kriging is used as a presampling filter to reconstruct the head distribution for a flow domain. Cokriging would be amenable to problems where the accuracy of estimation of a variable that is not sufficiently sampled can be improved by considering its spatial correlation with other variables that are better sampled.

While both Monte Carlo and Latin Hypercube methods do not require assumptions of stationarity and allow computer codes to be used for model simulations without major changes, they do not give direct insight into how the system response is controlled by different parameters. The other major limitation is the high computational effort usually involved.

8.3.6. FIRST-ORDER AND SECOND MOMENT METHOD

Non-sampling (or perturbation) methods are another alternative. The most common perturbation approaches are the first order and second moment method (e.g., Dettinger and Wilson, 1981) and the spectral methods (Gelhar, 1974). These methods assume that the simulation model solution and all uncertain parameters can be represented by the sum of a deterministic mean value and a random fluctuating part. The random fluctuations are assumed to be small, relative to the mean value, and have known statistical properties. Expressions for the statistical moments of the state variables are then found in terms of the statistical moments of the fluctuating quantities. The resulting expressions for the moments of the static variables are open in the sense that they always contain higher-order moments. Generally, products of fluctuating quantities higher than a certain order are assumed to be negligible to simplify acquiring closure and expressions for the state variable moments.

The main hypothesis implied by this perturbation procedure is that all important information about the stochastic nature of the system response can be expressed with the mean and the variance-covariance. The former depicts the expected trend of the system behavior while the latter describes the amount of spreading or variation about the mean. The ideal case to be portrayed by this method occurs when the probability density function of the response has a Gaussian shape.

It is the objective of this procedure to investigate the uncertainty of a selected system response function (or performance measure) R due to uncertainty in a system parameter set α. The choice of acceptable performance measure which identifies a region and time of importance is unlimited. The estimate of the mean and variance are obtained by utilizing a multidimensional Taylor series expansion of a system response about the mean values of the parameters.

The first-order and second moment method is not subject to restrictions of stationarity. Another important feature is that it reveals the parameters contributing the most to the uncertainty, using the adjoint operator method. The limitation in nonlinear problems with large parameter variances is the major drawback. Since the validity of the sensitivity coefficient is usually restricted to a neighborhood around the value α_0 for which the variation of R to a change of α remains linear, extrapolations outside this neighborhood may produce incorrect information.

8.4. Summary

Numerous different techniques exist for stochastic environmental modeling. The utility of some of the procedures is limited by the computational aspects. For others, the necessary assumptions indicate their restricted range of utility; in addition, essentially it is a model that is utilized many times in response to sampling from probability density functions. Much research and development work remains to be done before the full utility of stochastic environmental modeling will be realized.

References

Benjamin, J.R. and Cornell, C.A. (1970) *Probability, Statistics, and Decisions for Civil Engineers*, McGraw-Hill Inc., New York.

Bluestein, M.H. and Hendricks, D.W. (1975) Biota and water quality of the South Platte River, Past-Present-Projected, Env. Eng. Tech. Report, Dept. of Civil Engineering, Colorado State University, Fort Collins, Colorado.

Burges, S.J. and Lettenmaier, D.P. (1975) Probabilistic methods in stream quality management, *Water Resources Bulletin* **11(1)**, 115-130.

Chadderton, R.A., Miller, A.C., and McDonnell, A. J. (1982) Uncertainty analysis of dissolved oxygen model, *ASCE Journal of the Environmental Engineering Division* **108(5)**, 1003-1012.

Curi, W.F. and Unny, T.E. (1991) An automatic formulation of the first and second order moment equations for a set of linear Ito equations, *Proc. of IASTED*, International Conf. on Modeling and Simulation, Calgary, Canada, July 16-18.

Dettinger, M.D. and Wilson, J.L. (1981) First order analysis of uncertainty in numerical models of groundwater flow. Part I. Mathematical Development, *Water Resources Research* **17(1)**, 149-161.

Finney, B.A., Bowles, D.S., and Windham, M.P. (1982) Random differential equations in river water quality modeling, *Water Resources Research* **18(1)**, 122-134.

Gelhar, L.W. (1974) Stochastic analysis of Phreatic Aquifers, *Water Resources Research* **10(3)**, 539-545.

Imam, R.L. and Conover, W.J. (1980) Small sample sensitivity analysis techniques for computer models with an application to risk assessment, *Commun. Statist. Theor. Meth,* **A9(7)**, 1749-1842.

Jazwinski, A.H. (1970) *Stochastic Processes and Filtering Theory*, Academic Press, San Diego, California.

Journel, A.G. and Huijbrets, Ch.J. (1978) *Mining Geostatistics*, Academic Press, San Diego, California.

Leduc, R., Unny, T.E., and McBean, E. (1988) Stochastic models for first-order kinetics of biochemical oxygen demand with random initial conditions, inputs and coefficients, *Applied Mathematical Modeling* **12**, 565-572.

Leduc, R., Unny, T., and McBean, E. (1986) Stochastic Analysis of First-Order BOD Kinetics, *Water Research* **20(5)** , 625-632.

McLaughlin, D.B. (1983) Statistical analysis of uncertainty propagation and model accuracy, in M.B. Beck and G. van Straten (eds.) *Uncertainty and Forecasting of Water Quality*, Springer-Verlag, New York.

McWilliams, T.P. (1987) Sensitivity analysis of geologic computer models. A formal procedure based on Latin Hypercube sampling, *Math. Geology* **19(2)**.

Padgett, A. (1991) *Probability, Random Variables, and Stochastic Processes*, McGraw-Hill Inc., New York.

Ponnambalam, K., and Curi, W.F. (1992) Comment on the meaning of randomness in stochastic environmental models by P.A. Zielinski, *Water Resources Research* **28 (4)**, 1187-1189.

Snedecor, G.W. and Cochran, G.W. (1980) *Statistical Methods*, 7th edition, Iowa State University Press.

Song, Q. and Brown, L.C. (1990) DO model uncertainty with correlated inputs, *ASCE Journal of Environmental Engineering Division* **116(6)**, 1164-1180.

Soong, T.T. (1973) *Random Differential Equations in Science and Engineering*, Academic Press, New York.

Srinivasan, S.K. and Vasudevan, R. (1971) *Introduction to Random Differential Equations and their Applications*, American Elsevier, New York.

Streeter, H. W. and Phelps, E.P. (1925) A Study of the Pollution and Natural Purification of the Ohio River, U.S. Public Health Service, Publication Health Bulletin 146, February.

Tumeo, M.A. and Orlob, G.T. (1989) An analytic technique for stochastic analysis in environmental models, *Water Resources Research* **25(12)**, 2417-2422.

Tung, Y.K. and Hathhorn, W.E. (1988) Assessment of probability distribution of dissolved oxygen deficit, *ASCE Journal of Environmental Engineering Division* **114(6)**, 1421-1435.

Wallis, J.R., Matalas, N.C., and Slack, J.R. (1974) Just a moment!, *Water Resources Research* **10(2)**.

Yeh, W. W-G., Yoon, Y.S., and Lee, K.S. (1983) Aquifer parameter identification with kriging and optimum parameterization, *Water Resources Research* **19(1)**, 225-233.

Zielinski, P.A. (1988) Stochastic dissolved oxygen model, *ASCE Journal of Environmental Engineering Division* **114(1)**, 74-90.

Zielinski, P.A. (1992) Reply to the Comment on "On the meaning of randomness in stochastic environmental models", *Water Resources Research,* 1190-1191.

Zielinski, P.A. (1991) Meaning of randomness in stochastic environmental models, *Water Resources Research* **27(7)**, 1607-1611.

Zwilling, D. (1989) *Handbook of Differential Equations*, Academic Press, Toronto.

Appendix I - Ito's Differential Lemma

Let x_t be the unique solution of the Ito SDE

$$d x_t = f_t dt + g_t d \beta_t \tag{I.1}$$

where β_t is Brownian motion process as defined in the text of the chapter. Let $\phi(x_t, t)$ be a scalar-valued real function, continuously differentiable in t and having continuous second partial derivatives with respect to x. Then, the stochastic differential $d\phi$ of ϕ is:

$$d\phi = \frac{\partial \phi}{\partial t} dt + \frac{\partial \phi}{\partial x} dx + \frac{1}{2} g^2 E(d\beta^2) \frac{\partial^2 \phi}{\partial x^2} \tag{I.2}$$

If we want to calculate the second moment equation for the linear SDE in Eq. (8.4) in the text, then we use $\phi = x^2$ and hence $\frac{\partial \phi}{\partial t} = 0$, $\frac{\partial \phi}{\partial x} = 2x$ and $\frac{\partial^2 \phi}{\partial x^2} = 2$; therefore,

$$d x^2 = 0 + 2x[\mu_a x dt + x d\beta] + x^2 \sigma_a^2 dt \tag{I.3}$$

Expanding and taking expectations of Eq. (I.3) we get the second moment equation (8.8) in the text. For vector differential equations, the reader is referred to Jazwinski (1970).

Appendix II - First-Order Analysis Of Uncertainty

Let the random variable Y = g(x) where x is a set of N independent random variables and Y is the dependent random variable. Further, let g( ) be the function relating the dependent random variable with the independent variables. According to the FOA, the mean of Y is

$$\bar{Y} = g(\bar{x}) + \frac{1}{2} \sum_{i=1}^{N} \sum_{j=1}^{N} \left[\frac{\partial^2 g}{\partial x_i \partial x_j} \right]_{\bar{x}} Cov[x_i, x_j] \tag{II.1}$$

and the variance of Y is expressed as

$$V(Y) = \sum_{i=1}^{N} \sum_{j=1}^{N} \left[\frac{\partial g}{\partial x_i} \right]_{\bar{x}} \left[\frac{\partial g}{\partial x_j} \right]_{\bar{x}} Cov[x_i, x_j] \tag{II.2}$$

CHAPTER 9

MODELING OF ENVIRONMENTAL PROCESSES

N.B. Harmancioglu, O. Fistikoglu and V.P. Singh

Abstract. Modeling is the stage when data are transferred into information required for environmental management. Thus, it constitutes an essential component of the decision making process. In recent years, the adoption of integrated approaches to environmental management has also changed the way environmental models are structured and employed. A good example for this development is the case of hydrologic models which evolved from simple models simulating individual processes in a basin to the current comprehensive watershed models of an integrated nature. This chapter reviews the evolution of environmental models particularly for the case of hydrologic and watershed models. The currently used models are reviewed with respect to their purpose and content. Finally, the future of modeling is evaluated in consideration of current environmental problems and expected solutions.

9.1. Trends in Hydrologic Modeling

9.1.1. BASIC APPROACHES UNTIL THE 90'S

Hydrologic models, whether physical or mathematical, are simplified representations of real hydrologic systems. Physical models, which are based on a physical analogy with the operation of a system, are often inadequate when the purpose of modeling is to identify the basic processes and their interactions within a complex natural system. Thus, mathematical models are more widely used.

Basic features of mathematical modeling in hydrology have changed within the last three decades. The 70's may be considered as the decade of stochastic hydrology, which had then flourished and attained widespread usage. Two factors facilitated the use of statistical techniques in the analysis of hydrologic processes: availability of sufficient data that permit statistical analyses and developments in the capacity of computers.

For instance, until about the 70's, efforts towards modeling of sea levels and currents in the Black Sea were confined to hydrodynamic models which were run with a set of exact input values. The random variability of sea levels and flows could not be taken into account since the amount of available observations did not permit statistical analyses. It was only after the accumulation of monitored data that variables describing sea levels and currents were investigated as time series. In addition, as the monitoring of such data was realized at high frequencies in the order of minutes, both data storage and data analyses required computers with sufficient capacities.

N.B. Harmancioglu et al. (eds.), Environmental Data Management, 213-242.

The situation was similar for the majority of hydrologic variables so that the modeling practice of the 70's can be described as being statistically-based. Furthermore, these models were focused on the description of a hydrologic process at a particular point to reveal basically their temporal variability.

The 80's experienced the flourishing of another approach in modeling: optimization techniques. In those years, the problems associated with water resources planning and management, as well as those related to all environmental processes, had grown in extent and complexity. Objectives, demands and possible solutions became diverse and complicated. Thus, selection of the "best" alternative solution was deemed necessary in respect of both technological and socioeconomic objectives. Thus, optimization techniques became popular as significant tools to aid in the decision making process for environmental management in general, and for water resources management in particular. In parallel with the optimization methods, simulation techniques were also used to arrive at, not "the best", but better decisions.

In the broadest sense for environmental processes, models were classified into decision (management) models and technological models. Technological models were based on physical, chemical, and biological concepts (such as those of fluid dynamics and biochemistry) to delineate the behavior of natural environmental systems. These were basically scientific models which served to pinpoint representative variables and diverse processes that take place in the system. As such, technological models were considered as micro-level environmental models since they were concerned primarily with the identification of real phenomena. Decision or management models were macro-level models which served the purposes of decision making, long-term planning, testing and predicting the future effects of management and control strategies. These were essentially optimization and simulation models which supported decisions to be made or engineering practice to be realized in an environment where the true nature of the system is identified by a technological model (Spofford, 1975; Loucks *et al.*, 1981). In comparison with technological models, decision models had larger dimensions as they accounted for social, political, legal and economic factors (Beck, 1978; Uslu *et al.*, 1983). As shown in Fig. 9.1, technological models were then considered as components of decision models, incorporated into either the objective function or the constraints (Alpaslan and Harmancioglu, 1991).

To develop a decision model, the problem to be investigated had to be described first as a system. This could be an ecologic system in the broadest sense, consisting of economic, social, and finally, physical subsystems. Each subsystem constituted a component of a larger system, so that there were continuous interactions between the systems.

Once the system was specified, the decision model could be set up by determining the objective function and the constraints. In general, environmental decision models used to have such components as demand functions, pollution models, environmental development and quality models, hazard functions and management (decision) policies (Spofford, 1975).

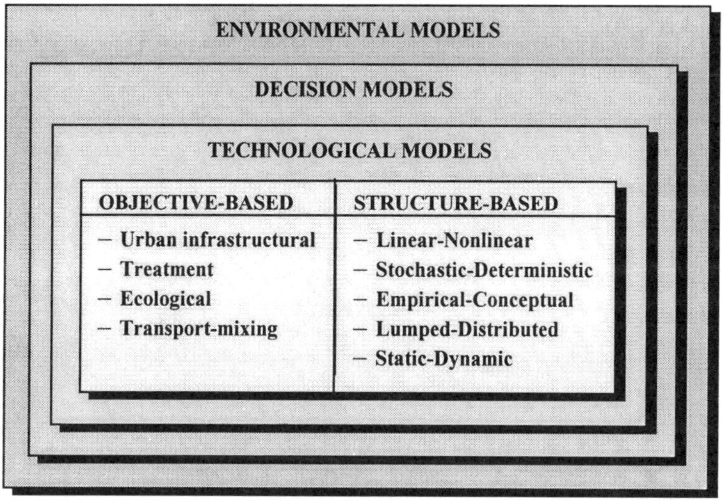

Figure 9.1. Classification of environmental models

Most decision models were formulated as optimization models where an optimum or the "best" solution could be obtained for the investigated problem. Often, the results of an optimization model were confirmed by simulation techniques, especially in case of complicated systems where the relationship between various system components could not be defined analytically. By means of simulation models, it was possible to observe system behavior in detail with respect to both physical and economic inputs. Furthermore, large numbers of variables could be considered to evaluate several alternative solutions (Alpaslan, 1983; Loucks *et al.*, 1981).

As discussed earlier, decision models were considered macro-level models covering the physical, economic, social, political and legal aspects of a problem. The physical behavior of an environmental system was then defined by a technological model, which used to be an ecological model at the broadest scale. Ecosystem models were evaluated in two ways for purposes of environmental management. First, such models could be used to predict the behavior of the environmental system for a set of given inputs. This is basically a simulation approach where the model plays a descriptive role. Second, when standards or specific management objectives were given, ecosystem models were evaluated jointly with a decision model (often an optimization model) where the optimal levels of inputs to the ecosystem model were determined by the decision model. In this case, system behavior was analyzed together with economic factors and standards so that the ecosystem model played a prescriptive role. Evidently, the mathematical formulation of such a model was much more difficult than it was in the first case.

To give an example of how such a model was incorporated into a decision model, the mathematical formulation of a typical problem, as described by Spofford (1975), is presented in the following. The environmental decision model is considered to be an optimization problem, where the approach is essentially deterministic. With respect to economic decision criteria, the model has a static character, although it is possible to convert it to a dynamic one. The objective function foresees the maximization of net benefits as (Spofford, 1975):

$$max \{F = f(x, R)\} \tag{9.1}$$

subject to constraints:

$$Ax \leq B \tag{9.2}$$

$$h_i(x) = R_i \tag{9.3}$$

$$R_i \leq S_i \tag{9.4}$$

$$R_i \geq 0 \tag{9.5}$$

$$x \geq 0 \tag{9.6}$$

In the above, x is the vector of decision variables that represent effluent discharges; R_i (i= 1, ..., q), the pollutant concentrations; S_i (i= 1, ..., q), quality standards; and $Ax \leq B$ the set of linear constraints related to effluent discharges. $h_i(x) = R_i$ (i= 1, ..., q) is the technological model (such as an ecosystem model) that describes the relationship between effluent discharges and pollutant concentrations.

The above formulated model is a standard linear programming problem when both the objective function and the constraints are linear. In general, the classical water quality management problems had been of this type. If nonlinearity were significant in any one of the above expressions (1) through (6), the problem was considered as a nonlinear one. In such cases, nonlinear simulation models were often preferred.

Technological models, which were incorporated into the decision model as constraints in the above formulation, could assume different forms depending on their objectives and structural properties. In essence, the selection of a particular type of model depended upon the nature of the problem to be investigated. The purposes of model application (or objectives) played a significant role in this selection (Vansteenkiste, 1978).

Several environmental models had then been developed at varying degrees of complexity. Essentially, modeling activities may be traced back to the Streeter-Phelps formulation of 1925, which is still valid with some modifications. Changes in objectives of environmental management had also affected modeling efforts so that not only specific, but also multidimensional problems were attacked to produce detailed formulations like the comprehensive ecological models. Actually, by the end of 70's, environmental modeling had almost become a custom. This was because decision makers tended to rely more on mathematical models, which, they thought, could be used at a significant advantage in making management decisions. Obviously, the advent of powerful computers was one of the major reasons to enhance this tendency (Alpaslan and Harmancioglu, 1991). The result was the development of highly complex models, mostly of a deterministic character (Whitehead, 1978; Beck, 1987)

It may be worthwhile to review the types of technological models as denoted in Fig. 9.1. The figure shows that four types of technological models can be identified with respect to model use. The first two groups, namely the urban infrastructure (water supply and sewerage systems) and the treatment (activated sludge, biofilm, etc.) models aid in both the design and operation of physical systems. Therefore, they constitute significant tools in engineering design and applications. When such mathematical models are calibrated and validated by using data of real systems, they can be used to determine optimum dimensions and operational policies for new systems.

Ecological and transport-mixing models are basically natural system models where the problem investigated is more complex. Ecological models consider the environment at a significantly large scale to delineate the complicated relationships between living things, matter, and energy in real systems. The other group of models mainly investigate the transport and mixing mechanisms (advection and diffusion) of pollutants within natural environments.

The above four types of models were developed more with a conceptual approach than an empirical one when physical processes were dominant in the investigated problem. Basically, infrastructure and transport-mixing models fall into this category. When chemical and biological processes were more significant as in ecological and treatment models, model characteristics were more on the empirical side (Uslu *et al.*, 1983).

Figure 9.1 also shows the classification of technological models with respect to model structures. The selection among these types basically depends on the dominant nature of the environmental system investigated.

One approach to modeling is of a deterministic nature, where the randomness in basic variables and parameters is ignored. The major difficulty lies in model calibration using observed data. If sufficient and reliable field data are not available, the predictive capability of the model is highly reduced no matter how well it is formulated.

Stochastic models account for the random nature of environmental variables and their complex interactions. They are more complicated in comparison with deterministic models; however, various uncertainties underlying environmental phenomena often necessitate a stochastic approach (Whitehead, 1978).

Similar comparisons exist between static (or steady-state) and dynamic model structures. The former is easier to handle although it contradicts the temporally varying nature of environmental phenomena and their interactions. Within this respect, dynamic models appear to be more favorable (Whitehead, 1978; Uslu and Turkman, 1987), yet they have a more complicated structure which limits the number of dynamic variables.

Environmental systems are nonlinear by nature; however, the assumption of linearity not only simplifies the model, but also provides the advantage of developing a linear programming formulation for the decision model. Otherwise, the incorporation of a nonlinear technological model into a management model complicates modeling procedures, as the latter must then be a nonlinear or a dynamic programming problem.

In essence, environmental processes constitute distributed parameter systems. In early models, this characteristic could not be accounted for, so that the real phenomena were identified by means of lumped parameters. It was with the advent of powerful computers that a detailed description of system variables and their interactions could be realized in the form of distributed parameter models.

9.1.2. RECENT TRENDS

By the end of 80's, water resources management became a much more complicated endevour as the problems experienced had grown both in "scope" and "scale". With respect to scope, previously, the basic question addressed was how much water was available at a particular point in a basin. Later, as water pollution problems grew more intense, it become imperative to consider the quality of water, as well as its quantity, for water resources development and planning. This meant that these two properties of water had to be evaluated with respect to all causative factors, including human-induced effects as well as natural causes.

With respect to water quality, the assessment of point polluting sources has been relatively more straightforward in comparison with that of nonpoint sources. The latter pollutants require an investigation into land processes, land use, human activities, soil and vegetation properties, agricultural practices, and the similar, which essentially constitute all components of the environment. Consequently, water resources have been recognized as part of the "environmental continuum". Consideration of the environment as a whole indicates a change in the scale of problems experienced; namely that, in recent years, environmental problems, their causative factors, and hence the solutions required have a significant spatial dimension.

Finally, recognition of the environment as a continuum of all natural processes, further, the concern over sustainable development of the environment, and several other environmental problems led to the adoption of integrated approaches to environmental management. These developments have previously been discussed in Chapter 1.

As shown in Fig. 9.2, water elements encompassing physical, chemical, and biological aspects of water quantity and quality, their interactions, and the effects on them of natural as well as external constraints, constitute the framework upon which integrated water management is to be built. External constraints such as economic, demographic, transportation and other forms of development directly influence one or the other water elements. Similarly, climatic factors, climate change, and climatic extremes, and a host of natural hazards are some of the natural causes that greatly influence the water elements and have significant impact on the integrated water management (Singh, 1995 a).

The need for water is derived from a variety of activities such as agriculture, water supply, industry, energy, transportation and recreation. Clearly, all uses of water cannot be supported to the fullest extent, and an integrated management policy has to be developed to incorporate all considerations shown in Fig. 9.3 (Singh, 1995 a).

The above mentioned evolution of environmental problems and approaches used in managing them have also led to changes in modeling procedures. Essentially, four major developments are observed:

a) Early models used in the analysis of hydrologic processes, and in that of watershed as the basic unit for water resources development, considered water quantity as the basic process. Thus, the majority of watershed models simulated watershed response either without consideration of water quality or with inadequate consideration thereof. Later, in recognition of water quality as an essential element for watershed management, models have been developed to bc more versatile and comprehensive to include physical, chemical and biological properties of water (Singh, 1995 a). Furthermore, these models have been developed into watershed-scale water quality models. Donigian *et al.* (1995) state that these models were needed basically to evaluate two issues: "nonpoint source pollution effects (i.e., exposure and/or loadings assessment), and chemical management plans such as agricultural best management plans for fertilizers and pesticides or, more recently, total maximum daily loads (TMDLs), including both point and nonpoint contaminant sources".

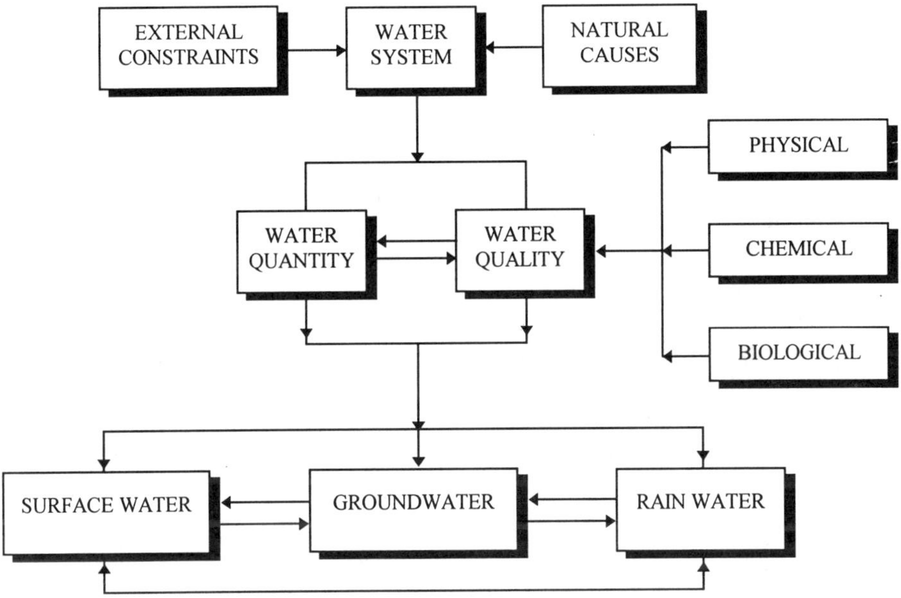

Figure 9.2. Water use and its interaction with the water system (Singh, 1995 a)

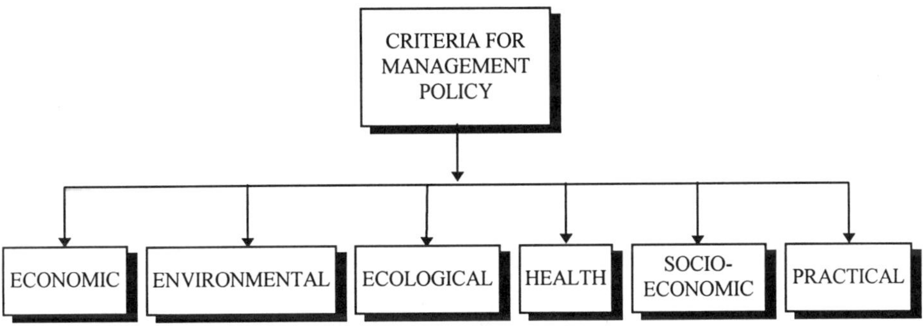

Figure 9.3. Criteria for an integrated management policy (Singh, 1995 a)

b) Previously, the traditional hydrological models, mostly of lumped conceptual types, had been adequate in solving problems related to water quantity, e.g., floods and droughts. However, in recent years, several diverse problems emerged in water resources planning and management, i.e., problems related to irrigation, land degradation, soil erosion, surface and ground water pollution, damages to aquatic ecosystems, possible climate changes, and the similar. These problems relate significantly to human-induced impacts and have the distinctive feature of being spatially distributed. Thus, to investigate management strategies required by such problems, the need has grown for advanced physically-based distributed type of models, mostly of a deterministic nature (Refsgaard and Abbott, 1996).

c) The adoption of integrated approaches to environmental management imposed new requirements on modeling techniques, as it follows from (a) and (b) above. The integrated environmental management requires the integration of various components such as physical, chemical, biological, social, economic, ecological, health, and other environmental elements. This can be accomplished through the development and application of comprehensive watershed models which are computer-based mathematical models simulating all processes and characteristics of a watershed. Eventually, models have been developed to perform comprehensive watershed analysis (water quality and quantity modeling) within a single framework. Thus, a number of advanced models have been formulated as "integrated watershed models", instead of externally linking models that perform individual analyses of particular watershed processes (Donigian *et al.*, 1995).

d) In the past few years, with revolution in information and communication technology, new vistas have unfolded that watershed models have not taken full advantage of. Examples include remote sensing and satellite technology (RSST), GIS, artificial intelligence, and risk and reliability analysis. The RSST is a powerful technology for data acquisition over extensive and even otherwise inaccesable areas. The GIS technology can be usefully employed for data management – data retrieval, manipulation, organization, and the like – that may be entirely suitable for development of distributed models. Artificial intelligence, or the so-called expert systems, allow one to make watershed models user-friendly and usable by nonhydrologists. Risk and reliability analysis allows for providing information on the accuracy of model results that the user wants to know about. Coupled with these developments, integrated watershed models have become significant management tools. The recent approach in developing and testing management strategies involves the integration of watershed models with GIS, RRST, and expert system technologies. Then, this system is run for a number of management schemes to evaluate the consequences of alternative management scenarios and to arrive at better decisions. This approach gives emphasis to simulation rather than optimization techniques which were discussed in section 9.1.1.

9.2. The Role of Models in Environmental Decision Making

9.2.1. PURPOSE OF MODELING

Today, mankind has reached the technological levels with which to both develop and destroy the natural environment. Despite the advances in available technologies, difficulties are still encountered in developing and managing natural resources due to the significantly complex nature of environmental systems and problems. The behavior of such systems entails various uncertainties; thus management decisions need to be realized under uncertain conditions. Furthermore, decision making for purposes of solving environmental problems becomes much more complicated when the objectives and priorities of the society are considered. Consequently, several diverse but interrelated factors of technological, social, political and economic nature need to be evaluated for planning and management of the environment. This actually is the essence of integrated environmental management.

Furthermore, natural resources have become limited both in quantity and quality so that the decision maker has to select the most efficient and effective policies, especially when he is also subject to budgetary constraints. At this point, mathematical models provide the basis for his decisions so that the most rational solutions can be attained. Evidently, this development puts significant demands on mathematical models if they are to be used as tools in search for effective decisions. These models are expected to realistically and reliably represent what occurs in the environment. The basic problem here is that real environmental systems are extraordinarily complex by nature, whereas mathematical models inevitably give way to simplifications and assumptions. However, a real system is understood and identified only as much as it can be described mathematically. Such a mathematical identification is the prerequisite for engineering judgment and decision, since information that cannot be described in mathematical terms does not constitute scientific information. This means that the interpretation of environmental systems and the eventual decision making process for management purposes need to be mathematically-based (Uslu *et al.*, 1983 and 1984; Alpaslan and Harmancioglu, 1991).

Any system that is described mathematically can be easily converted into a computer model. This constitutes a significant link between environmental problems and the advent of high-capacity computers. The complicated nature of environmental problems require complex, multifaceted and expensive technological solutions. Such solutions can be attained only by "information-intensive" approaches to evaluate all the available information and all the possible alternative plans. It is only by means of computers that such information can be processed; and consequently, the developments in computer technology have naturally affected the environmental modeling procedures.

It follows from the above that models serve two basic purposes:

a) identification of basic processes and their interactions in a real system, i.e., the environmental continuum in the case of integrated watershed models. This aspect of modeling helps both "to understand and reproduce how the modeled system works" (Donigian *et al.*, 1995);

b) testing of alternative management decisions and arriving at effective management policies. This aspect of modeling helps "to predict how the system will respond to future alternative conditions and actions" (Donigian *et al.*, 1995).

9.2.2. ENVIRONMENT-MODELING-DECISION MAKING RELATIONSHIPS

The environment-decision making relationship is established as the society manages the natural environment by means of the decision makers. Within the decision making process, modeling acts as a tool so that the environment-modeling-decision making relationship is developed as in Fig. 9.4. There are other factors that play significant roles in this triple relationship. Results of scientific research serve for a better understanding of the environment and thus affect the modeling efforts directly. Technological advances may produce the same effect, although they may also have adverse consequences upon the environment. Often, the social, political and economic circumstances where the decision maker functions are fairly different from those of the modeler so that a direct relationship between the modeler and the decision maker cannot be developed. In this case, investigations in the form of policy analyses provide the transition between modeling and decision making. Such analyses require expertise of both the modelers and the decision makers. The significant point here is the communication between the two groups; that is modelers must develop mathematical models that can be "used" by the decision makers. On the other hand, the extent up to which modeling contributes to the decision making process depends significantly on the decision makers' views and attitudes with respect to the problem encountered (Biswas, 1976; Alpaslan and Harmancioglu, 1991).

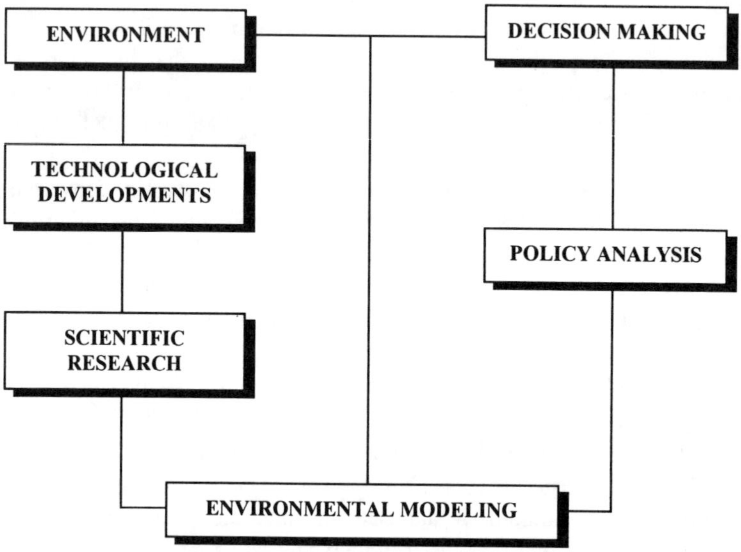

Figure 9.4. Environment-modeling-decision making relationships

The decision making process covers such activities as the identification of problem, collection of data, policy analysis, determination of objectives and their evaluation. In some cases, decisions are based on detailed analyses, while in some others the decision maker is contented to rely on results of past experiences. Often, the crucial factor here is the temporal (long- or short-term) and spatial (regional or wide-area) span over which the decisions will hold. In general, both long- and short-term decision making processes require a certain kind and level of analysis. Such analyses range from simple investigative methods to sophisticated mathematical formulations to be solved by the computer. The selection of analytical methods to be used depends on the technical nature of the environmental problem, the objectives foreseen, and the parameters of significance.

9.3. Classification of Hydrologic Simulation Models

9.3.1. GENERAL CLASSIFICATION

Hydrologic simulation models in general and watershed models in particular have been classified in a number of ways. (Singh, 1995 a) proposes a classification procedure as in Fig. 9.5, based on:

a) process description,
b) scale, and
c) method of solution.

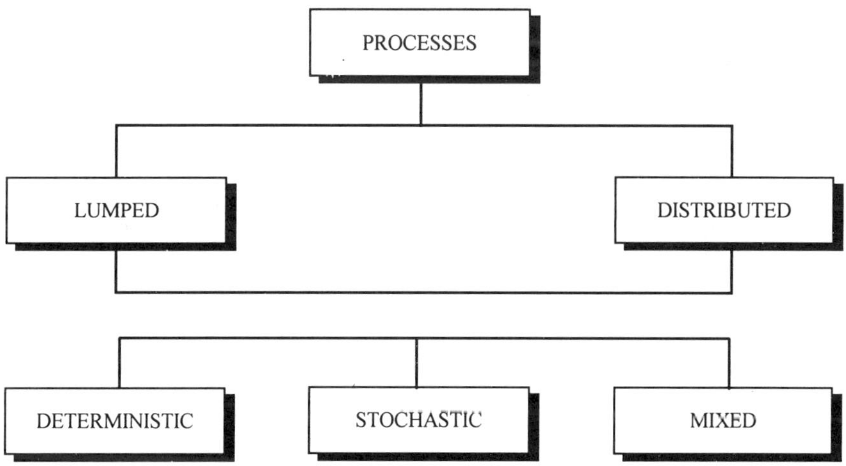

(a) Classification of models based on process description

Figure 9.5. Classification of hydrologic simulation models and watershed models

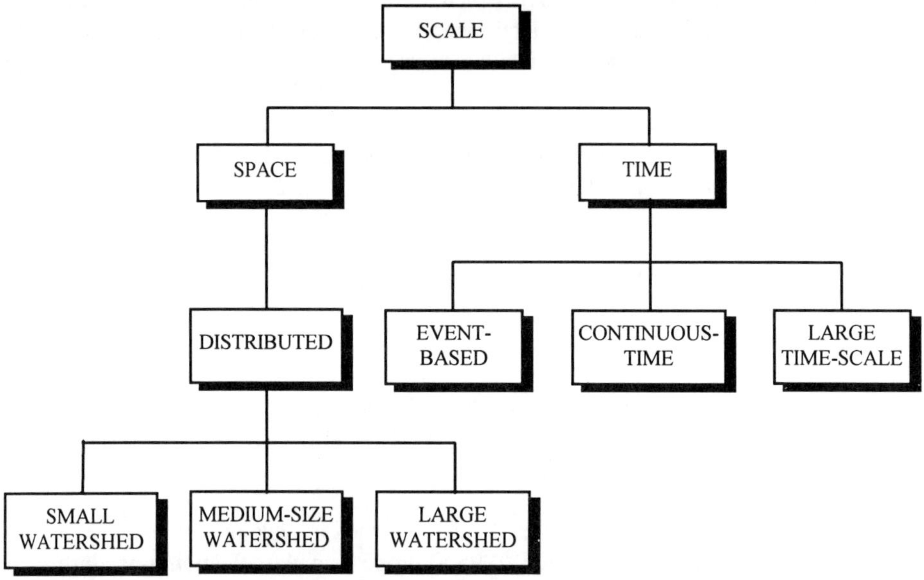

(b) Classification of models based on space and time scales

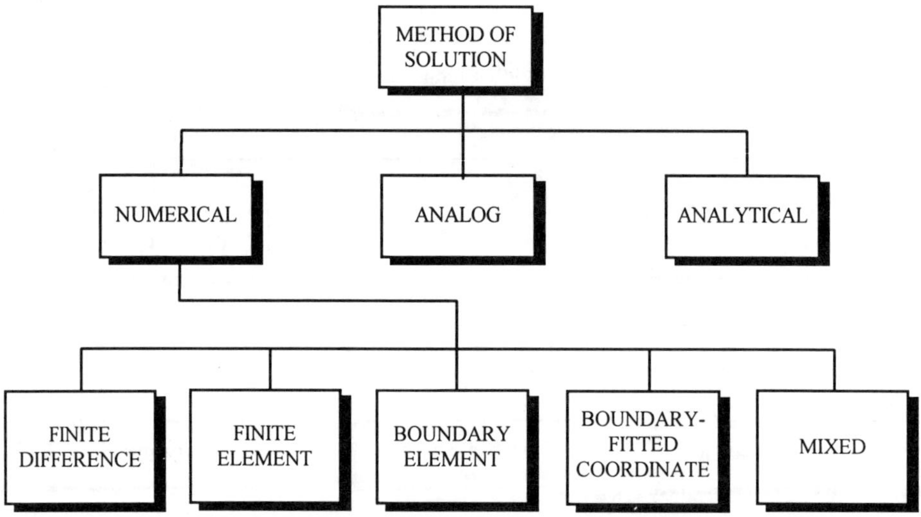

(c) Classification of models based on solution techniques

Figure 9.5. (continued)

9.3.2. PROCESS-BASED CLASSIFICATION

Based on the description of processes simulated, models can be described as lumped or distributed, deterministic or stochastic or mixed.

A lumped model describes the watershed as a whole unit. The variables and parameters used represent average values for the entire basin. A distributed model, on the contrary, takes into account the spatial variations of model variables and parameters (Refsgaard, 1996).

A lumped model, as shown in Table 9.1, is, in general, expressed by ordinary differential equations taking no account of spatial variability of processes, input, boundary conditions and system (watershed) geometric characteristics. In most lumped models, some processes are described by differential equations based on simplified hydraulic laws, and other processes are expressed by empirical algebraic equations. Examples of lumped models are HEC-1 (Hydrologic Engineering Center, 1981), HYMO (Williams and Hann, 1972), RORB (Laurenson and Mein, 1983), SSARR (U.S. Army Corps of Engineers, 1972), tank model (Sugawara *et al.*, 1984), amongst others (Singh, 1995 a).

TABLE 9.1. Lumped and distributed models (Singh, 1995 a)

Input	**System Characteristics**	**Component Processes**	**Governing Equations**	**Output**	**Model Type**
Lumped	Lumped	Lumped	ODE	Lumped	Lumped
Lumped	Lumped	Distributed	PDE	Distributed	Distributed
Distributed	Distributed	Distributed	PDE	Distributed	Distributed
Distributed	Lumped	Distributed	PDE	Distributed	Distributed

Distributed models, as shown in Table 9.1, take an explicit account of spatial variability of processes, input, boundary conditions, and/or system (watershed) characteristics. In practice, lack of data –field or experimental (laboratory)– prevents such a general formulation of distributed models. In a majority of cases, the system (watershed) characteristics are lumped, many of the processes are lumped, the input is lumped, and even some of the boundary conditions are lumped, but some of the processes that are directly linked to the output are distributed, as for example, rainfall-runoff process. These models are not fully distributed; rather they are quasi-distributed at best. Examples of distributed models are SHE (Abbott *et al.*, 1986a and b), IHM (Morris, 1980), SWMM (Metcalf and Eddy, Inc. *et al.*, 1971), NWSRFS (Hydrologic Research Laboratory, 1972), among a number of others.

9.3.3. DETERMINISTIC VERSUS STOCHASTIC MODELING

The classification of models into one of deterministic or stochastic types is essentially a process-based grouping. In the broadest sense, a model has five basic components as shown in Fig. 9.6.

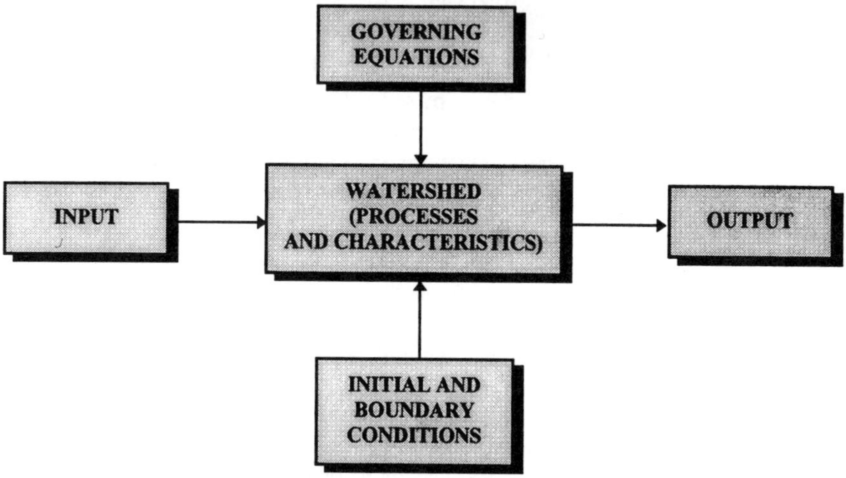

Figure 9.6. **Basic components of a watershed model**

The processes include all of the hydrologic processes that contribute to the watershed output. Depending upon the description of those processes, in conjunction with the system (watershed) characteristics, the models can be described as being deterministic, stochastic, or mixed (Singh, 1988 and 1995 a). If all of the components of a model are deterministic, the watershed model is deterministic. If all components are stochastic, the model is fully stochastic; if only some components are stochastic, then the model is quasi-stochastic. When the model components are described by a mixture of deterministic and stochastic components, the model is a stochastic-deterministic or hybrid model (Singh, 1995 a).

A model is basically a system description that comprises watershed processes and characteristics. The watershed system represents a transfer function by which inputs are transferred into outputs. In a deterministic model, the transfer function does not deal with the variability that naturally occurs in watershed processes and characteristics. On the contrary, stochastic models deal with the uncertainty associated with variable systems. Some deterministic models include stochastic processes to consider the spatial and temporal variability of some of the subprocesses.

With regard to the inputs and outputs of the system, a deterministic model always produces the same output when it is run under identical conditions, given two sets of equal inputs. In stochastic models, identical inputs generally lead to different outputs when the model is run under identical conditions. A similar situation occurs when the model inputs are described as stochastic processes (Refsgaard, 1996).

With respect to the outputs, deterministic models result in a predictable output for a given set of exactly known input values. Thus, the value of the output can be exactly determined. In contrast, stochastic models assign probabilities to the outputs, given a set of known inputs.

The majority of the watershed models are deterministic, and virtually no model is fully stochastic. Some deterministic models may include parts that are described by the laws of probability, in which case the model is characterized as quasi-deterministic or quasi-stochastic (Singh, 1995 a).

9.3.4. CLASSIFICATION BASED ON TIME SCALES

Watershed models can be classified based on the time scale of models. The time scale can be defined as a combination of two time-intervals (Diskin and Simon, 1979). One of the time intervals is used for input and internal computations. The second is the time interval used for the output and calibration of the model. Based on this description, the models can be distinguished as:

a) continuous-time or event-based,
b) daily,
c) monthly, and
d) yearly models.

This classification is based on the interval of computation. If the model components are available at shorter time intervals, such as hourly interval or shorter, the model will be a short-time interval model. Of course, the choice of a time interval is often a function of model's intended use (Singh, 1995 a).

An event-based model is one that represents a single runoff event occurring over a period of time ranging from about an hour to several days. The initial conditions in the watershed for each event are assumed or determined by other means and entered to the model as input data. The accuracy of the model output often depends on the reliability of these initial conditions.

A continuous watershed model is one that fuctions over an extended period of time, e.g., different periods of runoff conditions. The model keeps a continuous account of the basin input condition and therefore determines the initial conditions applicable to runoff events. At the beginning of the model run, the initial conditions must be known or assumed. However, the effect of the selection of those initial conditions decreases rapidly as the simulation advances. Most continuous watershed models utilize three runoff components: direct or surface runoff, shallow surface flow (interflow) and groundwater flow, while an event model may omit one or both of the subsurface components and also evapotranspiration.

9.3.5. CLASSIFICATION BASED ON SPACE SCALES

The spatial scale can be used as a criterion to classify models into small-watershed, medium-size watershed, and large watershed models. Certainly, the definition of a small watershed is somewhat vague. Usually, a watershed with an area of 100 km^2 or less can be called small; those with an area of 100 to 1,000 km^2, medium; and those with an area of larger than 1000 km^2, large. Such a classification is arbitrary, and is experimental rather than conceptual and is governed by data availability rather than physical meaning. The essential point is the concept of homogeneity and averaging of hydrological processes. For consideration of runoff generation on these watersheds, two phases can be considered: land phase and channel phase. Each phase has its own storage characteristics. Large watersheds have well-developed channel networks and channel phase; and thus, channel storage is dominant. Such watersheds are less sensitive to short duration, high intensity rainfalls. On the other hand, small watersheds have dominant land phase and overland flow, have relatively less conspicuous channel phase, and are highly sensitive to high-intensity, short-duration rainfalls (Singh, 1995 a).

If all three types of watersheds are spatially uniform, then they will behave hydrologically similarly. Under this condition, classification of watersheds serves no useful purpose. If a watershed is homogeneous with respect to its characteristics, then it is called a small watershed. In reality, though, watersheds are seldom homogeneous; therefore, homogeneity is defined in an average or statistical sense. For example, for a subsurface environment, if the average of logarithmic values of the hydraulic conductivity is the same, the aquifer may be considered homogeneous. Similarly, if the average system characteristics do not change, then the watershed may be considered small. For medium-size watersheds, the heterogeneity is present, but it is in the medium range (± 2 standard deviation around the mean), whereas it is quite large for large watersheds. Even if a watershed is homogeneous in one characteristic, it may not necessarily be homogeneous in another characteristic. In that sense, both the soil and land use characteristics are important. Strictly speaking, then, no watershed may be small; all will be either large or medium (Singh, 1995 a).

9.3.6. FURTHER TYPES OF CLASSIFICATION

Singh (1995 a) also classified watershed models on the basis of land-use and model-use. Based on land-use, the watersheds may be classified (Singh, 1992) as:

a) agricultural,
b) urban,
c) forest and range land,
d) desert,
e) mountainous,
f) coastal,
g) wetlands, and
h) mixed.

In many cases, large or even medium-size watersheds have mixed land use. These watersheds behave hydrologically differently, so that they have given rise to different branches of hydrology. For example, treatment of agricultural watersheds constitutes what is called agricultural hydrology. Similarly, there is urban hydrology for urban watersheds, coastal hydrology for coastal watersheds, forest hydrology for forest watersheds, desert hydrology for desert watersheds, wetland hydrology for marshes and wetlands, and mountain hydrology for mountainous watersheds. The soils, geology, and vegetation are different for these watersheds. As a result, the hydrologic processes are different in their evolution in these different watersheds, and hence their models are clearly different (Singh, 1995 a).

Frequently, watershed models are classified on the basis of their intended use:

a) planning models,

b) management models, and

c) prediction models.

A comprehensive watershed model can be employed to accomplish a considerable array of analytical tasks for planning and management of water resources. It offers the opportunity for systematic analysis of a management policy. On the other hand, it becomes an increasingly valuable predictor of river basin response to management strategies. For planning purposes, the model defines geographical and temporal trends that may not be accurate at specific sites in the river basin. Thus, the primary use of a planning model is for analyses of river system management strategies. The larger issues related to basin-wide trends and plans need to be analyzed to determine if there exist site interactions. Once these interactions are determined, a site-specific analysis can be performed with greater accuracy, perhaps with the use of a prediction model. Probably, the most powerful feature of a planning and management model is the ability it offers planners and managers to replace past management indices with improved economic performance measures of potential decisions.

Refsgaard (1996) gives the classification as in Fig. 9.7, which is applicable "both to catchment models and to simple component models such as groundwater models". According to this classification, empirical models are regarded as black box models where, not the physical processes in the catchment, but the concurrent inputs and outputs of the system are analyzed. Refsgaard further classified empirical models into *empirical hydrological* models (e.g., the use of unit hydrographs for streamflow routing and the linear reservoir concept used in modeling a groundwater system), *statistically based* methods (e.g., regression and correlation type models), and *hydroinformatics based* methods. The latter are a new group of models based on hydroinformatics; they include such techniques as neural networks and evolutionary algorithms as applied to rainfall-runoff modeling. These techniques have yet to develop further to be applied in practice (Refsgaard, 1996).

The second group of models in Refsgaard's classification (Refsgaard, 1996) comprise the lumped conceptual models which are described as grey box models. The Stanford Watershed Model of Crawford and Linsley (1966) is given as an example for this type of modeling.

For the third group of distributed physically-based models, Refsgaard (1996) states that these models describe "the natural system using the basic mathematical representations of the flows of mass, momentum and various forms of energy. For catchment models, a physically-based model in practice also has to be fully distributed". These models involve linked partial differential, integral-differential and integral equations. "together with parameters which, in principle, have direct physical significances and can be evaluated by independent measurements" (Refsgaard, 1996).

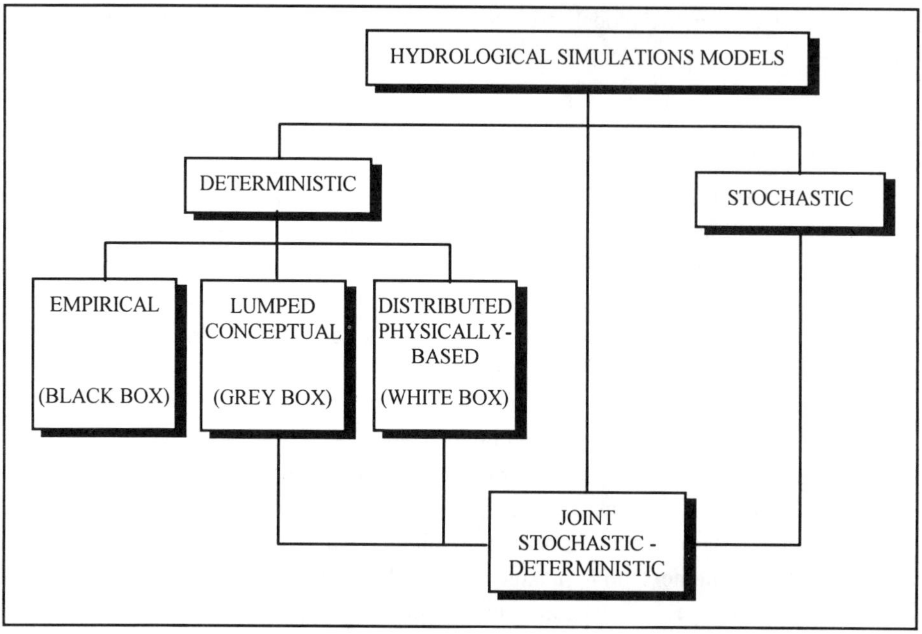

Figure 9.7. Classification of hydrologic models according to Refsgaard (1996)

9.3.7. MODEL DEVELOPMENT

Construction of a computer-based watershed model may be driven by the need to solve a particular problem or the scientific pursuit of model building which may or may not be used for problem solving. Frequently, the type of a model to be built is dictated by the availability of data. Different models have different data needs. In general, distributed models require much more data than do lumped models. In most cases, required data either do not exist or are not available in full. That is why regionalization and synthetic techniques are useful. Even if the needed data are available, problems remain with regard to incompleteness, inaccuracy, and inhomogeneity of data. Then, of course, storage, handling, retrieval, analysis, and manipulation of data have to be dealt with. If the volume of data required is large, data processing can be quite a sophisticated procedure (Singh, 1995 a).

A watershed model is an assemblage of component models corresponding to different components of the hydrologic cycle. The time scale of model output (e.g., streamflow) greatly influences the type of the model or the details to be included in the model. For example, a monthly watershed model is quite different in its architecture and construction from an hourly model. Thus, the elements of simulation differ with the model type. It remains an unsolved question as to the hydrologic laws operating at different time scales for different components of the hydrologic cycle. Solution of this question will greatly facilitate model construction and more clearly define the data needs (Singh, 1995 a).

After a model is developed, it needs calibration, verification, and an assessment of its reliability. Model calibration involves optimum parameter estimation subject to a specified error criterion. Usually, the available data is split into two halves, one for calibration and the other for verification. The calibration set should encompass the full range of data variability that may be expected in verification. A range of fairly sophisticated parameter estimation techniques are available. However, it is not clear how a model's parameter changes with variation in the objective function. Furthermore, an error analysis of calibration results must be performed. Through this analysis, it can be ascertained if the errors are due to errors in data, parameters, or model construction. Model verification and reliability of model results are essential to show if a model is any good (Singh, 1995 a).

9.4. Available Environmental Models

9.4.1. WATERSHED MODELS

As noted in section 9.1, the early problems encountered in water resources management related basically to water quantity. Along this line, hydrologic models have been developed since the early 70's at subbasin or basin scale to assess rainfall-runoff relationships, floods, water supply and drainage problems. One of the first products of such modeling efforts were the series of HEC models, developed by U.S. Army Corps of Engineers, Hydrologic Engineering Center.

Later, the watershed was considered to be made up of several components such as water, climate, soil and vegetation. Furthermore, runoff was evaluated in view of its interactions with other processes of the hydrologic cycle, i.e., precipitation, infiltration, and evapotranspiration. These considerations led to the development of more comprehensive watershed models comprising practically all processes of the hydrologic cycle and their interactions. Essentially, models of this type date back to 1966 when the Stanford Watershed Model (SWM) was developed (Crawford and Linsley, 1966) as the first example of a detailed watershed model. Essentially, SWM constituted the basis for most of the watershed models developed thereafter. One may refer to El-Kadi (1989) for an extensive review of such models. Among these are ANSWERS, CREAMS, DRAINMOD, HSP, HEC-1, NPS, SSARR, DMSM, and several others. All of these models simulate runoff, infiltration, and evapotranspiration (except ANSWERS); some take into account snowmelt, chemical transport and erosion (El-Kadi, 1989).

As described earlier in section 9.1, water pollution problems reflected an increasing trend in the 80's. The most important issue to be investigated was the problem of nonpoint source pollutants stemming primarily from land-use. Transport mechanisms of both point and nonpoint source pollutants were also significant in terms of identifying watershed processes and their interactions. These developments have led to more comprehensive watershed models which include water quality, as well as water quantity, as an essential element.

At present, watershed models have evolved into integrated type of models that incorporate all basin processes related to water, soil, and vegetation components and the interactions between these processes. This development has been concurrent with the adoption of integrated approaches to environmental management. Basically, an integrated watershed model comprises the elements shown in Fig. 9.8.

In essence, there are several watershed models developed for different purposes by different organizations. It may be observed that these comprehensive models, in general, have merged individual models previously developed for the analysis of particular catchment features. Many of the models reflect structural similarities since their underlying assumptions are the same. Some models, though, are distinctly different.

Donigian *et al.* (1995) classify watershed models into three groups as:

a) urban runoff models;
b) nonurban runoff models; and
c) surface receiving water quality models.

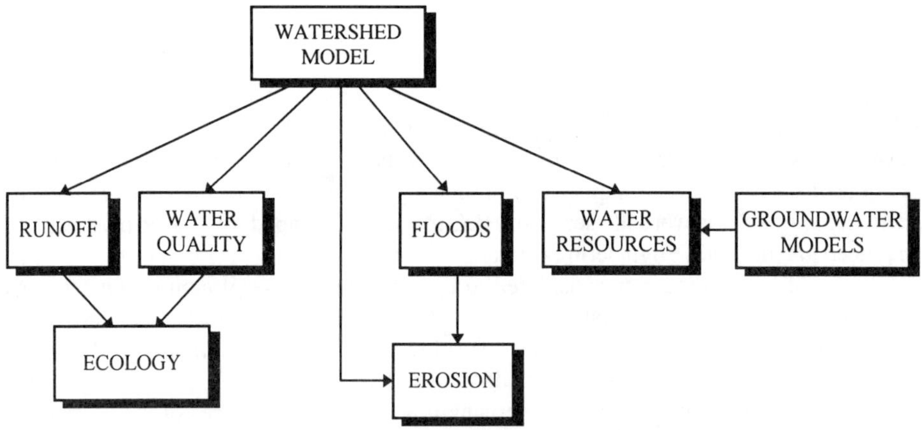

Figure 9.8. The elements of an integrated watershed model

Among the urban runoff models, Donigian *et al.* (1995) consider DR3M-QUAL (a version of USGS Distributed Routing Rainfall-Runoff Model), HSPF (Hydrological Simulation Program-Fortran, developed by USEPA), STORM (Storage, Treatment, Overflow, Runoff Model, developed by U.S. Army Corps of Engineers, HEC), and SWMM (Stormwater Management Model, developed by USEPA) as "the best choices" as full-scale simulation models for urban areas. Other available models have stemmed from SWMM; thus they are practically underlined by similar principles although they are given modified names. Apart from these U.S. models, two European models are worthy of note: MOUSE (Modeling of Urban Sewers developed by the Danish Hydraulic Institute) and the Wallingford model of HR Wallingford, UK, which includes a number of modules such as WASSP (simulation of runoff from rainfall), WALLRUS and SPIDA (sewer routing), and MOSQITO (quality simulation). In contrast to U.S. models, these European models have advanced graphics and "user-friendly" features which make use of the full capacity of microcomputers (Donigian *et al.*, 1995).

Some of the well-known nonurban runoff models include HSPF, CREAMS/GLEAMS (Chemicals, Runoff and Erosion from Agricultural Management Systems/Groundwater Loading Effects of Agricultural Management Systems, developed by the U.S. Department of Agriculture-Agricultural Research Service), ANSWERS (Areal Nonpoint Source Watershed Environment Response Simulation, developed by the Agricultural Engineering Department of Purdue University), AGNPS (Agricultural Nonpoint Source Pollution Model, developed by the U.S. Department of Agriculture-Agricultural Research Service), PRZM (Pesticide Root Zone Model developed by USEPA), and SWRRB (Simulator for Water Resources in Rural Basins, developed by Williams *et al.* (1985) and Arnold *et al.* (1989, 1991)).

On the European side, SHE (Systѐme Hydrologique Europѐen) was initiated in 1976 by a consortium of three organizations, the Institute of Hydrology (UK), the French consulting firm SOGREAH, and Delft Hydraulic Institute (Refsgaard and Storm, 1995; Refsgaard and Abbott, 1996). SHE was developed as a distributed hydrologic model. A current version of SHE is MIKE SHE, which is "a comprehensive deterministic, distributed and physically-based modeling system for the simulation of all major hydrological processes occurring in the land phase of the hydrological cycle" (Refsgaard and Storm, 1995). It simulates runoff, water quality, and sediment transport.

A suite of receiving media water quality models include (Donigian *et al.*, 1995): QUAL2E (the Enhanced Stream Water Quality Model), HSPF, WASPS (the Water Quality Analysis Simulation Program), EXAMSII (the Exposure Analysis Modeling System), CE-QUAL-RIV1, CE-QUAL-W2, and HEC-5Q. Two European models, SALMON-Q of HR Wallingford and MIKE-11 of the Danish Hyraulic Institute, are worthy of note.

A few of the above models incorporate urban runoff, nonurban runoff and receiving media modeling components into one comprehensive watershed simulation model. Essentially, these are the integrated type of watershed management models. Well-known models of this type are HSPF, SWMM, SWRRB, and MIKE SHE. If one attempts to use the other models for integrated watershed moanagement, he has to externally link them in an input-output framework, e.g., outputs of a nonurban runoff model providing inputs to a receiving media water quality model. As this approach is not convenient in terms of time and cost constraints, the use of comprehensive integrated models is preferred for watershed management (Donigian *et al.*, 1995).

For an extensive review of the above watershed models and their specifications, one may refer to recent publications like Singh (1995 b), Donigian *et al.* (1995), and Abbott and Refsgaard (1996).

9.4.2. GROUNDWATER CONTAMINATION MODELS

In the planning, management, and regulation of groundwater systems, groundwater contamination models are used. These models are of three types:

a) advection models which define the movement of contaminants as a result of groundwater flow only;

b) advection-dispersion models which add the concept of dispersion to advection models to consider dispersion related mixing and spreading and temporal changes in contaminant concentrations; and

c) advection-dispersion-chemical/biological reaction models which include the effects of reactions that change the concentration of transported contaminants (Bobba and Singh, 1995).

A large number of numerical models exist to investigate groundwater contamination. A survey of available models related to groundwater management resulted in a list of 138 flow models and 39 mass transport models (Bobba and Singh, 1995).

In their review of groundwater contamination models, Bobba and Singh (1995) have reached the following conclusions:

a) "Contaminant transport simulation techniques can be an important adjunct when dealing with:
   - the environmental acceptability of new and existing land sited facilities;
   - design of groundwater monitoring systems;
   - the need for or extent of remedial actions required to remediate an aquifer; and,
   - integration of subsurface and surface flow systems to provide a greater understanding of actual events that lead to environmental concerns."

b) "Further research and development is needed to improve contaminant transport models and increase confidence in their application. Areas for study include :
   - understanding the kinetics of processes occurring in groundwater systems, which is an important step to model improvement and utility;
   - improvements in data collection methods to decrease costs for obtaining information and to produce more representative model results; and
   - improvement in output formats to make modeling more useful for environmental management purposes".

9.4.3. SOIL EROSION MODELS

Soil erosion and degradation of land resources are significant problems in a large number of countries. Thus, substantial efforts have been spent on development of soil erosion models. These models aid in the evaluation of soil loss rates and of other problems associated with them, i.e., sediment transport and loss of nutrients. The eventual purpose of these models is to assess land use practices and land use planning. As in the case of water management, decision making for the management of land resources can be realized by developing a number of alternative land use scenarios and assessing their results by the use of soil erosion models.

Several soil erosion models exist with varying degrees of complexity. The simplest mathematical model is the Universal Soil Loss Equation, USLE, which has been used worldwide since the 60's. USLE is an empirical model which serves to estimate annual soil loss; it does not involve a spatial resolution; that is, it does not have a spatially distributed structure. With its revised (RUSLE) and modified (MUSLE) versions, USLE is still used in a large number of studies on soil loss. Other models range in various degrees of complexity. EUROSEM/MIKE SHE is a recently developed comprehensive soil erosion model with a distributed and physically-based character.

Lørup and Styczen (1996) classify soil erosion models into three groups:

a) empirical models;
b) conceptual (partly empirical/mixed);
c) physically-based.

Examples for the two first groups comprise the empirical USLE and its modifications, and more comprehensive models like ANSWERS and CREAMS mentioned earlier in section 9.4.1. The latter two are basically conceptual and event-based. The scale of their application extends to field and catchment levels. In particular, ANSWERS is a temporally and spatially (2-D) distributed model (Lørup and Styczen, 1996).

Since the 80's, a number of physically-based distributed soil erosion models have been developed at catchment or small subbasin scales. Among them are the most recently released European Soil Erosion Models, EUROSEM/KINEROS, EUROSEM/MIKE SHE and SHESED-UK, which are fully distributed in time and space scales (2-D). The latter two are continuous models, and EUROSEM/KINEROS is an event-based model.

Literature provides vast number of studies on the application of these models. One may refer to Lørup and Styczen (1996) for a fairly recent and extensive review of such models.

9.4.4. CLIMATE CHANGE MODELS

Global climatic change is one of the major issues addressed in UNCED of 1992, as discussed in section 9.1.1. Climatic change is expected due to the increasing atmospheric concentrations of carbon dioxide and other trace gases, which lead to changes in the radioactive balance of the atmosphere. Such changes propagate further to those in temperature and other climatic variables.

Hydrologic systems and water resources are likely to be seriously impacted by global climate change. Such processes as surface runoff, precipitation, soil moisture, groundwater, water quality, and sea levels will be significantly exposed to effects of climate change. Eventually, these effects will have to be considered in water resources planning and management.

Climatic changes are currently evaluated by various types of mathematical models which range from simple ones to the more complex, three-dimensional general circulation models. Mimikou (1995) summarizes these models in four groups:

a) *Energy-Balance Models (EBMs)*: these are simple models that focus on the radioactive balance of the earth/atmosphere system;

b) *Radioactive-Convective Models (RCMs)*: these models have a limited geographical and temporal coverage, and they overlook the effects of climatic change on hydrologic processes;

c) *Statistical-Dynamical Models (SDMs)*: these models are of intermediate complexity between the RCMs and the more complicated general circulation models;

d) *General Circulation Models (GCMs)*: these are the most complex, three-dimensional numerical simulation models that account for atmospheric motions, heat exchanges, and land-ice-ocean interactions.

Among all the others, GCM's are the most widely used models to simulate climatic change and its effects on various climatic and hydrological variables (e.g., temperature, water vapor distributions, radioactive fluxes, cloud and ice cover, precipitation, evapotranspiration, runoff, and soil moisture). They are considered to be more accurate than the other climate change models as they have a higher resolution and can provide better information on effects of climatic change at regional scales. On the other hand, their complexity and requirements for significant computer power are factors that limit their use. They are criticized for other characteristics such as "inadequate and limited representations of cloudiness, surface hydrology, soil moisture and turbulent heat transfer, and oversimplification in ocean modeling" (Mimikou, 1995). In addition, finer resolution is required to reliably assess the impacts of climatic change upon regional-scale processes.

Due to above limitations of GCM's, the predicted outputs of these models are considered rather as possible scenarios on future climatic conditions. Climate scenarios are also generated by other approaches. *Arbitrary (hypothetical) changes* in climatic characteristics may be formulated, and their effects on climatic and hydrologic variables may be assessed by sensitivity analyses. Some other studies use *temporal analogues*, based either on historical data of warm and cold climates in the past or on paleoclimatic data. *Spatial analogues* are also investigated by considering that the future climate of one region may be represented by the current climate of another region (Mimikou, 1995; Kaczmarek *et al.*, 1996).

The basic procedure used to investigate the effects of climate change on hydrologic processes and water resources involves the following steps (Kaczmarek *et al.*, 1996; Mimikou, 1995):

a) quantitative climatic scenarios are developed either by GCMs or other methods summarized above; these scenarios serve to quantitatively estimate the possible changes in climatic variables;

b) simulation of watershed hydrologic cycle and estimation of the impacts of climatic changes on hydrologic variables by using the scenarios of step (a);

c) assessing the impact of changes in hydrologic variables upon various components of water resources systems.

To evaluate the effects of climatic change on watershed response, either empirical or conceptual simulation models are used to model the watershed processes and characteristics. These simulation models range from simple empirical models, like the Turc formula or Thorntwaite's water balance model, to the more complex ones

(Kaczmarek *et al.*, 1996). More detailed analyses use the complex watershed models of a conceptual nature such as those discussed in section 9.4.1. Conceptual or deterministic models are preferred for this purpose since stochastic models are based on the assumption of stationarity of hydrologic variables, which validates the use of historical data in predicting the future (Mimikou, 1995; Kaczmarek *et al.*, 1996).

The above approaches to assessment of climate change upon watershed response and management have been applied to a number of test cases such as those presented by Holt and Jones (1996), Coker *et al.*, (1989), Lettenmaier and Gan (1990), Lettenmaier and Sheer (1991), Law (1989), Gleick (1987), Cohen (1986), Tung and Haith (1995), Chagnon and Kunkel (1995), Kirshen and Fennessey (1995), Seker and Harmancioglu (1997), to name but a few.

9.5. Future Expectations in Modeling

It is recognized at present that mathematical modeling is an essential tool for environmental management. Environmental models help to describe system behavior and to arrive at objective management decisions. However, it still seems difficult to admit that modeling of environmental processes has definitely reached a full level of usage. Indeed, models have become more sophisticated and more detailed in time; however, increased complexity does not necessarily mean increased efficiency and accuracy of a model. This basically appears to be a major problem in model development. A number of reasons may be cited to explain this problem.

First, there are conceptual difficulties associated with the development of model structures. Often, the concepts underlying real phenomena cannot be identified due to the large numbers of processes, variables, and interactions involved. In that case, trying to develop more comprehensive and detailed models does not look feasible. Such an approach increases the complexity of an already complicated problem. The solution may be to simplify the model by considering only those variables and interactions that dominate the system. For example, developing a complex ecological model may easily turn out to be a useless exercise if the uncertainties underlying the real phenomena do not permit a clear identification of the basic concepts.

Next, environmental problems are characterized not only by complexity but also by large numbers of processes, temporal points and spatial points to be analyzed. All of these factors require sufficient and reliable data so that they can be recognized. Yet, the case is often that we do not have such data. Then, no matter how elegantly a model is formulated, it cannot be calibrated or verified. Such a model can neither be used nor relied on. Under these conditions, one may either choose to develop his model according to the available information or may, more preferably, choose to put more demands on data monitoring systems.

The results of the two above problems is difficulty in the identification of model structure. Lack of suitable data and lack of an understanding of basic concepts leads to model structures that do not represent the true nature of environmental phenomena. For example, most processes of the environment are of a dynamic character; yet they are often considered in steady-state models (Alpaslan, 1983; Alpaslan and Harmancioglu, 1991).

The tendency among modelers to develop complex mathematical models was supported by the advent of powerful computers. However, the danger of this trend has been the development of new problems regarding the numerical analysis of highly complicated formulations. Numerical solutions solve only the mathematical equations but do not necessarily solve environmental problems. If modeling efforts are directed more to numerical analyses and solutions than to assessment of management strategies, the resulting models will be of no use to the decision maker.

As Whitehead (1978) pointed out earlier, losing sight of objectives in modeling precludes the development of efficient and useful models. The basic purpose of modeling is to simplify the complexity of a system by identifying the dominant variables and the significant variations. This objective should not be disregarded in attempts at developing more detailed and comprehensive models.

It may be interesting to note here the results of a survey carried out in the U.S.A. on assessment of modeling practice. Although this survey dates back to the 80's (Austin, 1986), most of the findings are still valid today.

The results of the survey has shown that the rate at which governmental and private agencies in the U.S.A. use mathematical models for purposes of water resources planning and management is in the order of 86%. The factors that limit a wider application of the models are listed as lack of appropriate data, costs of model development, and the failure of a large number of models in simulating the real behavior of natural systems.

It was further noted in the study by Austin (1986) that:

a) simulation models are more widely used in comparison with optimization models;
b) the most frequently used models are those that simulate surface runoff and water quality, reservoir systems, and hydraulic models;
c) statistical models are still widely used;
d) 40% of the available models are basically used by the organizations or agencies that developed them.

In several developing countries including Turkey, models are used, not by water resources planners and decision makers, but for research purposes by universities and research institutes. Often, these applications are limited particularly due to lack of adequate and reliable data (Harmancioglu and Ozkul, 1996).

The following questions need to be raised to increase the applicability of environmental models to real case problems (Harmancioglu and Ozkul, 1996):

a) How effective are the current models in guiding environmental planning and management practices?
b) What is the attitude of decision makers and managers towards employment of environmental models? What can be done to develop positive attitudes on the decision makers' side towards modeling so that it can be accepted by them as an effective management tool?
c) What factors limit the use of environmental models?

d) How well do the available models perform in simulating the behavior of real systems; in other words, how reliable are they?

The future of watershed models will be shaped by increasing societal demand for integrated environmental management; growing need for globalization by incorporation of biological, chemical, and physical aspects of the hydrologic cycle; rapid advances in remote sensing and satellite technology, geographical information systems (GIS) database management systems (DBMS), and expert systems; enhanced role of models in planning and decision making; mounting pressure on transinformation of models to user-friendly levels; and clearer statement of reliability and risk associated with model results.

Application of watershed models to environmental management will grow in the future. The models will be required to be practical tools readily usable in planning and decision making. They will have to be interfaced with economic, social, political, administrative, and legal models. Thus, watershed models will become a component in the larger management modeling strategy. Furthermore, these models will become more global, not in the sense of spatial scale but in the sense of hydrologic details.

Watershed models will have to embrace rapid advances occurring in remote sensing and satellite technology, geographical information and database management systems, and expert systems. With use of remote sensing and satellite technology, our ability to observe data over large and inaccessible areas is vastly improved, making it possible to develop truly distributed watershed models for both gaged and ungaged watersheds. Distributed models require large quantities of data which can be stored, retrieved, managed and manipulated with use of GIS and DBMS. This, of course, is possible because of the literally unlimited computing capability available at present and will be even more so in the future. If watershed models have to become practical tools, then they have to be simple, easy to use with a clear statement as to what they can and what they cannot do. They have to define the reliability with which they accomplish their intended functions, and require the user to posses the least of hydrologic training. The user is interested in what a model yields and how accurately it yields, not the biology, chemistry, physics and hydrology it is based on. The model results by themselves are useless unless they are associated with a clear statement of their reliability and accuracy.

The models will have to be described in simple terms such that interpretation of their results would not tax the ability of a user. They are designed to serve a practical end and their constituency is one of users. After all, computer models of watershed hydrology are to be used, not be confined to academic shelves. Thus, model building will have to gravitate around the central theme of their eventual practical use in integrated environmental management. Although much progress has been made in computer modeling of environmental processes and systems, there is still a long way to go before the models would be able to fully integrate rapidly occurring advances in information, computer, and space technology, and would become "household" tools.

References

Abbott, M.B. and Refsgaard, C. (eds.) (1996) *Distributed Hydrological Modeling*, Kluwer Academic Publishers, Dordrecht, 321 p.

Abbott, M.B., Bathurst, J.C., Cunge, J.A., O'Connel, P.E., and Rasmussen, J. (1986a) An introduction to the European Hydrological System - Systeme Hydrologique Europeen, "SHE", 1: History and philosophy of a physically-based, distributed modeling system, *Journal of Hydrology* **87**, 45-59.

Abbott, M.B., Bathurst, J.C., Cunge, J.A., O'Connel, P.E., and Rasmussen, J. (1986b) An introduction to the European Hydrological System - Systeme Hydrologique Europeen, "SHE", 2: Structure of a physically-based, distributed modeling system, *Journal of Hydrology* **87**, 61-77.

Alpaslan, N. (1983) *Modeling and Mathematical Simulation of Completely Mixed Activated Sludge Systems*, Ph. D. Thesis, Dokuz Eylul University, Institute of Science and Engineering, no. 7, Izmir, Turkey, (in Turkish).

Alpaslan, N. and Harmancioglu, N. (1991) Environmental decision models, in: G. Tsakiris (ed.), *Advances in Water Resources Technology*, A.A. Balkema, Rotterdam, pp. 339-347.

Arnold, J.G., Williams, J.R., Griggs, R.H., and Sammons, N.B. (1991) *SWRRBWQ, A Basin Scale Model for Assessing Management Impacts on Water Quality*, USDA-ARS, West Lafayette, IN.

Arnold, J.G., Williams, J.R., Nicks, A.D., and Sammons, N.B. (1989) *SWRRB, A Basin Scale Simulation Model for Soil and Water Resources Management*, Texas A & M Press.

Austin, T.A. (1986) Utilization of models in water resources, *Water Resources Bulletin* **22(1)**, 49-53.

Beck, M.B. (1987) Water quality modeling: A review of the analysis of uncertainty, *Water Resources Research* **23(8)**, 1393-1442.

Beck, M.B. (1978) Some observations on water quality modeling and simulation, in: G.C. Vansteenkiste (ed.), *Modeling, Identification and Control in Environmental Systems*, North Holland, pp. 775-786.

Biswas, A.K. (1976) *Systems Approach to Water Management*, McGraw-Hill, New York.

Bobba, A.G. and Singh, V.P. (1995) Groundwater contamination modeling, in: V.P. Singh (ed.), *Environmental Hydrology*, Kluwer Academic Publishers, Water Science and Technology Library, Dordrecht, pp. 225-319.

Chagnon, S.A. and Kunkel, K.E. (1995) Climate-related fluctuations in Midwestern floods during 1921-1985, *ASCE, Journal of Water Resources Planning and Management* **121(2)**, 326-334.

Cohen, S.J. (1986) Impacts of CO_2-induced climatic change on water resources in the Great Lakes basin, *Climatic Change* **8**, 135-153.

Coker, A.M., Thompson, P.M., Smith, D.I., and Penning-Rowsell, E.C. (1989) The impact of climate change on coastal zone management in Britain: A preliminary analysis, in: *Conference on Climate and Water*, The Publications of the Academy of Finland, Vol. 2, pp. 148-160.

Crawford, N.H. and Linsley, R.K. (1966) Digital simulation in hydrology, Stanford watershed model, IV. Department of Civil Engineering, Stanford University, Technical Report 39.

Diskin, M.H. and Simon, E. (1979) The relationship between the time bases of simulation models and their structure, *Water Resources Bulletin* **15(6)**, 1716-1732.

Donigian, A.S., Imhoff, J.C., and Ambrose, Jr., R.B. (1995) Modeling watershed water quality, in: V.P. Singh (ed.), *Environmental Hydrology*, Kluwer Academic Publishers, Water Science and Technology Library, Dordrecht, ch. 11, pp. 377-426.

El-Kadi, A. (1989) Watershed models and their applicability to conjunctive use management, *Water Resources Bulletin* **25(1)**, 125-137.

Gleick, P.H. (1987) The development and testing of a water balance model for climate impacts assessment: Modeling the Sacramento Basin, *Water Resources Research* **23**, 1049-1061.

Harmancioglu, N. and Ozkul, S. (1996) The use of computer-based mathematical models in water resources management (in Turkish), Istanbul Technical University , Faculty of Civil Engineering, V. Symposium on the Use of Computers in Civil Engineering, June 17-19, 1996, Istanbul.

Holt, C.P. and Jones, J.A.A. (1996) Equilibrium and transient global warming scenario implications for water resources in Wales, *Water Resources Bulletin* **32(4)**, 711-722.

Hydrologic Engineering Center (1981) *HEC-1, Flood Hydrograph Package - Users Manual*, U. S. Army Corps of Engineers, Davis, California.

Hydrologic Research Laboratory (1972) *National Weather Service River Forecast System: Forecast Procedures, Technical Memorandum NWS-HYDRO-14*, National Weather Service, National Oceanic and Atmospheric Administration, U. S. Department of Commerce, Washington, D.C.

Kaczmarek, Z., Strzepek, K.M., Somlyody, L., and Priazhinskaya, V. (eds.) (1996) *Water Resources Management in the Face of Climatic/Hydrologic Uncertainties*, Kluwer Academic Publishers, Water Science and Technology Library, 395 p.

Kirshen, P.H. and Fennessey, N.M. (1995) Possible climate-change impacts on water supply of metropolitan Boston, *ASCE, Journal of Water Resources Planning and Management* **121(1)**, 61-70.

Laurenson, E.M. and Mein, R.G. (1983) *RORB-Version 3: Runoff Routing Program - User Manual*, 2nd edition, Department of Civil Engineering, Monash University, Monash, Australia.

Law, F.M. (1989) Identifying the climate-sensitive segment of British reservoir yield, in: *Conference on Climate and Water*, The Publications of the Academy of Finland, Vol. 2, pp. 177-190.

Lettenmaier, D.P. and Sheer, D.P. (1991) Climatic sensitivity of California water resources, *Journal of Water Resources Planning and Management* **117(1)**, 108-125.

Lettenmaier, D.P. and Gan, T.Y. (1990) Hydrologic sensitivities of the Sacramento-San Joaquim River Basin, California, to global warming, *Water Resources Research* **26(1)**, 69-86.

Lørup, J.K. and Styczen, M. (1996) Soil erosion modeling, in: M.B. Abbott and J.C. Refsgaard (eds.), *Distributed Hydrological Modeling*, Kluwer Academic Publishers, Water Science and Technology Library, Dordrecht, pp. 93-120.

Loucks, D.P., Stedinger, J.R., and Haith, D.A. (1981) *Water Resources Systems Planning and Analysis*, Prentice Hall, Inc., Engelwood Cliffs, New Jersey.

Metcalf and Eddy Inc., University of Florida and Water Resources Engineers, Inc. (1971) *Storm Water Management Model, Vol. 1, Final Report*, Water Pollution Control Research Series 11024 DOC 07/71, U.S. Environmental Protection Agency, Washington, D.C.

Mimikou, M.A. (1995) Climatic change, in: V.P. Singh (ed.), *Environmental Hydrology*, Kluwer Academic Publishers, Water Science and Technology Library, Dordrecht, pp. 69-106.

Morris, E.M. (1980) Forecasting flood flows in grassy and forested basins using a deterministic distributed mathematical model, IAHS Publication 129, Wallingford, England, pp. 247-255.

Refsgaard, J.C. (1996) Terminology, modeling protocol and classification of hydrological model codes, in: M.B. Abbott and J.C. Refsgaard (eds.), *Distributed Hydrological Modeling*, Kluwer Academic Publishers, Water Science and Technology Library, Dordrecht, , pp. 17-40.

Refsgaard, J.C. and Abbott, M.B. (1996) The role of distributed hydrological modeling in water resources management, in: M.B. Abbott and J.C. Refsgaard (eds.), *Distributed Hydrological Modeling*, Kluwer Academic Publishers, Water Science and Technology Library, Dordrecht, pp. 1-16.

Refsgaard, J.C. and Storm, B. (1995) MIKE SHE, in: V.P. Singh (ed.), *Computer Models of Watershed Hydrology*, Water Resources Publications, Littleton, ch. 23, pp. 809-846.

Seker, S. and Harmancioglu, N. (1997) The effects of global climate change on management on the Gediz River Basin (in Turkish), Conference on Meteorologic Hazards, to be held in December 23-24, 1997, Ankara.

Singh, V.P. (1995a) Watershed modeling, in: V.P. Singh (ed.), *Computer Models of Watershed Hydrology*, Water Resources Publications, Littleton, ch. 1, pp. 1-22.

Singh, V.P. (ed.) (1995b) *Computer Models of Watershed Hydrology*, Water Resources Publications, Littleton.

Singh, V.P. (1992) *Elementary Hydrology*, Prentice-Hall, Inc., Englewood Cliffs, New Jersey.

Singh, V.P. (1988) *Hydrologic Systems, Vol. 1, Rainfall-Runoff Modeling*, Prentice-Hall, Inc., Englewood Cliffs, New Jersey.

Spofford, Jr., W.D. (1975) Ecological modeling in a resource management framework: An introduction, in: C.S. Russel (ed.), *Ecological Modeling in a Resource Management Framework: Resource for the Future*, Washington, D.C.

Sugawara, M., Watanabe, I., Ozaki, E., and Katsugama, Y. (1984) *Tank Model with Snow Component*, Research Notes of the National Research Center for Disaster Preventation, no. 65, Science and Technology, Ibaraki-Ken, Japan.

Tung, C.P. and Haith, D.A. (1995) Global-warming effects on New York streamflows, *ASCE, Journal of Water Resources Planning and Management* **121(2)**, 216-225.

U.S. Army Corps of Engineers (1972) *Program Description and User Manual for SSARR model-Streamflow Synthesis and Reservoir Regulation Model*, Corps of Engineers, North Pacific Division, Department of the Army, Portland, Oregon.

Uslu, O. and Turkman, A. (1987) *Water Pollution and Control*, Publications of the General Directorate of Environment, Education Series I, Ankara, (in Turkish).

Uslu, O., Harmancioglu, N., and Alpaslan, N. (1984) Decision models in environmental engineering, in: O. Uslu (ed.), *Environment'84: V. Turkish-German Symposium on Environmental Engineering*, pp. A/74-A/79, (in Turkish).

Uslu, O., Harmancioglu, N., Ozer, A., and Alpaslan, N. (1983) On environmental modeling, *Environment'83: II. National Environmental Engineering Symposium*, Izmir, Turkey, pp. CS.36-CS.44, (in Turkish).

Vansteenkiste, G.C. (1978) Environmental systems research: An interdisciplinary dedication, in: G.C. Vansteenkiste (ed.), Introduction to *Modeling, Identification and Control in Environmental Systems*, North Holland, pp. vii-xii.

Whitehead, P.G. (1978) Modeling and planning of environmental studies, in: G.C. Vansteenkiste (ed.), *Modeling, Identification and Control in Environmental Systems*, North Holland, pp. 527-556.

Williams, J.R. and Hann, R.W. (1972) HYMO-A Problem-oriented Computer Language for Building Hydrologic Models, Water Research Laboratory, vol. 8, no. 1, pp. 79-86.

Williams, J.R., Nicks, A.D., and Arnold, J.G. (1985) Simulator for water resources in rural basins, *ASCE, J. Hydraulic Engineering* **111(6)**, 970-986.

CHAPTER 10

DECISION MAKING FOR ENVIRONMENTAL MANAGEMENT

N.B. Harmancioglu, O. Fistikoglu, S.D. Ozkul and M.N. Alpaslan

Abstract. Modern technology has provided efficient tools such as advanced models, remote sensing and satellite imaging, GIS, and expert systems to facilitate decision making for environmental management. These tools are currently integrated to establish an environmental information system which permits testing and evaluating of alternative management scenarios. Thus, current decision making procedures are realized within a multimedia framework that is easily accessible by users at local or even global levels. The presented chapter summarizes these new developments along with difficulties of their implementation in developing countries.

10.1. Introduction

We live in an age of environmental alertness. Almost all natural resources are attacked by pollution at varying degrees of intensity. The quality of surface and ground waters is continuously degrading. The situation is similar for land resources with problems of soil erosion, deforestation, and desertification in many parts of the world. Air pollution has already reached life and health threatening levels in particular regions. These problems have eventually endangered physical habitat for biodiversity. Further difficulties are expected due to the possible effects of climate change on various components of the environment. All these adverse developments are induced by diverse human activities as well as by natural occurrences. The result is that environmental degradation not only endangers nature, but it also has serious social and economic implications.

Thus, we need urgent remedies, not short-term but long-term solutions, to preserve environmental quality for future generations as well as for the present. It was this consideration that led to the adoption of "sustainable development" as the basic policy in environmental management. The need for sustainability has put significant demands on the decision making process for management. We now need more efficient, more effective, and more reliable decisions with which to control and develop our environment.

Decision makers and planners are unfortunate in the sense that current problems have become multifold, multidimensional, and multifaceted. Similarly, there are numerous objectives, often of a conflicting nature, to be satisfied. Furthermore, technology has provided an abundant number of solutions that may be applied even though their consequences for a particular problem investigated are not known in

N.B. Harmancioglu et al. (eds.), Environmental Data Management, 243-288.

advance. Thus, the result is that decision makers have to perform in the realm of complexity and uncertainty. This is why the situation may be described as being "unfortunate" for them.

On the other hand, in the present age we live in, technology, although it has stimulated environmental pollution in a number of ways, has currently provided the most advanced and effective tools to facilitate decision making. Thus, decision makers can be considered "fortunate" as they are now better equipped in identifying, analyzing, and solving environmental problems.

The essential basis for decision making is information on the environment. This information is to be provided by available data on various components of the environmental continuum, as well as social, economic, and all types of demographic data.

The most significant issue in environmental management is, then, the transfer of data into information. In essence, Fig. 1.1 in Chapter 1 summarizes the basic steps in this transfer, i.e., all steps in the given sequential order contribute to the production of information required by the decision maker. Thus, information is the end product of a data management system.

Furthermore, effective and efficient decisions require information that is sufficient and reliable. This requirement explains why data have to be managed within a cohesive system as each step in the system affect the quality of the information produced.

On the other hand, to support the decision making process, information should not only be sufficient and reliable but must also satisfy three conditions:

a) it must be available *when* it is needed;
b) it must be *easily accessed* by the user; and,
c) it should be available in a form that is *easy to understand and use* for the decision maker or the planner.

10.2. Present Tools for Decision Making

10.2.1. BASIC TOOLS

As noted in section 10.1, modern technology has provided the most advanced and sophisticated tools for decision making. There are essentially four basic tools that support the decision making process:

a) data;
b) models (simulation or optimization);
c) Geographic Information Systems; and,
d) knowledge based decision support systems or expert systems.

Among these, data are generally collected by traditional methods, i.e., taking samples via field monitoring. Such data are available at varying degrees of completeness and accuracy. Often, particularly in developing countries, environmental data have short sample sizes and reflect incompleteness, inaccuracy, and inhomogeneity. Modern technology has provided means of alleviating these deficiencies. We now have automatic monitoring equipment to assure completeness. Most important of all, remote sensing and satellite technology (RRST) has been developed to provide a number of significant advantages as discussed in section 10.2.2.

Modeling, as the second important tool in decision making, has also experienced significant evolution within the last decade, moving from simple empirical models to simulate individual processes to physically-based distributed models of a conceptual nature, which model the environment in a multicomponent framework. These developments are discussed in detail in Chapter 9.

The two new tools are the Geographical Information Systems (GIS) and expert systems, which have developed into essential components of the decision making process, and they will remain so for the foreseeable future. The basic advantages of these tools are discussed in the following sections.

10.2.2. REMOTE SENSING AND SATELLITE TECHNOLOGY (RRST)

A major problem in environmental analysis is the lack of adequate data to quantitatively describe an environmental process accurately. To alleviate this problem, the remote sensing technology aids as follows (Schultz, 1988; Singh, 1995):

1) it produces areal measurements instead of point measurements;
2) all information is collected and stored at one place;
3) it offers high resolution in space and/or time;
4) data are available in digital form;
5) data acquisition does not interfere with data observation;
6) data can be gathered for remote areas that are otherwise inaccessible; and,
7) once the remote sensing networks are installed, data measurement is relatively inexpensive.

The main drawback in obtaining remotely sensed data is in calibrating the electromagnetic signals in terms of the processes observed.

Remote sensing and satellite data can be used for watershed modeling, and are useful, especially where data collection sites and cost-effectiveness of data collection are important on one hand, and increasing requirement for data on the other. Among the satellites with hydrologically useful data are NOAA series, TIROS N, SPOT, Landsat, GOES, GMS, and Meteosat.

Remote sensing allows to obtain data of a process from a location far away from the user or modeler. Sensing is taken to mean observation of the average value of a variable over some areal extent by examining the electromagnetic energy. Measurement techniques are either passive or active. Passive measurement techniques determine the amount of reflected sunlight or the amount of natural emissions from the target at various wavelengths. Active measurement techniques direct artificially generated signal at a target and measure the reflected signal. An example of a passive technique is air photography and that of an active technique is weather radar.

Several different remote sensing sensors provide hydrologically relevant information. These include aerial photography, scanning radiometers, spectrometers, and microwave radars. Since all information is obtained in the form of electromagnetic signals, it is important to determine the type of signal most suitable for the particular variable observed. Furthermore, the needed resolution in space and time is obtained with certain types of sensors only (Singh, 1995).

Remote sensing data from satellites are useful in watershed modeling in two ways. First, satellite data can be used to better define soils and land covers over a watershed, which are needed to determine infiltration, evapotranspiration, and runoff. Second, remote sensing measures data over space rather than at a point, and can, therefore, be used to correct errors in input data based on point measurements, as the case is with rainfall, evaporation, and the like. By far, the most important data produced by remote sensing are on soil moisture and evapotranspiration through satellite thermal infrared images. These data can be employed to model water exchange between land surface and atmosphere (Engman, 1986), and to couple land-surface hydrology and atmospheric models (Otte *et al*, 1989; Singh, 1995).

The remote sensing technology can be employed in a multitude of applications in hydrology. Radars are employed for rainfall measurements. Satellite data can be used for estimation of area and intensity of rainfall. Temperature estimates obtained from thermal infrared and soil moisture from microwave observations can be applied to develop mathematical models of evapotranspiration. Soil moisture is most adequately obtained through remote sensing. Inventories of water bodies such as lakes, dams, rivers, swamps, flooded areas, etc., are easily obtained by remote sensing imagery. Even groundwater resource evaluation can be assisted through the remote sensing information produced on geologic and geomorphologic characteristics. Snow and ice conditions are mapped with the use of satellites. Water quality parameters such as algae, chlorophyll, and aquatic life parameters, as well as thermal pollution, oil spill, etc., can be observed by remote sensing techniques.

With the vastly improved capability to observe hydrologic data, remote sensing technology has developed a significant role in a range of hydrologic applications, and these applications are becoming operational at an accelerated pace. Some of these applications are exemplified in real-time flood forecasting by a distributed model with radar rainfall measurements as input, computation of monthly runoff by a lumped model with NOAA infrared satellite data as input, model parameter estimation with multispectral Landsat satellite data, development of macroscale hydrological models coupled with atmospheric general circulation models with satellite data, and development of distributed hydrologic models based on a digital terrain model and a geographic information system with the use of Landsat satellite imagery. These applications permit operationalization of space technology in such areas as weather forecasting, forecasting seasonal/short-term snowmelt runoff, evolution of watershed management strategies for conservation planning, development of reporting services for drought assessment/forecasting, mapping of groundwater potential to support conjunctive use of surface and groundwaters, inventorying of coastal processes and marine process environment, environmental impact assessment of large-scale water resources projects, flood-damage assessment, and development of resource information matrix for irrigation development (Schultz, 1997; Singh, 1995).

10.2.3. THE USE OF GEOGRAPHIC INFORMATION SYSTEMS (GIS)

Geographic Information Systems (GIS) are computer-based tools that store, analyze, retrieve, manipulate and manage large amounts of spatial data. They are spatial database management techniques. Originally, these tools were developed to ease cartography, but are being used these days for inventory analysis, estimation, planning, and modeling.

Watershed modeling requires accurate and representative data. Collection, storage, and manipulation of the required data can be cumbersome for three reasons (Singh, 1995):

1) spatial nature of hydrologic data;
2) large volume of data; and,
3) organization of data.

However, through the use of a Geographic Information System and the associated software, such data can be compiled and processed with relative use. The watershed and hydrology are inherently spatial. The physical characteristics of the watershed such as soils, land use and topography vary spatially. In distributed watershed models, these characteristics can be used directly. However, the watershed will have to be spatially characterized into homogenous subareas. To that end, the traditional data file structure is less than troublefree. Clearly, a general-purpose spatial data analysis and manipulation tool is desirable.

Distributed watershed models, especially numerical models, require large volumes of data. A hydrologic unit is characterized by many data, including those on topography, land use, soils, geology, climate, and so on. On the other hand, the various application programs may require the same data stored in different files, thus creating storage redundancy with the traditional file structure system.

Hydrologic data are arranged in a variety of formats and stored on various media. It is often difficult and tedious to transform and manage these data. Furthermore, because many models have different data requirements, a collection program tailored to the demand of a particular model cannot be used for another with differing data requirements. Consequently, a separate collection program has to be developed and data transformation then has to be undertaken for each model. These data problems can be resolved through application of a GIS. Its ability to extract, overlay, and delineate watershed characteristics permits integration with watershed models. Design, calibration, modification, and comparison of these models can be significantly facilitated by the use of the GIS. However, application of GIS technology to watershed modeling requires careful planning and extensive data manipulation, involving three major steps (Singh, 1995):

1) spatial database construction;
2) integration of spatial model layers; and,
3) the GIS and model interface.

With ever increasing volume of data available from many sources, the first step is time consuming. The second step is becoming less tedious due to advancing GIS capabilities in that a series of spatial overlaying and/or projecting procedures can be derived yielding the final modeling layer. The last step involves actual programming for interfacing GIS with watershed models.

Although large quantities of data can be efficiently manipulated in time by using a GIS, certain limitations need to be recognized. First, building the database is expensive and time consuming. Also, appropriate data may be another problem, and making up for deficiencies in data is quite difficult with a GIS. Another deficiency of the GIS is related to real time applications, i.e., when processing rapidly varying dynamic variables (Harmancioglu *et al.*, 1997). Unless the same database can be used in a variety of analyses or models, the use of a GIS may not be cost-effective. However, with increasing complexity of investigations or models involving more data and more complex data processing, a GIS may indeed be cost effective. The PC-based GIS innovations have greatly cut down the cost and are more affordable now. A succesful GIS application calls for data accuracy, data integrity, and multiple attributes capability. A GIS package should have a relational database management system (RDBMS) that provides both logic and physical data independence.

A GIS may be either raster-based or vector-based. A raster-based GIS is more storage efficient than a vector-based GIS. The larger the coverage, the greater the storage space used by a vector-based GIS. The processing speed of a vector-based GIS is slower and can be an important consideration if extensive spatial processing is needed for a watershed model.

Construction of an appropriate GIS database is the most difficult and expensive procedure for watershed modeling. There exist large volumes of data on soils, geology, land use, vegetative cover, and so on in digitized form. However, there are problems with many of the existing digital databases:

1) they are in different formats;
2) they use different coordinate (projection systems);
3) they are too coarse for use in 2-D or 3-D watershed models.

Integration of GIS with watershed modeling accomplishes a number of significant hydrologic functions: designing, calibrating, modifying, and comparing watershed models. The use of a GIS permits subdiving a watershed spatially into hydrologically homogenous subareas. Depending upon the type of application requiring categorization of hydrologic properties, many combinations of spatial overlay can be performed. With this technique, it is possible to delineate areas of varying soil loss rates (Ventura, *et al*, 1988), identify potential areas of nonpoint source agricultural pollution (Schmidt, 1987), map groundwater contamination susceptibility (Whittenmore *et al*, 1982), and so on. GIS improves the ability to incorporate spatial details beyond the existing capability of watershed models. With much better resolution of terrain – streams and drainage areas – , the ability to delineate more appropriate grid layers for a finite element watershed model is enhanced (Singh, 1995).

The widely used applications of GIS in different fields of water resources research can be listed as follows, together with some relevant examples:

a) flood analysis (Warwick and Haness, 1993);
b) erosion and erosion-based pollution prediction (Tim *et al.*, 1992; Fraser *et al.*, 1996);
c) groundwater quality and monitoring (Baker *et al.*, 1993);
d) wetland analysis (Leipnik *et al.*, 1993);
e) monitoring applications (remote sensing) (Schultz, 1993);
f) basin morphology (Warwick and Haness, 1993);
g) geographic and nongeographic data queries (Heidke and Auer, 1993);
h) water quality and quantity model-GIS integrations (Engel *et al.*, 1993); and,
i) expert systems and decision making (Mc Kinney *et al.*, 1993).

10.2.4. EXPERT SYSTEMS IN DECISION MAKING

A knowledge-based expert system is an interactive computer program that encodes judgment, experience, rules of thumb, intuition, empirical knowledge, and other forms of information or expertise to give knowledgeable advice on a specified problem. Such a program is knowledge-based and is a domain-independent reasoning system. In other words, it can be applied to other numerical reasoning tasks in domains other than hydrology. The expertise about the problem area is specified in the form of a model of the decision making process. A model of expertise in the form of judgment or experience is constituted by a set of statements in a stylized language. The program of reasoning reads a model or models from a disk file and creates an internal data structure that drives the consultation. Thus, changing the expertise in a knowledge-based system simply involves replacing one model by another. In this sense the structure of such systems is fundamentally different from most "traditional" programs.

Perhaps the first attempt to develop a knowledge-based computer consultation system, called HYDRO, for water resources problems was by Gaschning *et al.* (1981). They applied the HYDRO system to determine appropriate numerical values for various parameters that describe the physical characteristics of a watershed. The HYDRO system was intended to provide advice comparable to that of an expert hydrologist in selecting parameter values characteristic of the watershed under consideration. The parameter values so obtained served as input to the Hydrocomp HSPF Simulation program for evaluating various hydrological aspects of the watershed under consideration.

Although expert systems have a vast potential, they have not yet found widespread usage in watershed hydrology. There is, however, every reason to believe that they will have commonplace application in water resources area in the not too distant future (Singh, 1995).

10.3. The Decision Making Process

10.3.1. A NEW APPROACH

Recent advances in information and communication technologies, coupled with the development of new tools such as RSST, GIS and expert systems, have also changed the framework of the decision making process. In the past, planners did also take advantage of models and data. A typical approach would be to develop a management model, often in the form of an optimization model, and incorporate into it a technological model (e.g., water quality, urban infrastructure, erosion, and the similar) as part of the objective function or the constraints. The solution to such an optimization problem would indicate the "best" decisions to be made. An example on how this procedure is accomplished has been given earlier in Chapter 9.

Nowadays, the increased complexity of the management problem on one hand, and the availability of advanced tools for management on the other hand, have changed the way the decision making process is accomplished. The current approach involves two basic steps:

a) development of an integrated information system to support the decision making;
b) building and testing alternative management scenarios by running each through the information system.

These two steps imply that the decision making process has become an interactive one, i.e., the planner communicates with the information system and tests the consequences of his decisions instead of arriving at a single fixed decision.

10.3.2. DEVELOPMENT OF AN INFORMATION SYSTEM FOR MANAGEMENT

Development of an information system for decision making involves three basic steps (Fedra, 1997):

a) preparation of data for analysis;
b) analysis of data and modeling to transfer data into information;
c) developing methods of communication for the user (planner) to interpret the results and interactively participate in the planning process.

Preparation of data by itself requires significant efforts to sort the available data into a usable format. Data comprise those on different processes. They come from numerous sources and different monitoring organizations so that they are often available in varying formats. Thus, incompatible data formats is a problem to be resolved.

Furthermore, if data are collected via automatic measuring equipment or are remotely sensed satellite data, they come in large volumes. If they are monitored by traditional sampling methods, problems of incompleteness, inadequacy and inhomogeneity of data, together with high sampling errors, must be handled. Accordingly, the data preparation step involves testing of available data for reliability and errors and integration of large volumes of multiformat, multisource and multimedia data into a database. Clearly, this step is essentially devoted to data management.

To effectively manage the available data and databases, GIS may be used as an essential element of the information system. Fedra (1997) states that GIS is the "backbone of data management". When GIS is integrated with the database system, the two elements together actually constitute the "backbone" of the whole management procedure. This is why current management practices include GIS as an essential tool.

When data are prepared via database-GIS integration, models can be run to interpret the data for production of information. This is another stage of integration where data + GIS + model, as primary tools of the information system, are integrated. GIS plays an active role here by preparing the input data for the model and further displaying the model results within a geographic reference frame.

A final stage of integration is realized between the model and an expert system. Expert systems function like a model as they define, for a given set of input variables, an output value for a variable. Unlike numerical models, expert systems accomplish this function in a rule-based, qualitative framework where not only physical, but also socioeconomic factors can be accounted for. A numerical model can be integrated with the expert system by substituting a set of rules (Fedra, 1997). Scenario development and testing, as the second step of the decision making process, can then be accomplished though the integrated model + expert system component of the information system.

As it follows from the above, the information system for environmental management is established by the integration of basic management tools: databases, GIS, models and expert systems. The integration of management tools constitutes a multimedia framework as a basis for interactive software systems, which are designed nowadays with an object oriented approach. These systems can be accessible to local users or even to global communities via networking systems such as the World Wide Web (Fedra, 1997). Thus, users can easily access to and interact with the system, test and interpret results of various scenarios, and make judgments.

The above procedure implies that the decision making process has now evolved into a dynamic procedure where the decision maker is actively involved in all aspects of the management problem, i.e., databases, modeling, GIS and expert systems.

10.4. Applications

Current literature provides several examples where the above management tools are integrated and used for environmental assessment and management. To refer to a few of the most recent ones, one may cite Queen *et al.* (1996) who integrated GIS and remote sensing data to provide inputs to linear regression models for assessment of the Nemadji River watershed in norteastern Minnesota. The results of their study indicated significant implications for the strategic maintenance and restoration of forest land in the basin, where substantial rates of sediment transport have led to adverse environmental impacts.

Xue *et al.* (1996) developed a mechanism-based Best Management Practices Assessment Model (BMPAM) to assess best management practices (BMPs) in agricultural and urban areas to minimize stormwater runoff and associated pollution problems. They demonstrated the integration of this model with a GIS.

Burks and Passmore (1996) have developed an integrated decision support system for the U. S. Army Corps of Engineers' environmental restoration projects. Their system integrates environmental planning processes, from design through incremental cost analysis, within a simple user interface. Burks and Passmore (1996) discuss the linkage of habitat evaluation models with cost matrices and spatial data taken from a GIS integrated database.

Yoon (1996) developed methods for directly linking the distributed parameter model AGNPS with a GIS and a relational database management system (RDBMS) to investigate a nonpoint source pollution problem. The purpose of this study was to identfy critical areas of nonpoint source pollution and further to investigate "what if" scenarios to support the decision making processes such as the Best Management Practices (BMPs) for the Sand Hill River subbasin in Minnesota. The results of the study has produced an optimal BMP scenario.

Ruggles *et al.* (1996) developed a system that uses a GIS as a platform to analyze information obtained from remotely sensed images, compare that information to existing coverages within the GIS and, with the capability of accessing an intelligent database, assist in characterization of the available information. Their system was tested on a small watershed in Pennsylvania and proved to be a useful tool in analyzing water resources information and relating this information to remotely sensed images.

Richards *et al.* (1996) used GIS for groundwater assessments to manage, analyze and display all types of spatial data, e.g., the spatial locations of wells, borings, sources of contamination, etc. Their study describes how ARC/INFO GIS and Oracle relational database management system (RDBMS) were used at the Northwest Florida Water Management District to conduct an analysis of data generated by a typical field investigation. They have employed this system in integration with a numerical groundwater model in support of an in-depth groundwater investigation.

Ford and Killen (1995) have demonstrated the integration of a personal computer (PC) based decision support system (DSS) with specialized versions of the well-known HEC-1 and HEC-5 models for forecasting runoff and simulating reservoir operation, respectively. The DSS they developed uses custom software for retrieving and processing rainfall and streamflow data and an industry-standard database management system (DBMS) for filing the data. The DBMS and a special program manager integrate the model components and greatly simplify forecasting and reservoir release selecting and evaluating.

Shepherd and Ortolano (1996) describe the development of a critiquing expert system to assist in operations of a water supply system of the San Francisco Water Department. They claim that the implementation of the critiquing expert system enhances the capabilities of a conventional expert system by refining an operator's proposed plan. The system evaluates the operating plan and provides feedback, which includes suggestions for improvement, warnings and alternatives. Shepherd and Ortolano (1996) note that the system is welcomed by the San Francisco Water Department although, historically, expert systems have not been widely adopted by water utilities in the U.S.A.

Fedra (1997) describes two integrated environmental information systems with easy-to-use and easy-to-understand multimedia user interface. One of them serves for local air quality assessment and management, and the other focuses on regional water resources management. Both include interactive information systems equipped with GIS, simulation models, and expert systems.

The information systems for air pollution assessment and management, AirWare is presented by Fedra (1997) to include:

a) a geographical information system with background maps including a DEM (Digital Elevation Model) and satellite imagery;
b) object databases (air quality observation time series, meteorological data, emission inventories, and model scenarios);
c) an expert system for the estimation of point and area source emissions;
d) simulation models (short-term and long-term Gaussian models, a pseudo 3D (three layer) dynamic PUFF model, a photochemical box model), and an optimization model.

The water resources management information system introduced by Fedra (1997) is called WaterWare which provides advanced tools to analyze environmental impacts and constraints of water resources management options. The system uses and integrates a set of databases, GIS, simulation models, and expert systems. It is developed in a modular, easy-to-use, and interactive framework and has an open, object oriented structure. This structure involves a hybrid GIS with hierarchical map layers, object databases, time series analysis, expert system, reporting functions and a multimedia user interface with Internet access. The models integrated to the system perform management scenario analysis for water quantity and quality problems.

The system includes both simulation and optimization models and related tools such as:

a) a rainfall-runoff and water budget model;
b) an irrigation water demand estimation model;
c) a graphical river network editor;
d) dynamic and stochastic water quality models;
e) a groundwater flow and transport model;
f) a water resources allocation model; and,
g) an expert system for environmental impact assessment.

The component models and tools of the system relate to objectives, criteria and constraints of decision problems. The system has been developed through a series of applications to the River Thames in UK, the Lerma-Chapala basin in Mexico, and the West Bank and Gaza in Palestine.

Young *et al.* (1995) describe the Catchment Management Support System (CMSS), developed in support of the development of Nutrient Management Plans for many Australian Watersheds. CMSS is a PC-based program that calculates average annual nutrient loads and allows the effects of land use changes and land management changes to be investigated with ease. The system can be integrated with GIS to prepare and edit the required databases. Young *et al.* (1995) claim that the CMSS approach stresses the importance of wide community involvement in the decision making process for watershed management. The system has been used in at least a dozen major catchments in Australia.

In all studies cited above, the integration of DBMS, GIS, models and expert systems has been described to be a flexible, efficient and versatile approach to environmental management problems.

10.5. Applications in Developing Countries - Two Case Studies in Turkey

10.5.1. DIFFICULTIES IN DEVELOPMENT OF ENVIRONMENTAL INFORMATION SYSTEMS

In the previous section, examples on development and application of advanced environmental information systems for environmental management have been presented. The majority of these developments are experienced in developed countries. The management problems of the developing countries are at least as severe as those of the developed ones. Thus, advanced tools to support decision making are definitely needed. However, the development and implementation of an integrated system of tools are often hindered by such problems as lack of adequate and reliable data, planners' attitudes towards sophisticated tools, lack of communication between decision makers and scientists/engineers who develop the tools, lack of training and education, lack of agreement among different organizations who are responsible for various stages of the management process, etc. These problems are discussed in the following sections on two case studies from Turkey: management of the Yesilirmak River basin for maximization of fishery production and assessment of the sensitivity of a subbasin to soil and nutrient loss. The first case is a three-year project realized by the Dokuz Eylul University (DEU) research group, the Ministry of Agriculture of Turkey (TKB), and the Scientific and Technical Research Council of Turkey (TUBITAK). The other case is carried out solely for research purposes at DEU.

10.5.2. MANAGEMENT OF THE YESILIRMAK RIVER BASIN TO MAXIMIZE FISHERY PRODUCTION

The primary objectives of water resources management in Turkey have been water supply, irrigation, energy production and flood control for more than sixty years. These objectives have achieved the priority among all other alternative goals in natural resources development plans to meet the demands of the growing population, agriculture, and industry. Accordingly, the majority of large river basins have been engineered into systems of numerous hydraulic structures that impound water and regulate runoff for the above basic purposes.

Within the last decade, the governmental institution on Planning and National Policy Development has stressed the need for enhancement of freshwater productivity and management of fishery resources of inland waters. Investigations on the current status of freshwater fisheries in Turkey revealed that inland fish production increased from 7,000 tons/yr to 49,000 tons/yr within the 25 years between 1963 and 1988 (Ministry of Environment, 1991). Studies also disclosed the presence of 192 species of freshwater fish living in riverine systems. However, as of 1988, the proportion of inland fishing in the total fishery production of the country is stated to be as low as 7.2%. There are basically two factors underlying this rather poor development. First, freshwater fish production has not been properly addressed in natural resource development plans as other objectives (i.e., irrigation, water supply, energy production, and flood control) were given the first priority. Second, the recently emerging issue of water pollution has reduced the fishery value of water courses. Intensive industrialization and agricultural practice have significantly degraded the quality of rivers which, in turn, led to decreases in freshwater productivity. On the other hand, no extensive study has been carried out to define the ultimate goals of fishery management and to establish a scientific basis for multifaceted guidelines to be followed in achieving these goals (TKB, 1992; Alpaslan *et al.*, 1994 a).

Upon the apparent need for management of freshwater fishery, TKB (the Ministry of Agriculture) and TUBITAK (The Scientific and Technical Research Council of Turkey), with the contribution of Istanbul Technical University (ITU) and later of Dokuz Eylul University (DEU), started a research program in 1991 to collect, analyze and synthesize the scientific information needed to conserve and manage fish resources of the country. The Yesilirmak River basin along the Black Sea coast was selected as the pilot study due to its potential for development of fishery production. This potential is indicated by the quality of river runoff and by the possibility of river regulation via series of existing and planned reservoirs. The management plan proposed for the basin has essentially been an integrated scheme that covers scientific, technical, institutional, social, economic and legal aspects of decision making to define the optimum fishery yield. Thus, the pilot study is a multifaceted large scale project to be realized by a number of stages, each of which is undertaken as a separate subproject.

Objectives of Management in the Yesilirmak River Basin

The Study Area. Yesilirmak River basin in the Black Sea region has a drainage area of 36114 km^2 and an annual average flow of 5.80 km^3. The basin permits a storage of 3464 hm^3 of water with the establishment of 57 dam reservoirs which serve energy production, irrigation and flood control purposes. In addition, the existing reservoirs along the river, together with those planned, constitute significant potential for development of fishery production. At the stage of full development, the systems planned along Yesilirmak and its tributaries will irrigate 1,326,046 ha of land, provide flood control over a total area of 73,000 ha, and will supply domestic water in the order of 144 hm^3/yr (DSI, 1991). With these characteristics, Yesilirmak ranks to be the third largest basin in Turkey.

Yesilirmak River and its tributaries receive pollutant discharges from domestic and industrial sources. A wide range of pollutant constituents are discharged into the river by different types of industry including leather, chemicals, sugar, cigarettes, dairy, and food industries. Since no significant treatment is applied to domestic and industrial wastewaters, the river is gradually degraded in its quality. Nonpoint pollution due to intensive agricultural activities contribute significantly to this degradation. The annual loads of water constituents are estimated to be 1532 μmhos/cm of EC, $5.67x10^6$ kg of total dissolved solids, $13.8x10^8$ kg of Cl^-, $3.96x10^6$ kg of NH_3-N, $3.08x10^5$ kg of NO_2-N, $4.5x10^8$ kg of NO_3-N, $18.46x10^8$ kg of SO_4, and $2.31x10^8$ kg of Mg (Ministry of Environment, 1991).

Two agencies, DSI (State Hydraulic Works Authority) and TKB run the major water quality monitoring stations in the basin where the longest period of observation is approximately 15 years. EIE (Electrical Works Authority) is basically involved in streamflow gaging although water quality is also sampled at some of these gages, particularly in areas of significant pollution. At the time when the DEU group started the research program, DSI had been collecting data at 7 and TKB at 49 sampling points, the majority of which were located downstream of significant industrial discharges. The data collected by then had not yet been extensively evaluated to disclose the spatial and temporal distribution of water quality constituents.

Objectives of Management. The pilot study initiated earlier by TKB and TUBITAK had foreseen the following basic objectives for the management of the Yesilirmak basin (TKB, 1992):

a) determination of the basin-wide potential for fishery production and identification of the optimum fishery yield (in tons/year);
b) delineation of potential sites and appropriate fish species for economic freshwater production;
c) development, conservation and proper management of fishery resources;
d) development of a regulatory system to control and maintain a high quality natural environment that will not have adverse effects upon fishery production; and,
e) development of the required scientific and technical basis to realize the above four objectives.

Essentially, the stated goals of the study necessitated a management scheme to define appropriate policies for maximization of fishery production in the basin. Such a scheme was also expected to provide realistic answers to the following specific questions:

a) What is the potential (in tons/yr) of the Yesilirmak basin with respect to fishery production?
b) Which regions of the basin constitute potential sites for economic production?
c) Which species can be harvested in different regions selected by (b)?
d) How can a regulatory system be established to control and protect the quality of the environment for purposes of maintaining favorable conditions required by fishery production?
e) Which polluting sources must treat their wastewaters before discharging into the riverine system?
f) What are the scientific, technical, administrative, financial, and legal requirements in defining a national water quality monitoring system?

g) What kinds of databases are required for application of GIS (Geographic Information Systems) and expert system technologies in management of the drainage basin?

Furthermore, three secondary objectives of the pilot study were implied by these questions, which foresaw:

a) development of a basin-wide database (or information system) that includes not only hydrologic, but also geographic and demographic characteristics of the basin;
b) establishment of coordination between various institutions that are involved in basin management;
c) application of new technologies (e.g., GIS, expert systems, etc.) in management.

The above objectives were to be achieved by separate subprojects incorporated into the macro level management plan. In essence, the pilot study planned for the Yesilirmak basin has been an integrated river basin scheme which would eventually assume the form of a regional development plan. Similar schemes would be applied to other basins in Turkey once the above objectives could be efficiently realized in the case of Yesilirmak. The results of the pilot study were expected to delineate basic guidelines for management of the nation's natural resources using new technologies (e.g., GIS and expert system methodologies) and to initiate cooperation between various institutions related to management. Realization of the latter issue is particularly significant as the water resources and related agencies in Turkey have long lacked interactive communication and collaboration which are essential for development of sound management policies and actions (Alpaslan *et al.*, 1994 b).

Proposed Pilot Study

Various levels of activities in the form of subprojects have been planned for Yesilirmak basin as outlined in Fig. 10.1. Among these, two stages, i.e., the statistical analysis of river runoff (step 1) and development of a mathematical model for plankton (step 6) were completed in 1991 (Alpaslan *et al*, 1995).

Steps 4 and 7 are stages undertaken by DEU to satisfy one of the specific objectives of the project, i.e., the establishment of a regulatory system to control and protect the quality of the environment for purposes of maintaining favorable conditions required by fishery poduction. Such a system has to be developed by resolving the following issues (Harmancioglu *et al.*, 1994):

a) determination of seasonal variations in the physical, chemical and biological properties of the river using water quality simulation program packages and delineation of required databases;
b) establishment of coordination among various institutions to decide upon "who" should monitor "which" variables;
c) determination of compliance with receiving media standards in the case of each polluting source;

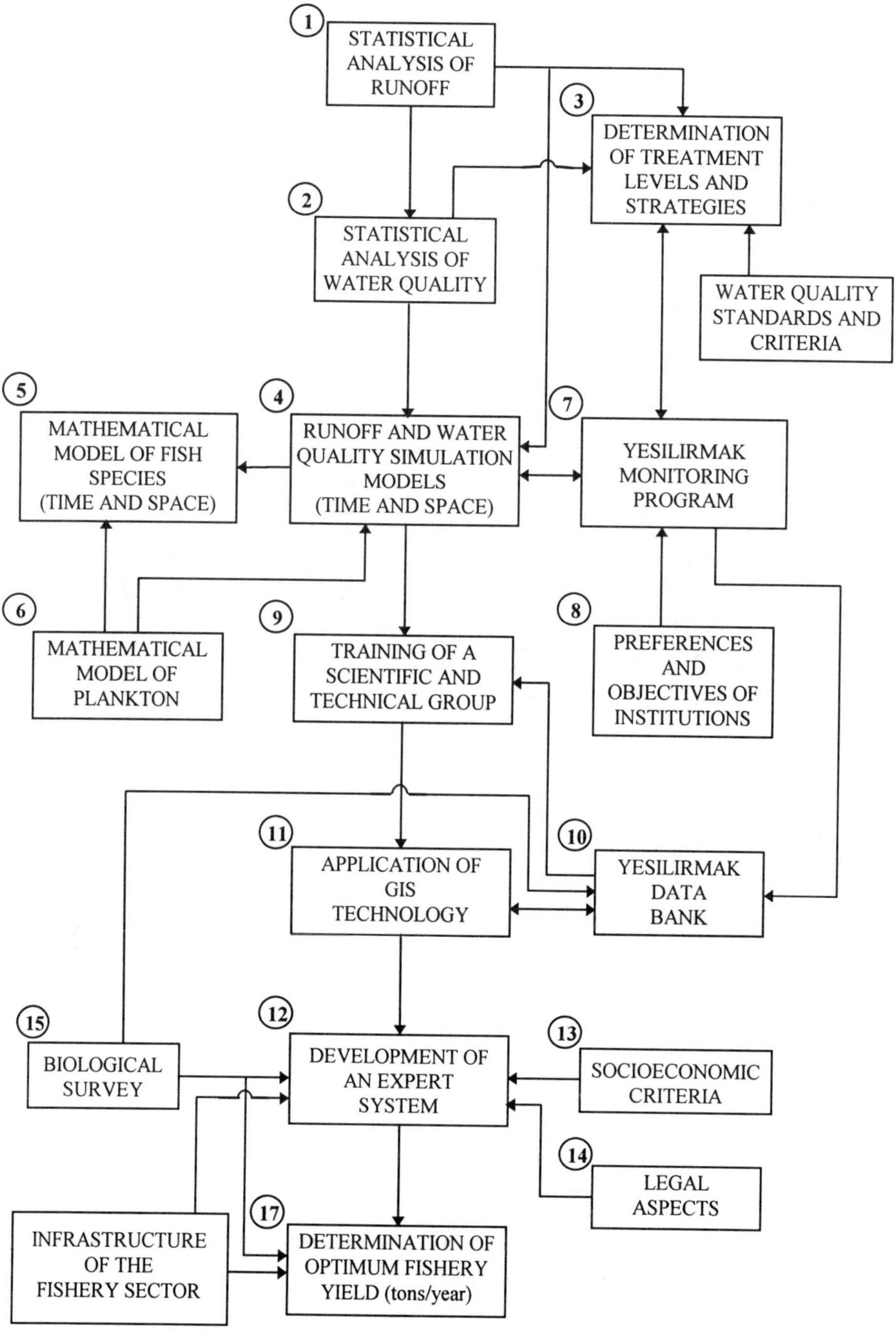

Fgure 10.1. Stages of the pilot study proposed for the Yesilirmak Basin

d) assessment of the share of each polluting source in the variation of water quality;

e) determination of the assimilation capacity of the river and those water quality variables the discharge of which must be prevented.

After a thorough survey of available models developed by various international agencies and research institutes, two water quality simulation models, SALMON-Q of HR Wallingford and QUAL2E-UNCAS of EPA, have been selected to realize (a) and (b) above. The results of model applications were expected to delineate space-time distributions of various water quality variables.

However, it was found at this stage that available data both on the river system itself and on water quality were incomplete, inadequate, and incompatible. The major diffculties observed were (Harmancioglu *et al.*, 1993):

a) data on riverflows did not have a sufficient coverage of the river system; in addition, adequate information on the hydraulic, topographic, and geometric properties of the system was not available so that the hydrodynamic structure of the river could not be identified into the selected models;

b) data on water quality were quite messy; they had short sample sizes, significant numbers of missing values and gaps within the data series, and irregular sampling frequencies; furthermore, they did not have an adequate spatial coverage;

c) water quality and quantity data were collected by three different agencies that monitored the river independently of each other. Thus, the available data were not compatible in terms of units used, data formats, sampling frequencies, and water quality variables observed. Thus, the data from the three agencies could not be merged into a complementary framework to develop a database for the basin. There were several cases where two agencies monitored at the same spatial points, but different variables at different time frequencies and with different units of measurement;

d) there were no simultaneous streamflow and water quality observations; in addition, some water quality variables required by the selected models were not observed.

In view of the above difficulties associated with available data, a major revision and modification of the monitoring system in the basin was foreseen particularly in respect of model requirements. These modifications included the selection of new sampling sites, appropriate temporal frequencies, and new water quality variables to permit efficient model applications (Harmancioglu *et al.*, 1993).

Essentially, the new sampling program foresaw the monitoring of those water quality variables that could be simulated by both SALMON-Q and QUAL2E. The formerly irregular patterns of sampling frequencies were revised into systematic programs. Thus, water quality would be observed on a regular monthly basis to permit an evaluation of seasonal changes. The new sampling locations were identified on the basis of model requirements to provide data on initial conditions for particular simulation runs. Furthermore, additional sampling sites were selected to observe the effects of significant pollutant discharges.

A significant progress realized at this stage was the agreement between three monitoring agencies, DSI, TKB, and EIE, to support a regional cooperative approach toward Yesilirmak basin management. EIE has undertaken the work of measuring streamflow and river cross sections at required points. DSI and TKB have started to evaluate their sampling and laboratory facilities on a joint basis to eventually decide upon "who" should sample "which" variables. They have also initiated field trips to investigate the variability of hydraulic conditions along the river. As a result of these studies, the initially defined river reaches would be specified more precisely to permit effective model applications together with model calibration and verification processes.

In the development of the new data collection program, preferences and monitoring objectives of various agencies in Turkey have been investigated and evaluated by the university on the basis of a questionnaire. In view of the current lack of coordination among these agencies, this study has constituted a significant initial step towards establishment of cooperation. The questionnaire has shown that most agencies have developed the tendency towards cooperative approaches to monitoring practices. Such tendency indicates possibilities for development of a national data bank for management of river basins, and the Yesilirmak pilot study is expected to initiate the efforts for such a progress as foreseen in step 10 of Fig. 10.1.

In the meantime, the laboratory and computer facilities of each monitoring agency in the Yesilirmak basin have been updated and modified. The DEU research group started training courses for DSI and TKB regional staff so that they could apply SALMON-Q and QUAL2E programs and further evaluate model simulation results. These courses have been supported by reports and user manuals of program packages prepared by the university (Harmancioglu *et al.*, 1993). Such activities constitute the basis for step 9 of Fig. 10.1, i.e., the development of a scientific and technical working group that will actively contribute to basin management.

On another line, investigations have been carried out by TKB and DEU to identify the pollutant loads in the basin. Numerous types of industries, settlements, and their discharges have been analyzed for the amounts and types of water quality constituents they involve. Such information was then transferred onto discharge maps which were used in revising and modifying the water quality monitoring scheme planned by concurrent studies.

The investigations described have been designed to eventually constitute inputs to the basin management model which would be developed at further stages of the pilot study. This model was to help identify the required levels and locations of treatment as a part of the regulatory system to control and protect the quality Yesilirmak River.

Application of Water Quality Simulation Models

Application of water quality simulation models constituted an integral part of the pilot study as shown in step 4 of Fig. 10.1. These models simulate water quality processes in space-time dimensions, thereby making it possible to estimate the contribution of each polluting source to the variability of water quality along the river. By using such models, it is also possible to plan a monitoring system to identify nonpoint discharges.

Information provided by simulation studies would then constitute inputs to the management model at the final stage.

As mentioned earlier, two program packages have been selected for modeling water quality in the Yesilirmak basin. SALMON-Q has been used only by the university for basic research and decision-making purposes while developing the framework for Yesilirmak management plan. It seemed highly costly in 1993 to distribute the package among regional offices of TKB and DSI in the basin. However, plans were made to obtain an additional copy of the package from HR Wallingford and to install it at one of the offices which would eventually serve as a reference laboratory. On the other hand, QUAL2E-UNCAS program package was supplied to every laboratory in the basin.

QUAL2E is a program package that permits the simulation of basically 15 water quality constituents in a branching river system. It applies a finite difference solution to the one-dimensional advective-dispersive mass transport and reaction equation (Brown and Barnwell, 1987). The program can be operated either as a steady-state or a dynamic model. In the former case, the model can be used to investigate the effects of pollutant loads upon river quality and to identify the magnitude and structure of nonpoint discharges. As a dynamic model, QUAL2E makes it possible to analyze the effects of diurnal variations in meteorological data on water quality. Diurnal dissolved oxygen (DO) variations caused by algal growth and respiration can also be studied by the dynamic model. Recently, the program has been enhanced to cover:

a) addition of the QUAL2E-UNCAS model which provides extensive capabilities for uncertainty and risk analyses;
b) inclusion of an option for reach-variable climatology input for steady-state temperature simulation; and
c) capability to plot observed DO data on the line printer plots of estimated DO concentrations.

In particular, the development of the QUAL2E-UNCAS model has provided the capability to assess the effect of model sensitivities and of uncertain input data on model forecasts. As a result, the risk of a water quality variable being above or below an acceptable level can be evaluated.

QUAL2E is a highly comprehensive and flexible program to suit a wide spectrum of physical conditions. However, this flexibility may be a disadvantage to users unfamiliar with the structure of the package since several decisions have to be made to adjust this structure to the specific problem investigated. Nevertheless, QUAL2E has been applied for more than 20 years both in the United States and worldwide and is considered to be a "proven, effective modeling tool" in the analysis of DO balance in streams (Barnwell *et al.*, 1989). Since sufficient experience and knowledge about QUAL2E has been developed through the years, the program is also claimed by Barnwell *et al.* (1989) to be suitable for incorporation into expert system methodologies. Recently, Melching and Yoon (1996) have demonstrated a method to handle key sources of uncertainty in QUAL2E model for the case of Passaic River in New Jersey. They claim that application of stream water quality models in decision making has been hampered by a lack of data appropriate for minimization of model-simulation uncertainty. Essentially, a similar problem of data deficiencies were encountered in the present case of the Yesilirmak basin.

SALMON-Q is a one-dimensional mathematical model of flow, mud transport, water quality, and saline intrusion in tidal and riverine channel networks. It can simulate the oxygen and nutrient balance of looped and dendritic networks using an implicit finite difference approach to the solution of advection-dispersion equations. The model can be used to assess:

a) impacts of effluent discharges;
b) effects of water abstraction and flow regulation schemes on water quality, sediment transport and saline intrusion;
c) effects of engineering schemes on hydraulic and water quality processes; and
d) alternative pollution control schemes (Wallingford Software, 1993).

Although SALMON-Q is a fairly new release, it comprises process modules which are developed as a result of extensive studies by HR Wallingford over many years. Thus, it is supported by a long history of experience and applications. Its capabilities are still being developed and enhanced in the light of new applications and requirements suggested by its users. At present, it is highly flexible to model a wide spectrum of physical conditions. Thus, it permits the evaluation of particular river management plans to control whether or not specified water quality objectives can be met by proposed actions.

One of the most powerful aspects of the SALMON-Q program is its user-friendly graphical interface which comprises data capture and edit, data visualization, model run management, and model results review. The menu system is easy to use as it permits access to all sections of the program and controls their operation. The interface also has a comprehensive help system which guides the user at various levels of modeling. The context sensitive spreadsheet type editor is again fairly easy to use since it operates like an expert system checking input errors and parameter suitability.

The setting up module is another advantageous feature of SALMON-Q, where input cross-section and topography data are used to generate the geometric data required by the model. When this module runs, the river network comprising reaches, nodes and junctions is schematized on the screen, allowing the user to check whether the physical system is correctly defined in the program. For complex systems such as the Yesilirmak river, particular reaches and nodes can be easily identified and accessed through the labeling and zoom options provided by the module. These features are not available in QUAL2E where it is difficult to visualize the network system defined.

Furthermore, the graphics module of SALMON-Q is much more advanced than that of QUAL2E as it includes colored time history and longitudinal profile plots of any parameter simulated. In essence, this module involves several options to produce output plots of any kind and quality. Evaluation of outputs is further enhanced by the statistical post-processor which again is not available in QUAL2E.

On the other hand, the hardware requirements of SALMON-Q are more extensive than those of QUAL2E. This is in a way related to the dimensional limitations imposed on each model. In SALMON-Q, the number of reaches that can be defined is approximately three times as much as that allowed by QUAL2E. Similarly, the maximum number of computational elements is twice as that specified for QUAL2E. Thus, the user can describe the actual physical system more precisely by SALMON-Q provided that sufficient data are available. This advantage is enhanced by the capability of SALMON-Q to accept variable lengths of computational elements in different reaches, whereas QUAL2E requires uniform lengths throughout all reaches. Furthermore, SALMON-Q introduces a calibration file which guides the user with respect to the information required for calibration. QUAL2E involves only an observed DO data file to assist in calibration and verification procedures (Alpaslan *et al.*, 1994 a).

On the other hand, SALMON-Q input data files appear to be too comprehensive as they are described in the user manual. For example, although the program is a depth averaged model, the vertical structure of a river cross-section is divided into four sub-compartments comprising water column, bed, fluffy layer, and pore water. The definition of such compartments is essentially useful when simulating mud transport in estuaries or when modeling the processes in the pore water. In particular, the latter simulation approach enables the model to represent the full impact of re-erosion of polluted bed deposits. On the other hand, a significant number and type of data are specified for each compartment in the initial conditions file of the program. Such data are to be provided at the downstream and upstream ends of each reach. At this point, it is difficult for the user to comprehend how these data can possibly be supplied. Practically, it seems pretty difficult and costly to survey the specified parameters within each vertical compartment. A similar difficulty occurs in the simulation of water quality parameters such as biochemical oxygen demand (BOD) and nitrogen. BOD is considered as four fractions: dissolved and particulate at fast and slow decay rates. Organic nitrogen is also considered as both fast and slow decaying forms. This feature of the model makes it easier to simulate a wide range of organic effluents. However, it is not clear how these different fractions are to be measured and introduced to the model. The monitoring agencies in the Yesilirmak basin have actually been confused with such data requirements and have expressed the need for a precise definition of how such parameters may be specified for different vertical compartments.

A full application of the above two models to the Yesilirmak basin has been hindered by data availability. Although water quality observations were found to be sufficient for preliminary simulations, data on hydrodynamic variables did not meet the requirements of either of the models. Accordingly, a survey program has been started to define flow conditions (flow, cross-sections, slope, roughness, etc.) at necessary points along the river for a realistic description of river reaches. At present, both models can be run with the available data; however, the physical system of the basin cannot be introduced to the computer models precisely.

A particular difficulty encountered in applications of SALMON-Q to Yesilirmak river is the incorporation of a large number of existing reservoirs into the model. At present, it has not been clarified how such large structural elements should be treated except to apply the model at discrete groups of river reaches. Such an approach increases the number of reaches to be investigated, leading to further data requirements and surveying efforts.

At further stages of the Yesilirmak pilot study, an expert system will be developed for management of water quality. Accordingly, possibilities and necessary conditions of incorporating SALMON-Q into this system will have to be investigated. Such an analysis has already been realized for QUAL2E and the results have shown the applicability of this model within an expert system methodology (Barnwell *et al.*, 1989). Similarly, elements of an expert advisor will need to be developed for SALMON-Q as part of future studies.

Further Stages of the Yesilirmak Management Plan

In parallel with water quality modeling and simulation studies, further stages of the Yesilirmak management plan have been initiated in 1994 to cover (Alpaslan *et al.*, 1994 a and b):

a) biological survey: this survey is planned to collect data on existing fish species, to analyze the current fishery practices, to define the fishery potential of the basin, and to investigate possibilities of increasing the current freshwater production to an optimum yield level;
b) development of a geographical information system (GIS): all types of information (e.g., geographical, topographic, meteorological, hydrologic, demographic, land use, etc.) relevant to the basin will be incorporated into a GIS system; for this purpose, a GIS database will be developed;
c) development of an expert system: an expert system will be developed to incorporate the GIS system and the simulation models together with other relevant information regarding the basin; this system will then be used for management purposes.

Conclusion

Upon the need in Turkey for enhancement of freshwater productivity, a pilot study has been introduced in one of the largest basins, the Yesilirmak basin, for purposes of management of fishery resources. The study presented above summarizes the developments achieved so far and future investigations foreseen for this pilot study. An integral part of the plan is devoted to modeling of water quality and quantity inputs to the basin as fishery production is significantly hindered by domestic and industrial pollutant discharges. To this end, two simulation models, QUAL2E and SALMON-Q, have been applied, and concurrent monitoring activities have been carried out to collect data required by the models. The features of the pilot study described indicate that the Yesilirmak plan has been developed into an integrated river basin scheme. It is essentially a regional development plan which is foreseen for other basins in the country.

Although significant accomplishments were expected from the Yesilirmak project, the status of the pilot study as of 1994 remains practically the same. This is due to budgetary constraints which emerged as a consequence of a severe economic crisis experienced in the country since 1994. Thus, further stages of the project could not be initiated, including the finalization of the modeling procedure, biological survey, establishment of GIS and expert systems, and finally the integration of these tools into a complete integrated environmental information system. This means that the initial phase for the development of an integrated information system discussed in section 10.3.2, i.e., preparation of data and databases, could not be realized. Thus, the further stages of the system cannot be established since the problem of data availability, as the essential initial step of basin management, have not been solved at full scale.

It must be noted here that one of the major problems developing countries are faced with are the limited funds that can be reserved for water management programs (Frederiksen, 1996). Thus, the primary constraint is the availability of sufficient funds, as the case is with the Yesilirmak management plan. In Turkey as a developing country, funds are allocated more to programs that meet urgent demands such as water supply, irrigation, energy production, and flood control. In this respect, the objective of the Yesilirmak plan, i.e., maximization of fishery production, happens to be a secondary concern for the planners.

However, the studies carried out so far in the pilot project did realize a few significant accomplishments:

a) the deficiencies of the monitoring program have been recognized by the responsible governmental organizations; thus, it is expected that the design of monitoring programs for the future will receive due consideration of these deficiencies;
b) coordination between different governmental organizations has been established; otherwise, the traditionally independent practices of the responsible agencies hinder the development of efficient management policies;
c) the personnel of the responsible agencies who are involved in basin management have been trained to use models and interpret the results;
d) the basic requirements for the revision and modification of the current monitoring system have been identified.

It is expected that the pilot project will be restarted in the near future to accomplish its predefined objectives.

10.5.3. INTEGRATION OF GIS WITH USLE IN ASSESSMENT OF SOIL EROSION - A CASE STUDY FOR THE GEDIZ RIVER BASIN IN TURKEY

A Geographic Information System (GIS) has been integrated with the USLE (Universal Soil Loss Equation) model in identification of rainfall-based erosion and the transport of nonpoint source pollution loads to the Gediz River, which discharges into the Aegean Sea along the western coast of Turkey (Fistikoglu, 1996). The purpose of the study is to identify the gross erosion, sediment loads, and organic N loads within a small region of the Gediz River basin. Similar studies are available in literature, ranging from those that use a simple model such as USLE to others of a more sophisticated nature (Fraser *et al.*, 1996). The reason why this study is presented here is to reflect the difficulties in applying the methodology when the required data on soil properties, land use and vegetation are not available.

An effective investigation of soil loss within a watershed by using GIS - USLE integration requires spatially distributed data on several parameters describing the basin. Such parameters include topography, rainfall characteristics, soil types, vegetation, land use, and the similar. In Turkey, data on most of these parameters are collected often on a local basis, and therefore, a well-organized regional or basin-wide database is not available. Considering this lack of information, the study by Fistikoglu, (1996) is carried out by assumed values of regional parameters although a topographically known region is selected. The results have shown that GIS can be effectively used to investigate critical regions within a basin with respect to erosion and to estimate nonpoint source pollutants carried by sediment loads. However, realistic results can only be obtained if adequate data on soil types, vegetation, land use, etc., are available to the user.

Some limitations encountered in the application stem also from the limitations of the specific GIS software used. The study has shown that the use of GIS at the most efficient level requires sufficient databases and an understanding of GIS system capabilities.

Nonpoint Source (NPS) Pollution and GIS

In order to monitor and report on NPS pollution and its relationship with land use activities, a substantial environmental database and a diverse analytical capability to evaluate that information has to be established. There is a universal need to integrate catchment-based environmental and land use information with stream ecosystem information over both space and time in order to define these relationships.

Geographic Information Systems are useful tools to assist with this task. GIS provide a generic knowledge-based management tool for the analysis and interpretation of primary data being collected on land use activities and water quality. Moore (1990) has reviewed the functional advantages of using GIS for identifying and targeting areas susceptible to NPS pollution. The ability to develop knowledge-based watershed models of pollutant transport, using the analytical subsystems of GIS, is seen as particularly useful in evaluating the performance of environmental protection strategies (Moore, 1990).

The GIS-based modeling approaches commonly use geographic attributes of the watershed that affect NPS pollution. The basic procedure in NPS pollution modeling is to estimate soil erosion and pollution transported by eroded soil. In order to obtain a quick and effective solution for determining NPS pollution carried by eroded soil, a GIS-Universal Soil Loss Equation (USLE) integration is often used. NPS identification by USLE is based on a number of spatially distributed parameters which can be effectively described and evaluated by GIS.

Soil Loss Estimation by USLE

The Universal Soil Loss Equation is the most widely used erosion equation as discussed in Chapter 9. The equation and its development utilized more than 40 years of experimental field observations gathered by the Agricultural Research Service of the USDA (Novotny and Olem, 1994). USLE estimates rainstorm based sheet and rill erosion which occur in small watersheds (Wischmeier and Smith, 1965). The USLE equation is described as;

$$A = (R)\,(K)\,(LS)\,(C)\,(P) \tag{10.1}$$

where;

A : calculated soil loss in ton/ha/year
R : rainfall runoff factor
K : soil erodibility factor
LS : topographic factor
C : cover and management factor
P : support practice factor

With appropriate selection of its factor values, the equation computes the average soil loss for a multicrop system, for a particular crop year in a rotation, or for a particular cropstage period within a crop year. It computes the soil loss for a given site as the product of five major factors whose most likely values at a particular location can be expressed numerically. Erosion variables reflected by these factors vary considerably about their means from storm to storm, but effects of the random fluctuations tend to average out over extended periods.

The equation basically expresses soil loss per unit area due to rain. It does not include wind erosion, and it does not give direct sediment yield estimates (Novotny and Olem, 1994).

Since all factors in the USLE equation have a spatial distribution in a watershed, a GIS based evaluation of the equation gives more accurate results. In the GIS assisted approach, each factor of the equation is described in the form of digital maps, and five digital layers of the equation are overlaid in order to obtain spatially distributed soil loss in a watershed as shown in Fig. 10.2.

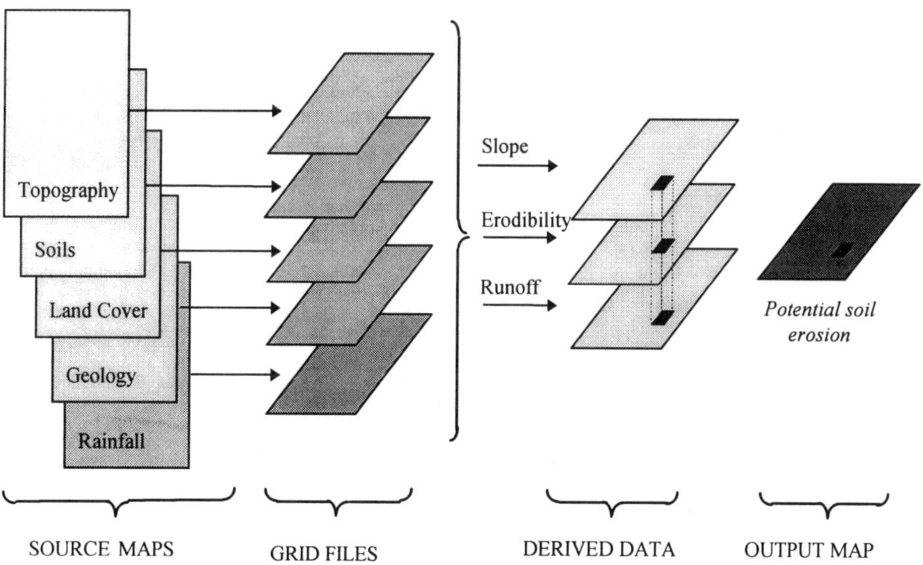

Figure 10.2. The use of GIS in estimation of potential erosion

An effective application of USLE for a small watershed requires several types of information about the study area. In order to get accurate results, the data required by USLE should be available to reflect the watershed characteristics.

The data needs for an effective USLE application on erosion-based pollution can be defined as:

1) rainfall records;
2) soil types and their properties;
3) land use and management practices;
4) topographic information;
5) water quality observations.

The USLE-GIS integration can be established by converting all parameters of USLE into a raster-based format and evaluating these digital parameter layers as in Fig. 10.2. Each parameter (R, K, C, P) and topography (LS) are digitized from the associated maps. LS factor of the watershed is derived from digital elevation model (DEM) obtained from topography. Then, the digital maps in vector format are converted into raster format in which each parameter of a specific pixel is known (Fig. 10.3). The USLE equation (Eq. 10.1) is applied to five digital parameter-layers by overlaying them.

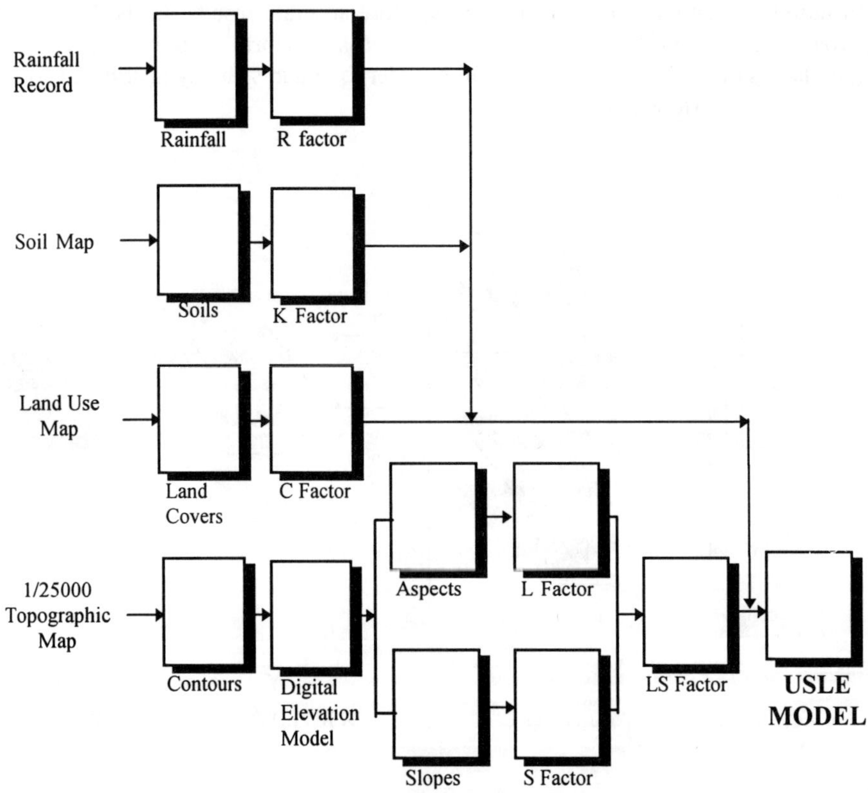

Figure 10.3. Steps in estimation of soil loss via USLE

Enrichment Ratio and Organic N Transport by Sediment

Organic N pollution, which is transported by clay particles in sediment, is estimated by applying the enrichment concept of contaminants to sediment yield computed by USLE.

Sediment yield refers to the amount of sediment measured at a watershed outlet or a point on the waterway. Basically, sediment yield is not equal to the upland erosion (Novotny and Olem, 1994).

The ratio of sediment delivered at a given area in the stream system to the gross erosion is the sediment delivery ratio for that drainage area. Thus, the annual sediment yield of a watershed is defined as follows:

$$SY = (A)(DR) \tag{10.2}$$

where;

A : total gross erosion computed from USLE (Eq. 10.1)
DR : sediment delivery ratio

A general equation for computing watershed delivery ratios is not yet available since they depend on several properties of the watershed like infiltration, roughness, vegetation cover, hydrograph or runoff drainage, etc. But the ratios for some specific drainage areas have been computed directly from local data obtained from sediment gauging stations. As a rough approach, Roehl (1992) established a relationship between the delivery ratio and watershed size (Novotny and Olem, 1994).

Tim *et al.* (1992) have proposed an equation for deriving sediment delivery ratio of a specified area in a watershed. The point value of delivery ratios can be calculated from topography (relief and slope) of watershed, water body, and land use data maps according to the expression;

$$DR_i = \exp(-k\, S_{fi}\, L_{fi}) \tag{10.3}$$

where;

k: coefficient that varies with land cover;
S_{fi}: slope function;
L_{fi}: length of the flow path between the point i and channel outlet.

By simplification, DR_i can be assumed as;

$$DR_i = 10\,(R/L) \tag{10.4}$$

where; R is the relief which is the difference in elevation between a land point and the point which surface flow reaches to stream channel (Hession and Schanholtz, 1988).

Enrichment ratio is the concentration of contaminant in the sediment divided by that of the soil. Most soil contaminants are adsorbed by clay and organic matter because of their high surface area, leading to strong adsorption bonds. When rainwater reaches the surface horizon of the soil, some contaminants are desorbed and go into solution; others remain adsorbed and move with soil particles. The contaminant content of runoff sediment is then higher than that in the parent soil. The difference is called enrichment ratio and defined as follows:

$$ER = Cr / Cs \qquad (10.5)$$

where;

Cr: contaminant concentration of the runoff per gram sediment;
Cs: contaminant content of the parent soil per gram.

The enrichment concept can be applied to clay, organic matter, and all contaminants adsorbed by soil particles, which include phosphates, ammonium, metals, and pesticides. It is not appropriate to apply the enrichment ratio to contaminants that are mobile in soils, such as nitrates or soluble pesticides (Novotny and Olem, 1994).

The enrichment ratio refers to the difference in particle size distribution and associated or adsorbed contaminant content of washload particles and soils from which the sediment originates. Enrichment of sediments by clay is a two-step process: enrichment during particle pick up and enrichment during redeposition of the coarser particles during delivery in overland and channel flows. Thus, as the delivery ratio decreases with the increasing watershed area or time of overland flow, the enrichment ratio of the washload increases.

Organic N transport by sediment can be calculated by using sediment yield and enrichment ratio of each subbasin. A loading function developed by McElroy *et al.* (1976) is used to estimate organic N loss for each subbasin. The loading function is;

$$ONY = 0.001\,(SY)\,(CON)\,(ER) \qquad (10.6)$$

where;

ONY : the organic N runoff loss at the subbasin outlet in kg/ha,
SY : sediment yield in ton/ha from Eq.(4.13),
CON : concentration of organic N in the top soil layer in g/ton.

Application to the Gediz River Basin

In Turkey, within the last 20 years, there also have been some studies to determine the local values of USLE parameters. However, these studies have remained on a local basis and the evaluated parameters have yet not been established in widely usable formats (e.g., maps, tables). Due to the lack of information about the spatially distributed values of the USLE parameters in Turkey, in the presented study, most of the USLE parameters, except for R and LS, had to be assigned assumed values.

As a demonstrative case, a small region in the Gediz River basin around the city of Usak is selected. The topography of this area is known via available maps. The other required data (soil types, land use and management, sediment observations, etc.) are not available in organized and spatially distributed formats. Therefore, these characteristics are assumed in as much a realistic manner as possible.

The study is carried out by the specific GIS software, IDRISI, which is an educational version of a raster based geographic information system. Thus, some applications that need vector formats are not applied in the study. However, IDRISI can generally be used in water resources research by considering basic raster procedures such as producing digital elevation models (DEM), basin boundaries determination, slopes determination, image overlaying, etc.

The initial step of the study has been the vector identification of the topographic map by digitizing a 1:25000 contour map. This map is then converted into raster format and used to obtain the digital elevation model by operating INTERCON module in IDRISI. WATERSHED and ASPECT modules in IDRISI are operated to generate basin boundaries and river network of the watershed. The raster images generated by IDRISI have 45000 (225x200) pixels, each one being a square with 30x30 m dimensions.

The generated images of the elevation model for the selected region have considerable accurate results in comparison with the real case. Figures 10.4 through 10.6 show the digital elevation model with river network, subbasins, and 3-D topography of the demonstration basin, respectively. In Fig. 10.4, the elevation increases from light-gray-zones to the black-zones. The demonstration watershed is divided into 15 subbasins. Figure 10.5 shows 15 subbasins of the watershed with respect to their basin number indicated with the gray scale.

The USLE parameters are obtained by digitizing associated maps or spatial distribution information described below. USLE calculations that are a multiplication of 5 parameters (R, K, LS, C, P) are derived by overlaying each five digital-parameter-layer in raster format. The computations are applied to each subbasin as:

Gross Erosion (Eq. 10.1); $A_i = (R_i)(K_i)(Ls_i)(C_i)(P_i)$

Sediment Yield (Eq. 10.2); $SY_i = (A_i)(DR_i)$

Organic N Yield (Eq. 10.6); $ONY_i = 0.001\,(SY_i)(CON_i)(ER_i)$

where i represents the number of subbasin.

The rainfall-runoff factor, R, is derived by evaluating available rainfall records of a raingage in the region. As the selected region is small (23 km^2), the spatial distribution of the R factor was assumed to be uniform.

In order to get an accurate result, K factor has to be determined by field measurements, which were not available for the selected region. Thus, K values were selected as literature values given by Novotny and Olem (1994), where estimated values of the soil eridibility factor are given for technical soil classes. Figure 10.7 shows the spatial distribution of K in the demonstration basin.

The topographic factor (LS) which depends on the mean slope and the mean slope-length of the subbasins is derived from the values obtained from images of both the mean slope and the mean slope-length of the demonstrated basin. Figures 10.8 and 10.9 show the mean slopes and slope-lengths of the subbasins.

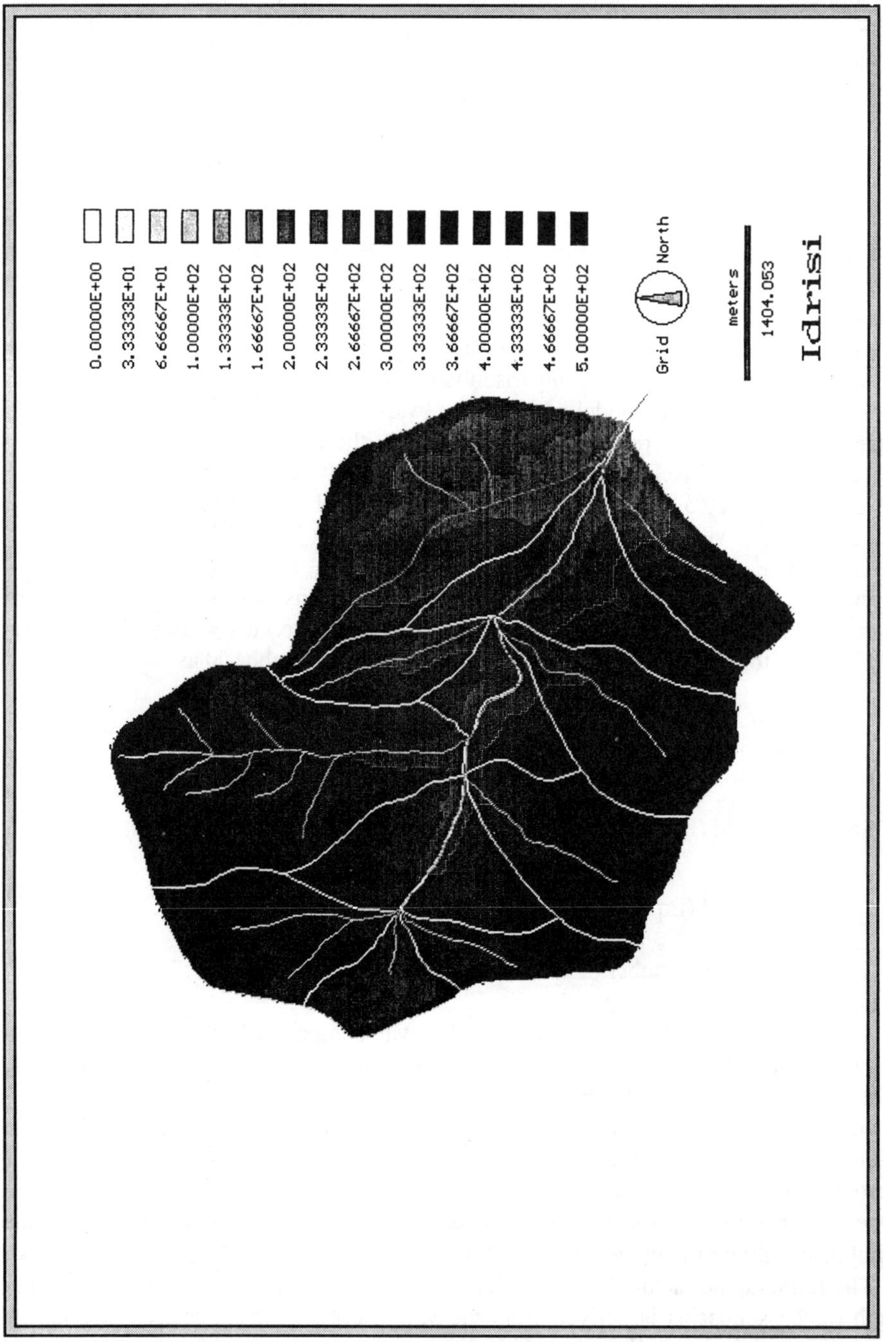

Figure 10.4. Digital Elevation Model and river network of the demonstration basin

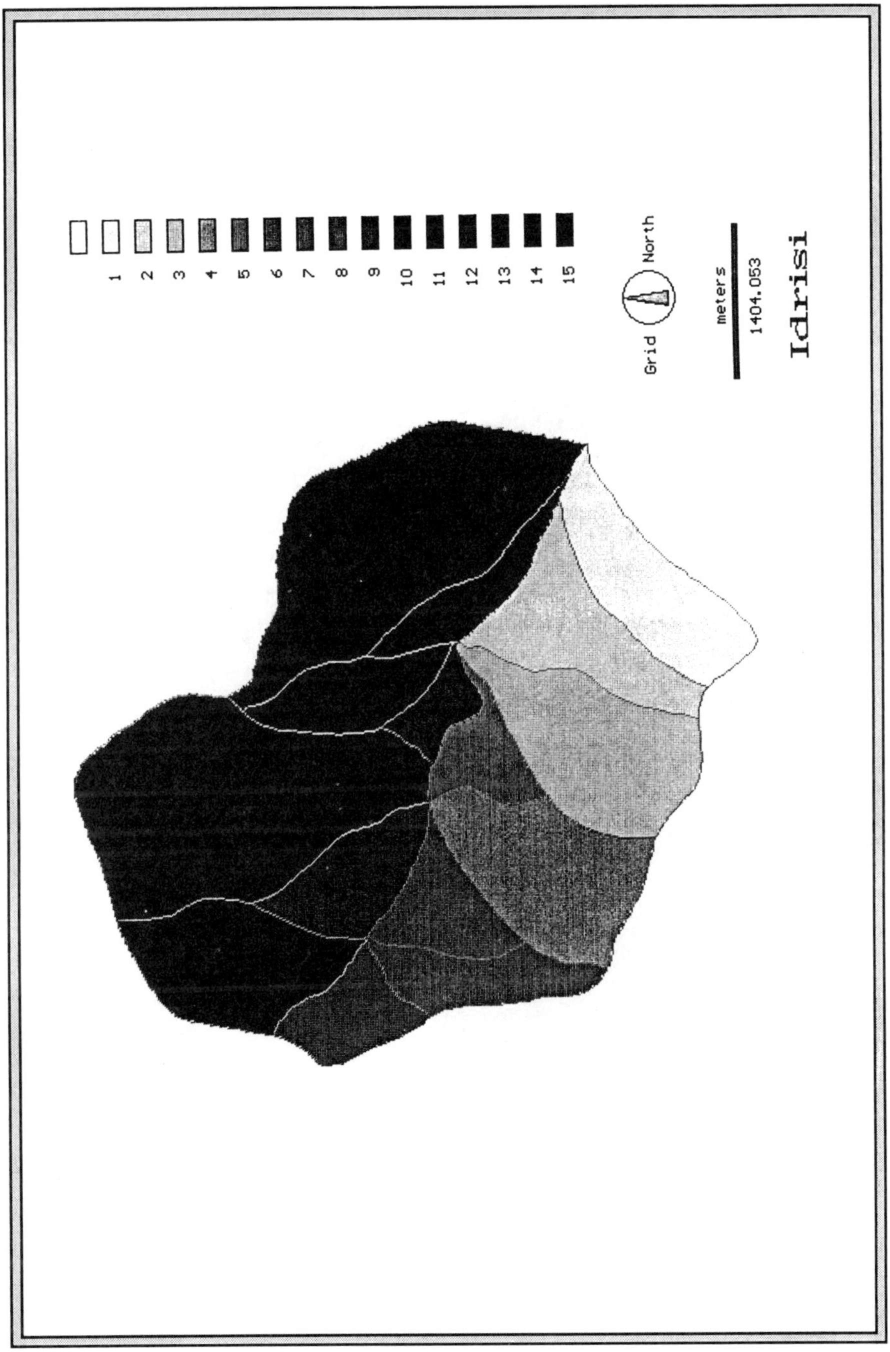

Figure 10.5. Subbasins of the demonstration basin

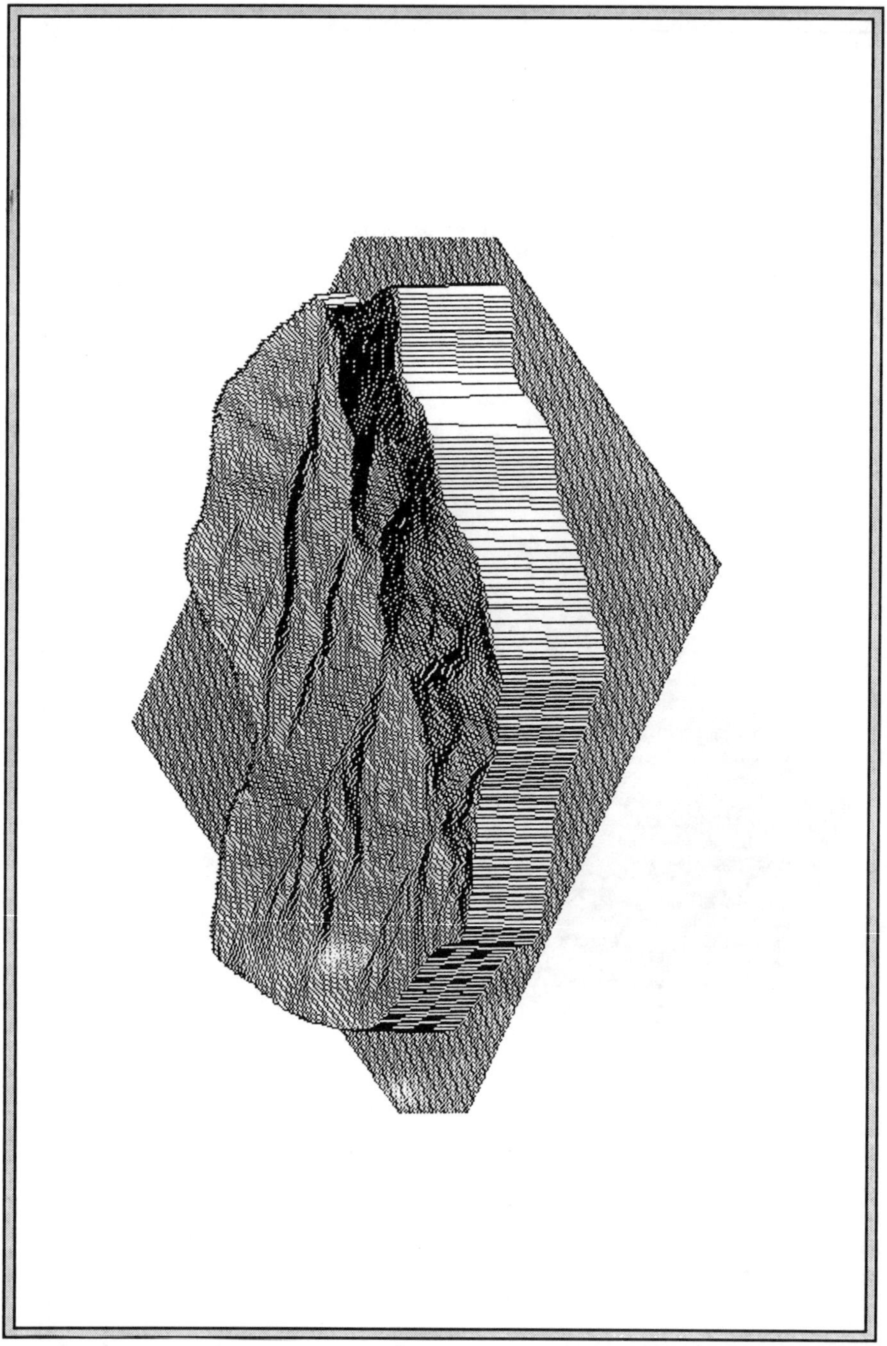

Figure 10.6. 3-D topography of the demonstration basin

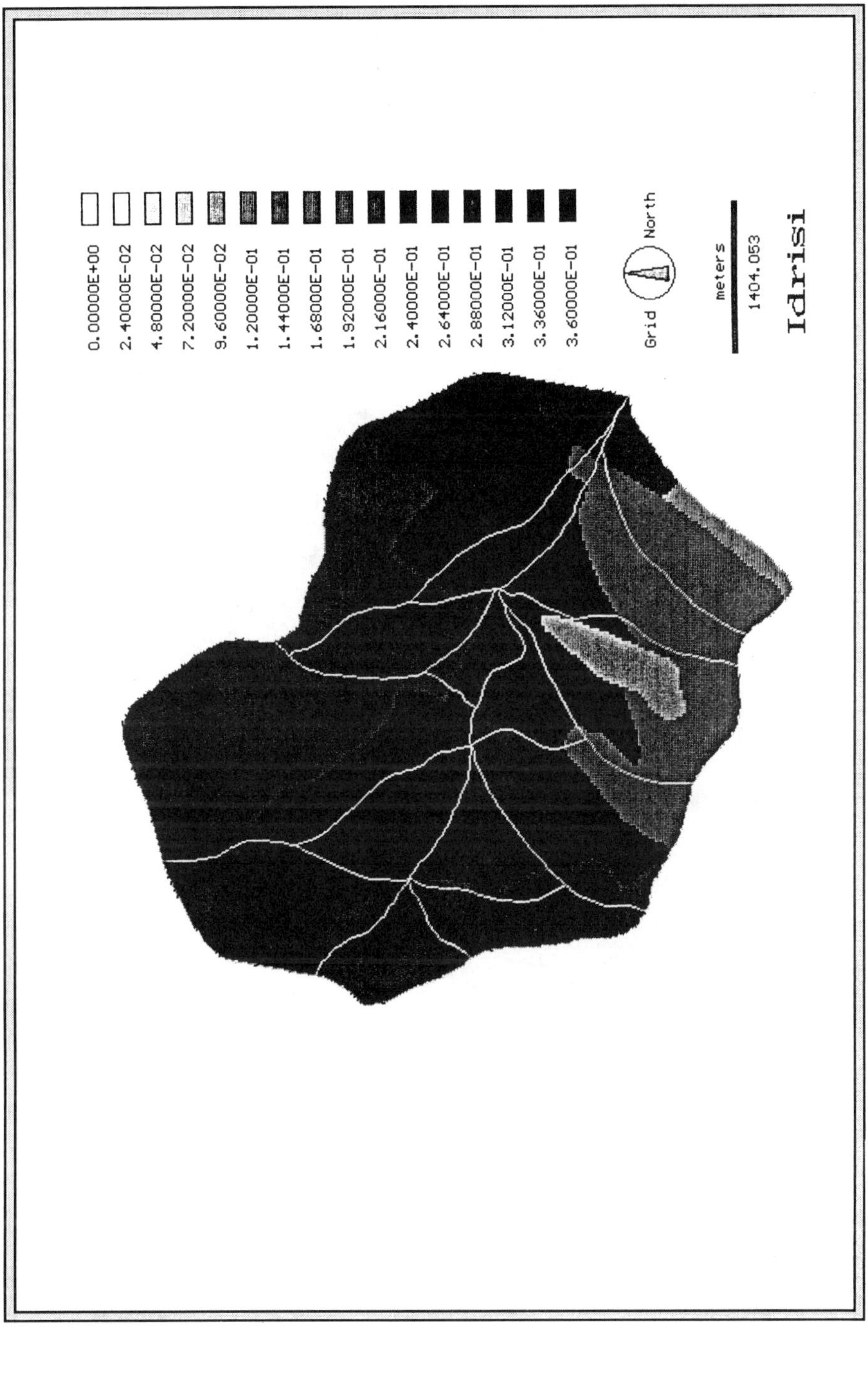

Figure 10.7. Spatial distribution of the K factor in the demonstration basin

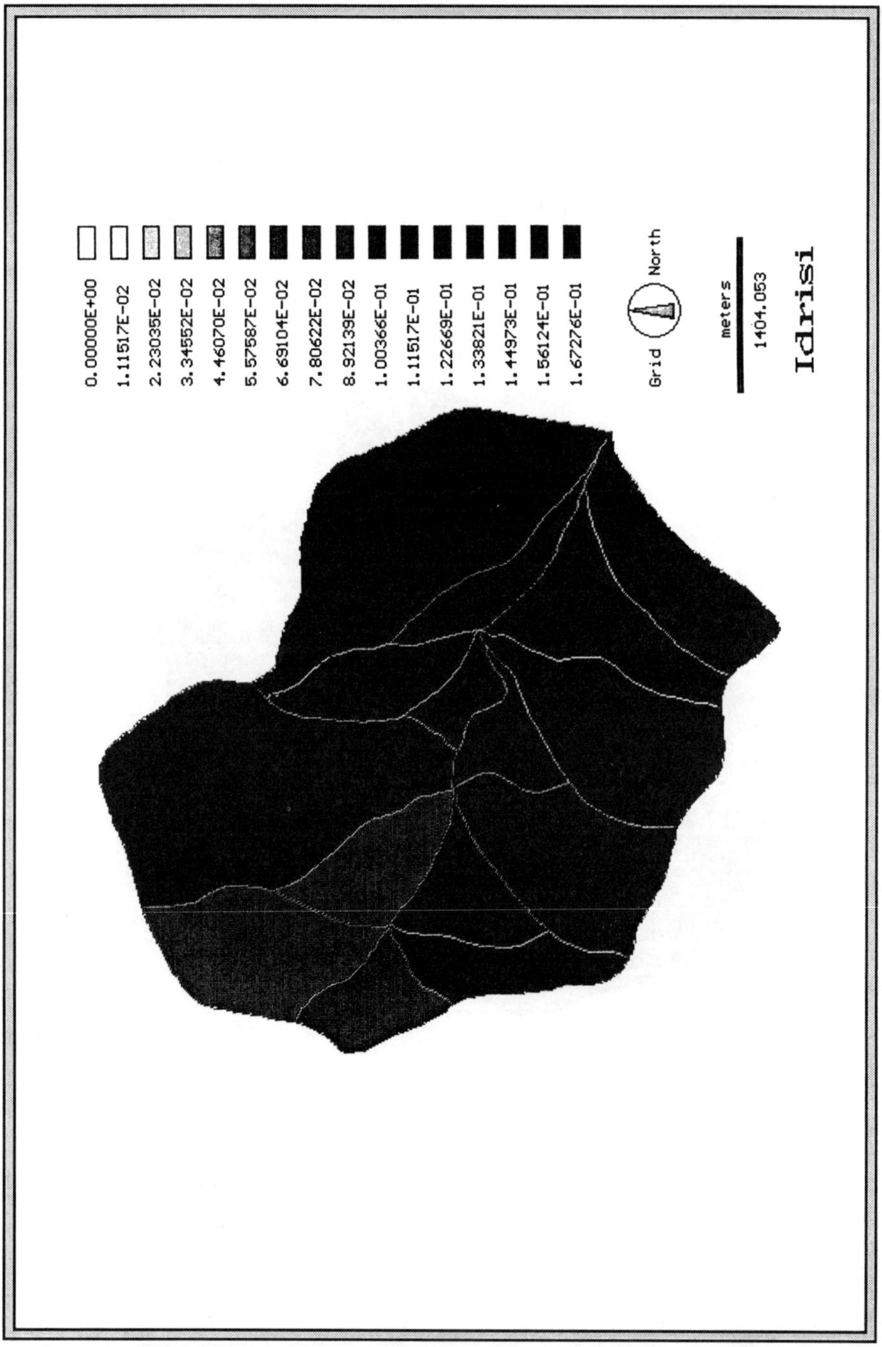

Figure 10.8. Mean slopes of the subbasins

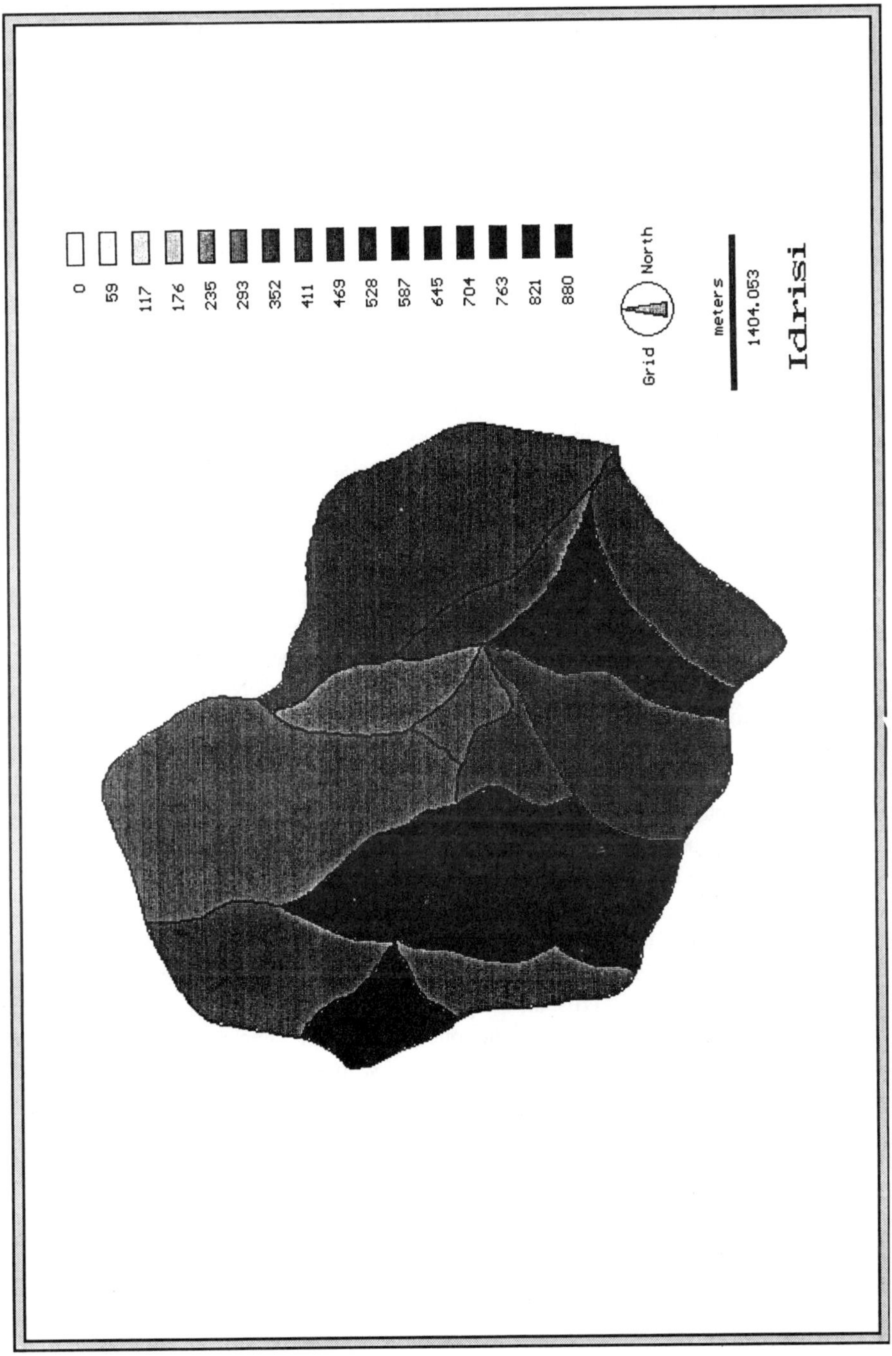

Figure 10.9. Me ı slope-lengnts of the subbasins

The mean slope image of the demonstrated basin is obtained by averaging the cell slopes associated with each subbasin. Fig. 10.10 indicates the cell slope distributions for the demonstrated basin, where the bright zones show low slopes and dark zones show the highest slopes.

The mean slope-length of a subbasin is derived by the module DISTANCE in IDRISI, which calculates distances from each cell to the stream. In Fig. 10.9, the slope length of subbasin 2 is the longest and it is indicated with a dark gray scale.

In order to calculate the topographic factor (LS) of a subbasin, mean slope and slope-length of that subbasin are entered on the slope-length chart provided in literature by Novotny and Olem (1994). Fig. 10.11 shows the topographic factors obtained from this slope-length chart.

Consequently, LS factor is considered as the value of LS obtained from the graphical solution in Fig. 10.11. As seen from Fig. 10.11, LS factor of the 2nd subbasin is the highest since its mean slope and slope length are also the highest. The other subbasins which have high slopes show moderate magnitudes of LS as their slope lengths are relatively shorter than those of subbasin 2.

To calculate the cover and management factor, C, land use and management practices in each subbasin are assumed, again by evaluating values given in literature. Figure 10.12 shows the distribution of C factors in the demonstration basin. Similarly, support practice factor, P, is also assigned assumed values to result in the distribution shown in Fig. 10.13.

Determination of sediment delivery ratios, DR, of a specific basin requires sediment data. In the present case, since such data were not available for the selected subbasins, the relationship between DR and watershed area, again given in literature by Novotny and Olem (1994), is used to roughly estimate the DR values. Next, enrichment ratios were calculated for the subbasins, considering the delivery ratios of each and assuming a constant sediment concentration of $C_a = 100$ gr/m^3.

To determine the critical soil loss regions in the basin, the spatial distribution of cell-based USLE parameters is obtained by multiplying the five USLE parameters in the specified 30x30m cells. Here, each cell is assumed as a closed plot where surface flow cannot enter a cell from another cell. The LS factor is computed by using the cell-length (30m) and the slope of each cell. The USLE equation is applied to each five digital parameter layers by the OVERLAY module in IDRISI, and the resulting image is given in Fig.10.14. Figure 10.14 does not indicate soil loss of each cell but gives an information about cell sensitivity with respect to parameters that affect erosion; thus it provides roughly some information about the critical zones of soil loss. In the figure, light zones adsorb rainfall energy more than the dark zones do; thus the southern part of the demonstration basin and also the majority of subbasin 14, which has great slopes, are more sensitive to erosion.

Gross erosion of the demonstrated basin is also computed by USLE. The USLE equation is applied to each subbasin by overlaying the 5 parameters. The resulting multiplication image is in Fig. 10.15 where all parameter patterns can be observed.

The mean gross erosion of the subbasins is computed by MEAN module of IDRISI, for each subbasin, and the spatial distribution of these means is given in Fig. 10.16. As seen in Fig. 10.16, the southern part of the demonstration basin has more erosion productivity than the other parts. Finally, the sediment yield and organic N yield of the basin are computed by Eqs. (10.2) and (10.6).

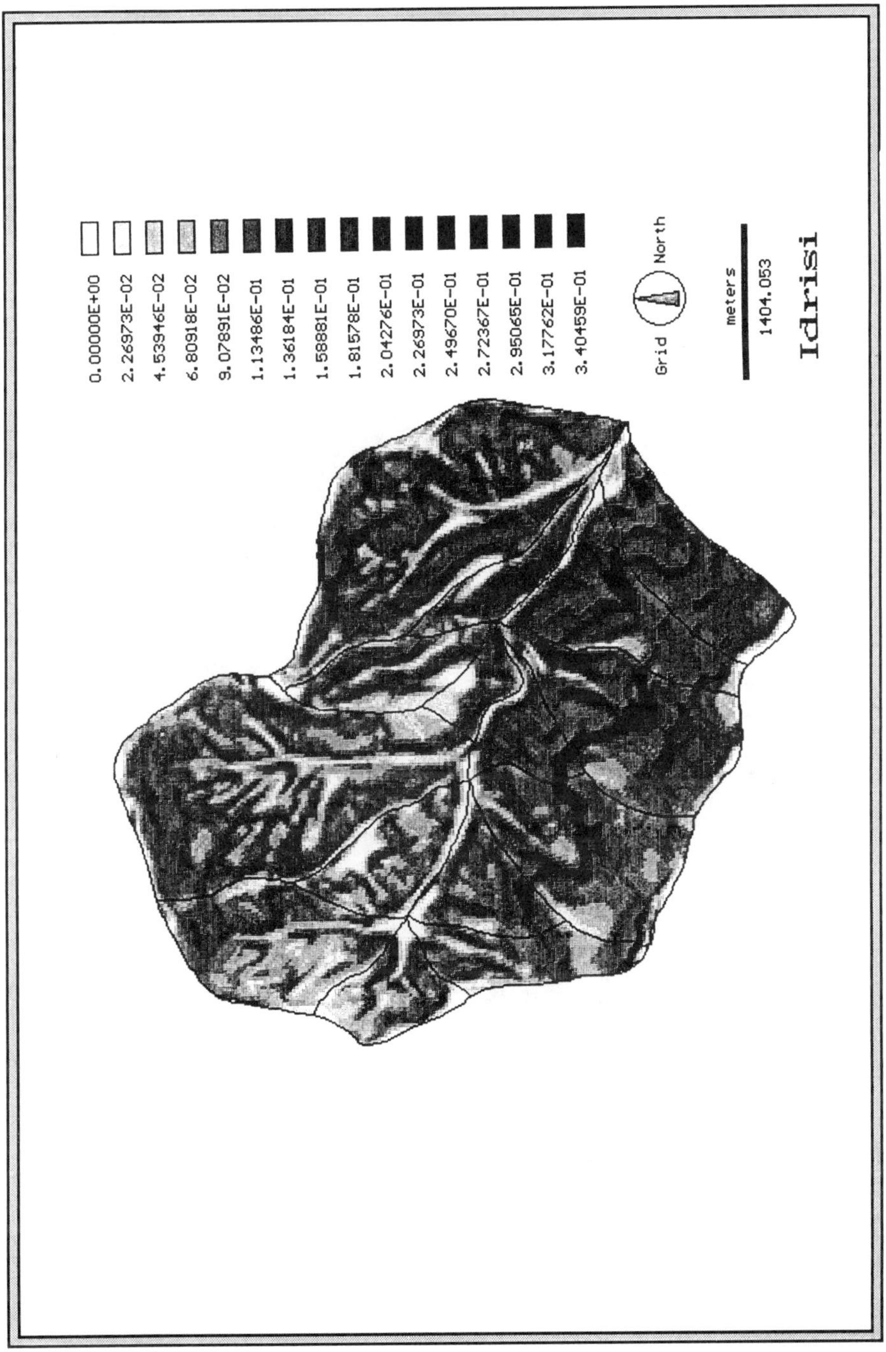

Figure 10.10. Cell-based slopes in the demonstrated basin

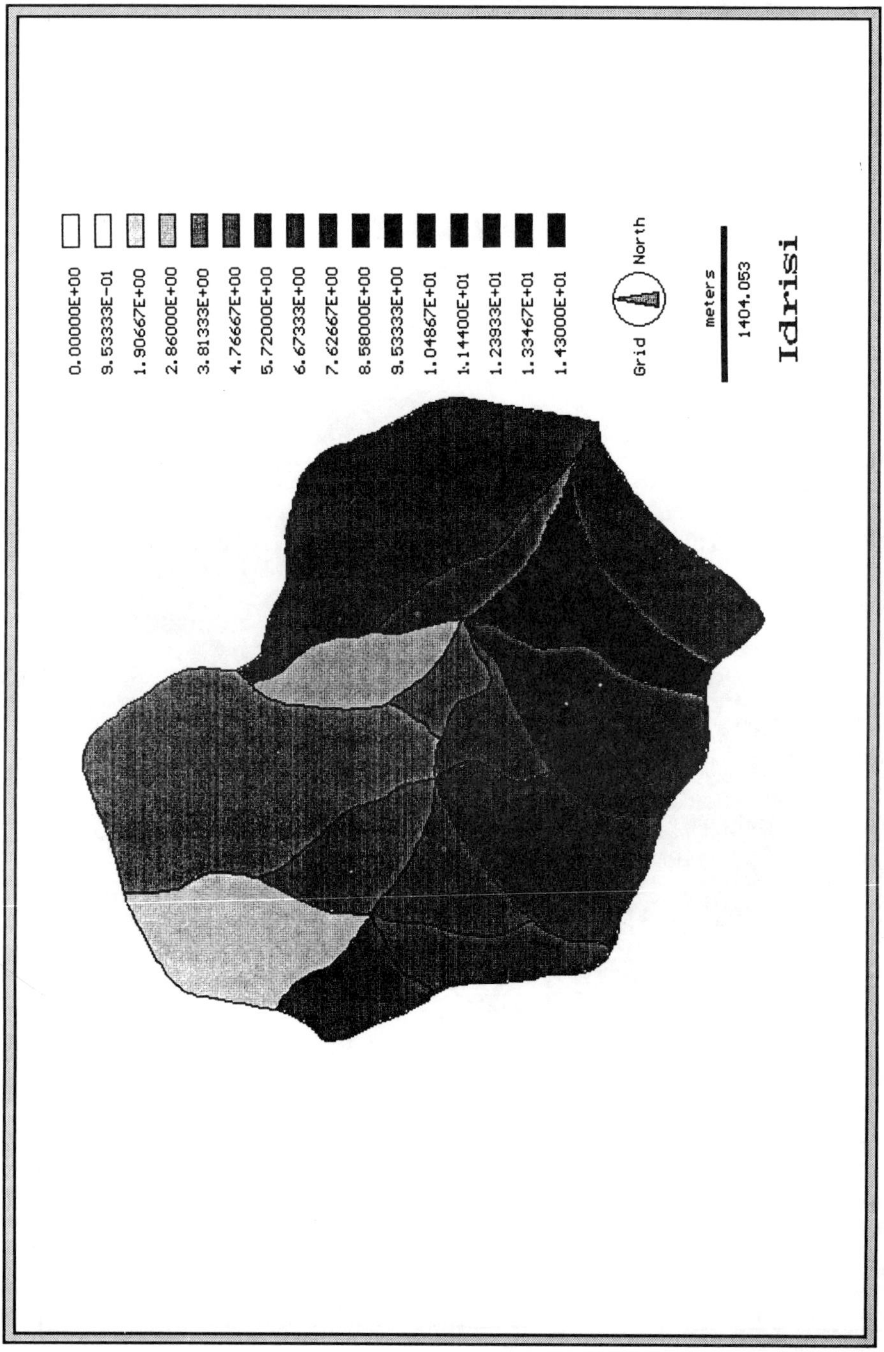

Figure 10.11. Topographic factor obtained from slope-lenght chart of the demonstrated basin

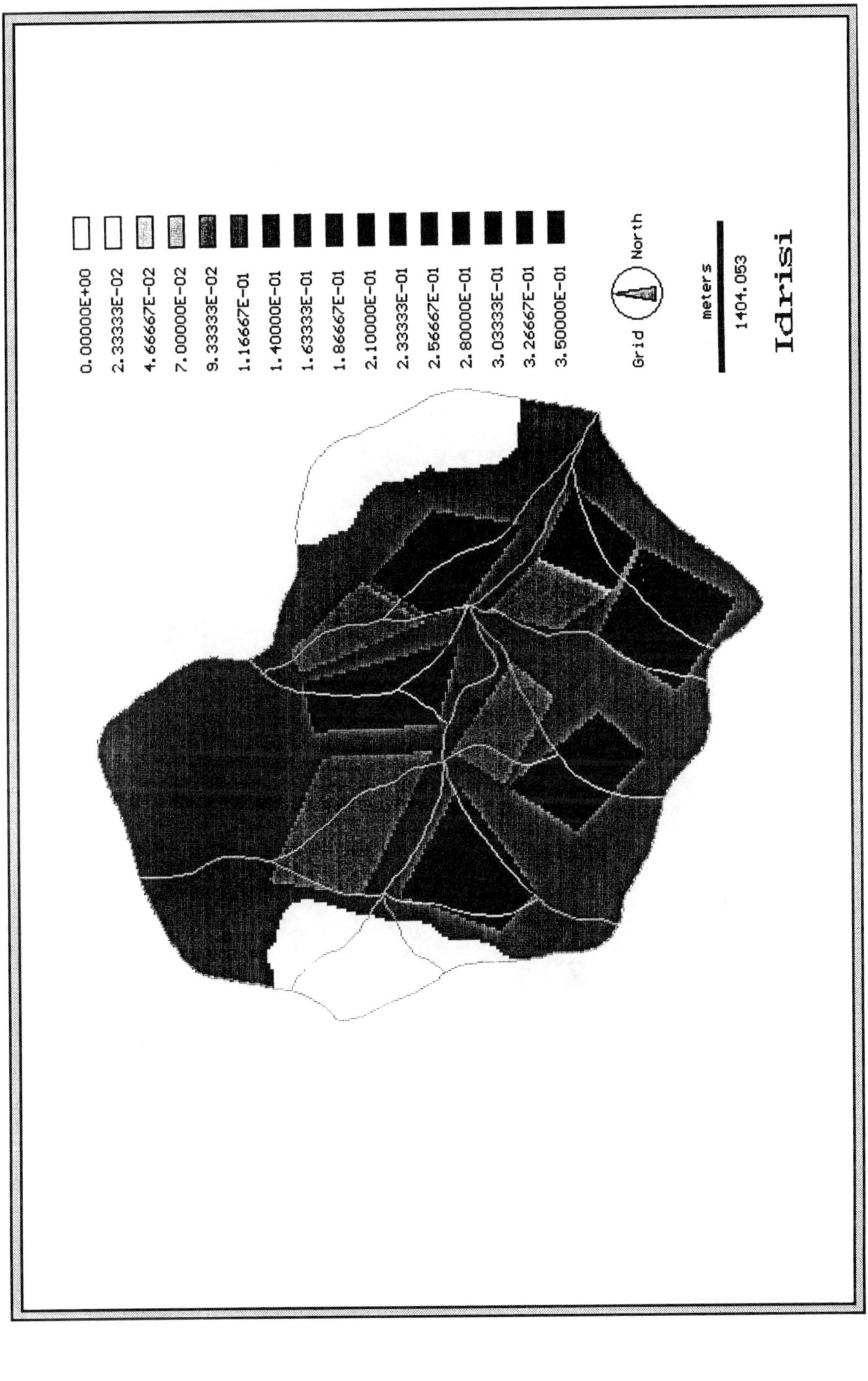

Figure 10.12. Spatial distribution of C factors of the demonstration basin

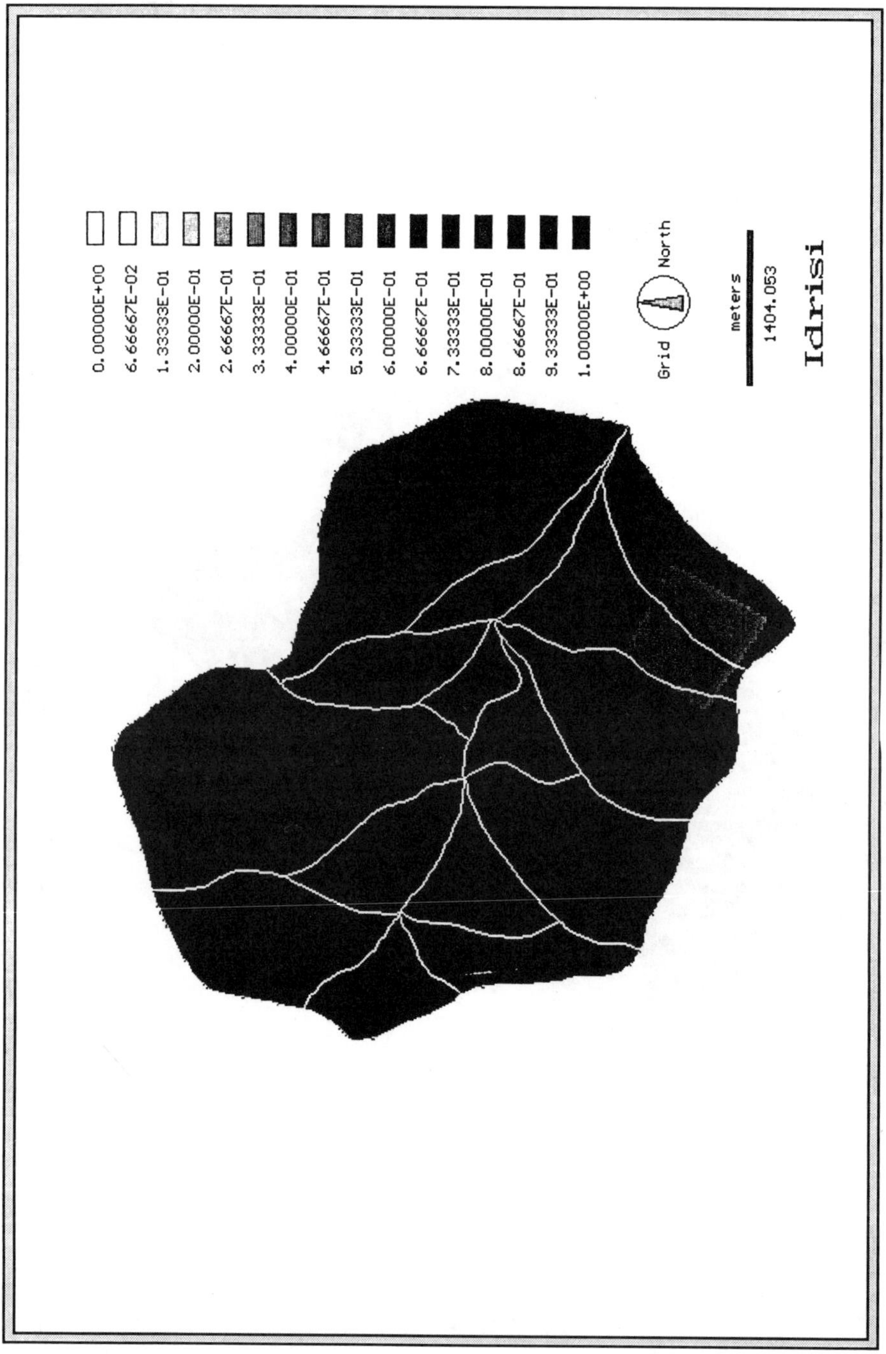

Figure 10.13. Spatial distribution of P factor in the demonstration basin

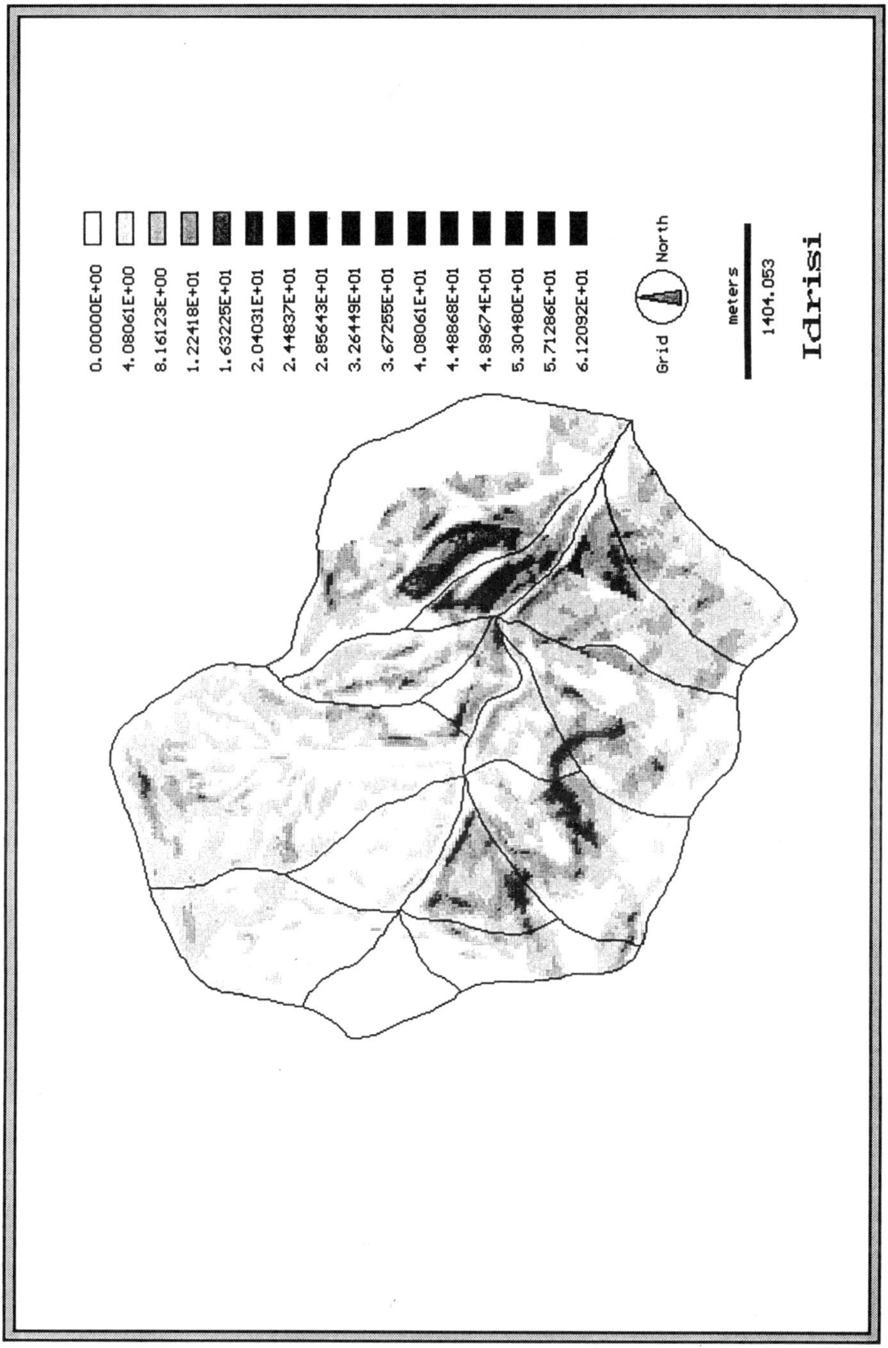

Figure 10.14. Cell-based description of basin sensitivity to erosion in terms of USLE parameters

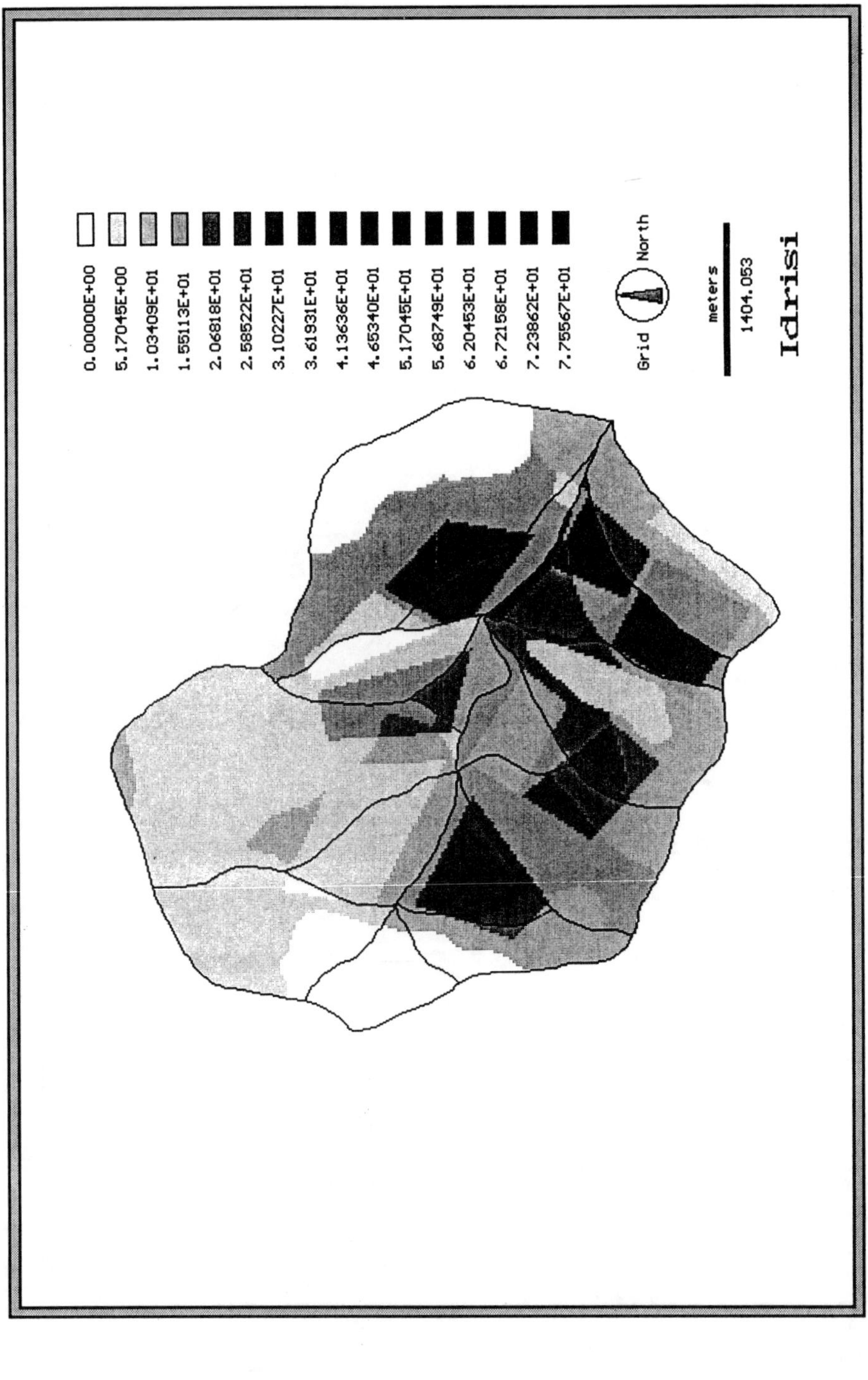

Figure 10.15. Patterns of the USLE parameters

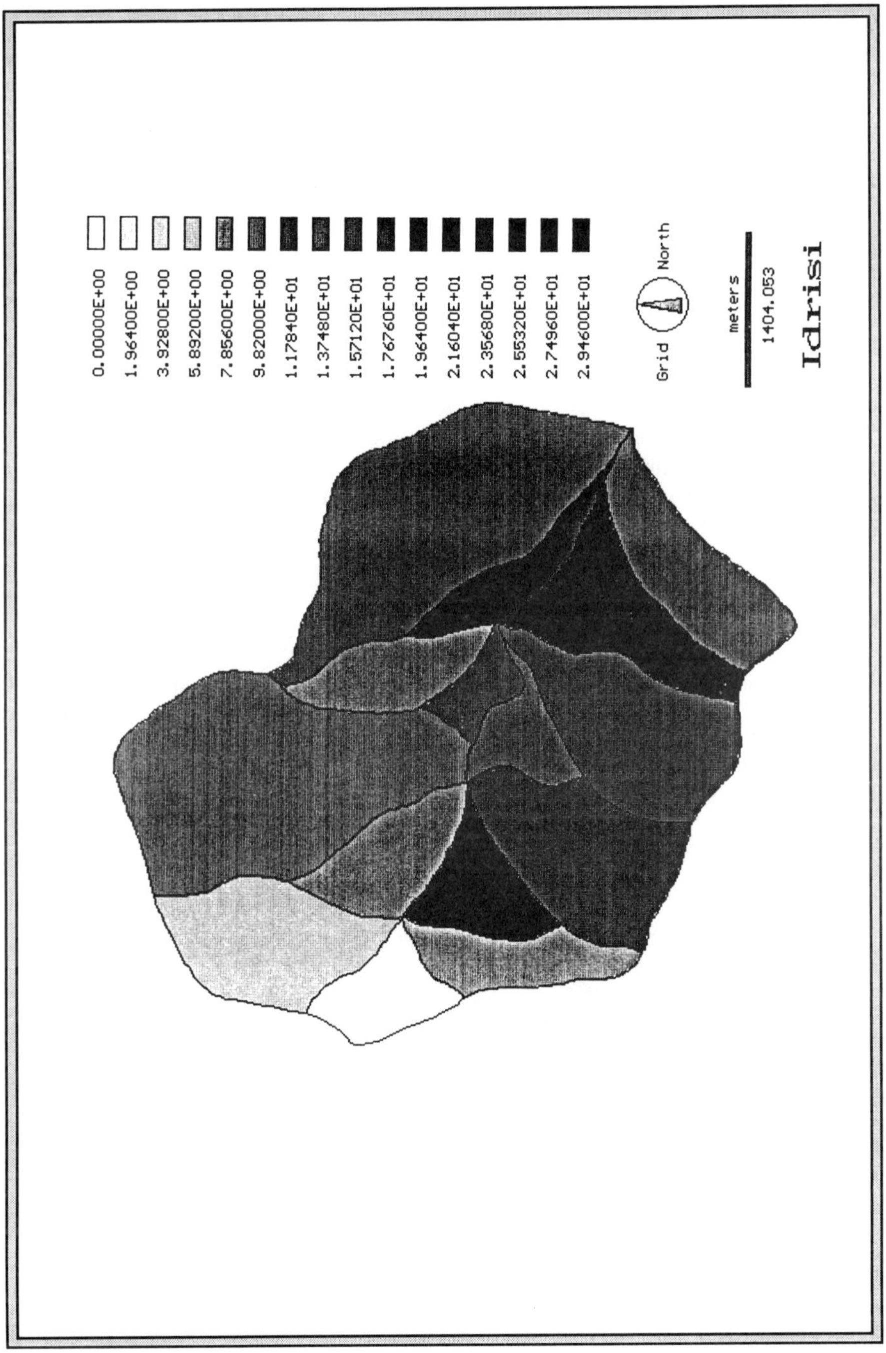

Figure 10.16. Spatial distribution of gross erosion in the demonstrated basin

Conclusion
The results of the study have shown that GIS permits more effective and accurate applications of the USLE model for small watersheds provided that sufficient spatial data are available.

As for the presented case, data availability has hindered the application of the methodology to reflect the real erosion potential of the basin; thus the results of the study have remained on an experimental basis.

Essentially, this situation may be generalized for attempts at applying the same procedure in other basins of Turkey. At present, there are no organized databases covering spatial data on soil properties, land use and vegetation. Such measurements are taken often on a local basis for purposes of specific surveys and projects. However, they are not readily available to the general user. This implies that every investigation directed towards assessment of soil loss and land degradation must involve an initial step of data collection and processing, which impose a significant cost component on the planned investigation program.

The above considerations reflect the major feature of developing countries in planning and implementing environmental management schemes; namely that, the tools for effective management, i.e., models, GIS, and expert systems, are available, but their application is significantly hindered by availability of adequate, reliable and processed data.

References

Alpaslan, N., Harmancioglu, N.B., and Saner, E. (1994 a):Proposed management of the Yesilirmak River Basin, in: C. Kirby and W.R. White (eds.) *Integrated River Basin Development*, John Wiley & Sons Ltd., England, pp. 97-107.

Alpaslan, N., Harmancioglu, N.B., and Saner, E. (1994 b) Management of the Yesilirmak Basin to maximize fishery production, in: G. Tsakiris and M. Santos (eds.) *Advances in Water Resources Technology and Management,* A.A. Balkema, Rotterdam, pp. 343-350.

Alpaslan, N.;, Harmancioglu, N.B., Saner, E., Ozkul, S., and Fistikoglu, O. (1995) Maximization of freshwater productivity: Case of the Yesilirmak River basin in Turkey, *Ninth World Productivity Congress,* (June 4-7, 1995, Istanbul), Proceedings, Vol. 2, pp. 1295-1308.

Baker, C.P., Bradley, M.D., and Bobiak, S.M.K. (1993) Wellhead protection area delineation: Linking flow model with GIS, *ASCE, Journal of Water Resources Planning and Management* **119(2)**, 275-287.

Barnwell, T.O., Brown, L.C., and Marek, W. (1989) Application of expert systems technology in water quality modeling, *Water Science and Technology* **21**, 1045-1056.

Brown, L.C. and Barnwell, T.O. (1987) *The Enhanced Stream Water Quality Models QUAL2E and QUAL2E-UNCAS: Documentation and User Manual,* EPA-600/3-87/007, Env. Research Lab., US EPA, Athens, GA, 30613.

Burks, K.A. and Passmore, M.F. (1996) An integrated decision support system for Corps of Engineers' environmental restoration projects, in: C.A. Hallam, K.J. Lanfear, J.M. Salisbury, and W.A. Battaglin (eds.), *GIS and Water Resources*, AWRA Technical Publication Series, TPS-96-3, pp. 76-88.

DSI (1991) *Statistical Bulletin for the Year 1991* (In Turkish), General Directorate of the State Hydraulic Works Authority of Turkey, Ankara.

Engel, B.A., Srinivasan, R., Arnold, J., Rewerts, C., and Brown, S.J. (1993) Nonpoint source (NPS) pollution modeling using models integrated with geographic information system (GIS), *Water Science and Technology* **28(3-5)**, 685-690.

Engman, E.T. (1986) Hydrolic research before and after AgRISTARS, *IEEE Transactions, Geoscience and Remote Sensing* **GE-24**, 5-11.

Fedra, K. (1997) Integrated environmental information systems: from data to information, in N.B. Harmancioglu, M.N. Alpaslan, S.D. Ozkul and V.P. Singh (eds.), *Integrated Approach to Environmental Data Management Systems*, Kluwer Academic Publishers, NATO ASI Series, 2. Environment, vol. 31, pp. 367-378.

Fistikoglu, O. (1996) *The Use of Geographic Information Systems (GIS) in Water Resources Research and Manegement*, Dokuz Eylul University, Graduate School of Natural and Applied Sciences, M. Sc. Thesis in Civil Engineering (Advisor: Prof. Dr. N. Harmancioglu).

Ford, D.T. and Killen, J.R. (1995) PC-based decision support system for Trinity River, Texas, *Journal of Water Resources Planning and Management* **121(5)**, 375-381.

Fraser, R.H., Barten, P.K., and Tomlin, C.D. (1996) SEDMOD: A GIS-based method for estimating distributed sediment delivery ratios, in: C.A. Hallam, K.J. Lanfear, J.M. Salisbury, and W.A. Battaglin (eds.), *GIS and Water Resources*, AWRA Technical Publication Series, TPS-96-3, pp. 137-146.

Frederiksen, H.D. (1996) Water crisis in developing world: misconceptions about solutions, *Journal of Water Resources Planning and Management* **122(2)**, 79-87.

Gaschning, J., Reboh, R., and Reiter, J. (1981) *Development of a knowledge-based expert system for water resource problems*, Final Report, SR1 Project 1619, SRI International, California.

Harmancioglu, N.B., Alpaslan, M.N., and Ozkul, S.D. (1997) Conclusions and recommendatios, in N.B. Harmancioglu, M.N. Alpaslan, S.D. Ozkul and V.P. Singh (eds.), *Integrated Approach to Environmental Data Management Systems*, Kluwer Academic Publishers, NATO ASI Series, 2. Environment, vol. 31, pp. 423-436.

Harmancioglu, N.B., Alpaslan, N., Ozkul, S., Saner, E., and Fistikoglu, O. (1994) Application of simulation models in integrated river basin management, *Fifth International Conference ENVIROSOFT 94,* San Fransisco, USA, Nov. 16-18, 8p.

Harmancioglu, N., Alpaslan, N., Ozkul, S., Saner, E., and Fistikoglu, O. (1993) *Project on Determination of Water Pollution in Yesilirmak Basin: Water Quality Simulation Program Packages (in Turkish)*, Report prepared by Dokuz Eylul University (DEVAK) for TKB as part of TUBITAK project no.71G, vol. I: General Guidelines, vol II: QUAL2E User Manual, vol III: SALMON-Q User Manual.

Heidke, T.M. and Auer, M.T. (1993) Application of a GIS-based nonpoint source nutrient loading model for assessment of land development scenarios and water quality in Owasco Lake, New York, *Water Science and Technology* **28(3-5)**, 595-604.

Hession, C.W. and Shanholtz, V.O. (1988) Geographic information system for targeting nonpoint source agricultural pollution, *Journal of Soil and Water Conservation* **43(3)**, 264-266.

Leipnik, M.R., Kemp, K.K., and Loaiciga, H.A. (1993) Implementation of GIS for water resources planning and management, *ASCE, Journal of Water Resources Planning and Management* **119(2)**, 184-205.

McElroy, A.D., Chiu, S.Y., and Nebgen, J.G. (1976) *Loading Functions for Assessment of Water Pollution from Nonpoint Sources*, Environmental Protection Technical Service, EPA 600/2-76-151.

McKinney, D.C., Maidment, D.R., and Tanriverdi, M. (1993) Expert geographic information system for Texas water planning, *ASCE, Journal of Water Resources Planning and Management* **119(2)**, 170-183.

Melching, C.S. and Yoon, C.G. (1996) Key sources of uncertainty in QUAL2E model of Passaic River, *Journal of Water Resources Planning and Management* **122(2)**, 105-113.

Ministry of Environment (1991) *Turkey: National Report to UNCED 1992.*, Ministry of Environment, United Nations Conference on Environment and Development, Ankara.

Moore, I.D. (1990) Geographic information systems for land and water management, *Proceedings of Soil and Water Conservation Association Australia Annual Conference*, November 5-6, 1990, Canberra.

Novotny, V. and Olem, H. (1994) *Water Quality Preventation, Identifiction, and Management of Diffuse Pollution*, Van Nostramd Reinhold, New York, USA.

Otte, C., Vidal-Madjar, D., and Girard, G. (1989) Remote sensing applications to hydrological modeling, *Journal of Hydrology* **105**, 369-384.

Queen, L.P., Wold, W.L., and Brooks, K.N. (1996) Application of GIS and remote sensing for watershed assessment, in: C.A. Hallam, K.J. Lanfear, J.M. Salisbury, and W.A. Battaglin (eds.), *GIS and Water Resources*, AWRA Technical Publication Series, TPS-96-3, pp. 337-346.

Richards, C.J., Roaza, H.P., and Pratt, T.R. (1996) Applying geographical information systems to groundwater assessments, in: C.A. Hallam, K.J. Lanfear, J.M. Salisbury, and W.A. Battaglin (eds.), *GIS and Water Resources*, AWRA Technical Publication Series, TPS-96-3, pp. 183-192.

Roehl, J.W. (1992) Sediment source areas, delivery ratios and influencing morphological factors, *International Association Hydrological Science* **59**, 202-213.

Ruggles, R., Lindquist, L., Rutt, K., and Szekalski, J. (1996) Using a GIS to analyze remotely sensed images and provide information to an intelligent database, in: C.A. Hallam, K.J. Lanfear, J.M. Salisbury, and W.A. Battaglin (eds.), *GIS and Water Resources*, AWRA Technical Publication Series, TPS-96-3, pp. 19-28.

Schmidt, R.R. (1987) *Groundwater Contamination Susceptibility in Wisconsin*, Report 5, Wisconsin's Groundwater Management Plan, Wisconsin Department of Natural Resources, Madison.

Schultz, G.A. (1997) Integration of remote sensing information, digital elevation models and digital maps within a GIS to generate new spatial environmental data sets for water management purposes, in N.B. Harmancioglu, M.N. Alpaslan, S.D. Ozkul and V.P. Singh (eds.), *Integrated Approach to Environmental Data Management Systems*, Kluwer Academic Publishers, NATO ASI Series, 2. Environment, vol. 31, pp. 153-170.

Schultz, G.A. (1993) Application of GIS and remote sensing in hydrology, *HydroGIS'93: Application of Geographic Information Systems in Hydrology and Water Resources, Proceedings of the Vienna Conference*, April 1993, IAHS Publ. No. 221.

Schultz, G.A. (1988) Remote sensing in hydrology, *Journal of Hydrology* **100**, 239-265.

Shepherd, A. and Ortolano, L. (1996) Water supply system operations: critiquing expert system approach, *Journal of Water Resources Planning and Management* **122(5)**, 348-355.

Singh, V.P. (1995) Watershed modeling, in: V.P. Singh (ed.), *Computer Models of Watershed Hydrology*, Water Resources Publications, Littleton, ch. 1, pp. 1-22.

Tim, U.S., Mostaghimi, S., and Shanholtz, V.O. (1992) Identification of critical nonpoint pollution source areas using GIS and water quality modeling, *AWRA, Water Resources Bulletin* **28(5)**, 877-887.

TKB (1992) *Project on Determination of Water Pollution in Yesilirmak Basin: Final Report for the Year 1991* (in Turkish), Ministry of Agriculture (TKB) and The Scientific and Technical Research Center of Turkey (TUBITAK), Project No. DEBAG-11G, Samsun, Turkey.

Ventura, S.J., Chrisman, N.R., Conners, K., Gurda, R.F., and Martin, R.W. (1988) A land information system for erosion control planning, *Journal of soil and Water Conservation* **43(3)**, 230-233.

Wallingford Software (1993) *SALMON-Q User Documentation Version 1.0,* HR Wallingford Ltd., Wallingford, Oxfordshire, Feb. 1993.

Warwick, J.J. and Haness, S.J. (1993) Efficacy of ARC/INFO GIS application to hydrologic modeling, ASCE, *Journal of Water Resources Planning and Management* **119(2)**, 366-381.

Whittenmore, D.O., Merchant, J.W., Whistler, J., McElwee, C.D., and Woods, J.J. (1987) Groundwater protection planning using the ERDAS GIS: Automation of DRASTIC and time-related capture zone, *Proceedings, National Water and Well Association Focus Conference on Midwest Ground Water Issues*, Dublin, Ohio.

Wischeimer, W.H. and Smith, D.D. (1965) *Predicting Rainfall Erosion Losses from cropland East of Rocky Mountains*, U.S. Department of Agriculture, Agricultural Handbook, No. 282, Washington, D.C.

Xue, R.Z., Bechtel, T..J., and Chen, Z. (1996) Developing a user-friendly tool for BMP assessment model using a GIS, in: C.A. Hallam, K.J. Lanfear, J.M. Salisbury, and W.A. Battaglin (eds.), *GIS and Water Resources*, AWRA Technical Publication Series, TPS-96-3, pp. 285-294.

Yoon, J. (1996) Watershed-scale nonpoint source pollution modeling and decision support system based on a model-GIS-RDBMS linkage, in: C.A. Hallam, K.J. Lanfear, J.M. Salisbury, and W.A. Battaglin (eds.), *GIS and Water Resources*, AWRA Technical Publication Series, TPS-96-3, pp. 99-108.

Young, W.J., Farley, T.F., and Davis, J.R. (1995) Nutrient management at the catchment scale using a decision support system, *Water Science and Technology* **32(5-6)**, 277-282.

SUBJECT INDEX

— G —

— H —

— I —

— K —

— L —

— M —

— N —

— O —

— P —

— Q —

— R —

— S —

— T —

— U —

— V —

— W —